New Research in Dietary Supplements and Healthy Foods

New Research in Dietary Supplements and Healthy Foods

Editor

Antonella Amato

MDPI • Basel • Beijing • Wuhan • Barcelona • Belgrade • Manchester • Tokyo • Cluj • Tianjin

Editor
Antonella Amato
University of Palermo
Italy

Editorial Office
MDPI
St. Alban-Anlage 66
4052 Basel, Switzerland

This is a reprint of articles from the Special Issue published online in the open access journal *Nutrients* (ISSN 2072-6643) (available at: https://www.mdpi.com/journal/nutrients/special_issues/Dietary_Supplements_Healthy).

For citation purposes, cite each article independently as indicated on the article page online and as indicated below:

LastName, A.A.; LastName, B.B.; LastName, C.C. Article Title. *Journal Name* **Year**, *Volume Number*, Page Range.

ISBN 978-3-0365-7020-4 (Hbk)
ISBN 978-3-0365-7021-1 (PDF)

Contents

About the Editor

Antonella Amato

Antonella Amato is Associate Professor of Physiology in the Department of Biological, Chemical, and Pharmaceutical Science and Technology (STEBICEF), University of Palermo, Italy. She teaches Nutrition Physiology for Master Degree courses. Prof. Amato is a guest editor of the journal Nutrients and director of the Advanced Course on "Nutrition and Health" at the University of Palermo. Her main research interest is analyzing the effects of the natural bioactive compounds contained in dietary supplements and functional foods on the metabolism dysfunctions related to obesity, including insuline resistance, adiposity and neurodegeneration. Her research has highlithed the beneficial effects induced by the chronic consumption of Sicilian pistachio and honey, or indicaxanthin or clorogenic acid, in dysmetabolisms occurring in obese mice. Moreover, the mechanisms of action underlying these beneficial effects have been explicated. Prof. Amato has published over 60 papers and she is co-inventor of the data submitted for patent applications. Much of her research has been funded by companies specializing in the synthesis of dietary and food supplements. Orcid ID: 0000-0002-3343-9656...

Preface to "New Research in Dietary Supplements and Healthy Foods "

Dear Readers,

This book comprises 14 scientific manuscripts belonging to the Special Issue in Nutrients titled "New Research in Dietary Supplements and Healthy Foods", edited by me.

The subject of the book is the role of natural bioactive compounds contained in dietary supplements and functional foods in the control of good health. The scientific manuscripts colllected here constitute the state of art regarding the beneficial effects induced by the long-term comsuption of functional foods or natural supplements, rich in antioxidant and anti-inflammatory compounds, against the onset and progression of chronic diseases such as cardiovascular diseases, obesity, type 2 diabetes, insuline-resistance, dyslipidaemia, hepatic steatosis, neurodegeneration, and cancer. Moreover, much of the data clarify the molecular mechanisms by which these compounds exert their beneficial effects.

This book is addressed to researchers involved in the field of nutrition or food science, providing an update to the most recent data in the field; to clinicians for integration in their therapies and management; and also to academic professors to update the theoretical contents of their lectures.

I would like to thank all the eminent authors that contribuited to this book and also to the editorial assistants for their valuable support.

Antonella Amato
Editor

nutrients

MDPI

Article

Discovery of Nobiletin from Citrus Peel as a Potent Inhibitor of β-Amyloid Peptide Toxicity

Kumju Youn [1,†], Seonah Lee [1,2,†] and Mira Jun [1,2,*]

1 Department of Food Science and Nutrition, College of Health Sciences, Dong-A University, 37, Nakdong-daero 550 beon-gil, Saha-gu, Busan 49315, Korea; kjyoun@dau.ac.kr (K.Y.); seonah_lee@daum.net (S.L.)

2 Center for Silver-Targeted Biomaterials, Brain Busan 21 Plus Program, Graduate School, Dong-A University, 37, Nakdong-daero 550 beon-gil, Saha-gu, Busan 49315, Korea

* Correspondence: mjun@dau.ac.kr; Tel.: +82-51-200-7323; Fax: +82-51-200-7505

† Contributed equally.

Received: 7 October 2019; Accepted: 29 October 2019; Published: 4 November 2019

Abstract: Increasing evidence has demonstrated that amyloid-β peptide (Aβ), the hallmark of Alzheimer's disease (AD), evokes oxidative and inflammatory cascades, which ultimately lead to the death of neurons. The purpose of the present study is to demonstrate the effect of nobiletin, a representative compound of citrus peel, in preventive and therapeutic approaches against neuronal damage by exposure to $A\beta_{25-35}$. Nobiletin significantly ameliorated $A\beta_{25-35}$-mediated cell death by restoring abnormal changes in intracellular oxidative stress, cell cycle, nuclear morphology, and activity of apoptotic caspase. Regarding anti-inflammatory responses, nobiletin significantly suppressed interleukin-1β, tumor necrosis factor-α, nitric oxide (NO), and prostaglandin E_2 production in response to Aβ stimulation. Moreover, nobiletin inhibited Aβ-stimulated inducible NO synthase and cyclooxygenase-2 expression, which was attributed to the blockade of nuclear factor-κB p65 and phosphorylation of its inhibitor, IκB-α. Interestingly, nobiletin decreased expression of c-Jun N-terminal kinase and p38 without affecting extracellular signal-regulated kinase 1/2 activation. Taken together, the novel data implicate nobiletin as a potential candidate for the prevention of AD through the inhibition of oxidative stress, apoptosis, and inflammation.

Keywords: Alzheimer's disease; amyloid-β peptide; oxidative stress; inflammation; nobiletin

1. Introduction

Amyloid-β peptide (Aβ) is the major component of amyloid plaque. Aβ is a characteristic marker of Alzheimer's disease (AD) and is believed to initiate the pathological cascade of the disease. Aβ is a 39 to 43 amino acid peptide formed by β- and γ-secretases that catalyze the splitting of amyloid precursor protein (APP) [1]. Aβ accumulation triggers the cascade of events, such as reactive oxygen species (ROS) production, cell cycle dysregulation, tau phosphorylation, and inflammation, which, ultimately, lead to the death of neurons [2,3].

The detrimental role of Aβ in the stimulation of oxidative stress has been reported in various studies. In a transgenic mouse model, Aβ increased the levels of protein and lipid oxidation markers, including protein carbonyls, 4-hydroxy-2-nonenal, and 3-nitrotyrosine [4,5]. The observed oxidative injury appeared to be dependent on the methionine 35 of Aβ peptide [6]. Notably, Aβ can regulate ROS generation and ROS can reciprocally promote overproduction of Aβ in a vicious cycle [7].

Nuclear factor-kappa B (NF-κB) regulates Aβ homeostasis through transcriptional upregulation of various related enzymes and proteins [8]. Under physiological conditions, NF-κB regulates the expression rates of genes for APP, β-secretase 1 (BACE1) and several components of the γ-secretase complex. However, activation of NF-κB by Aβ regulates the transcription of all these genes and over-stimulates Aβ production [9]. Mitogen-activated protein kinases (MAPKs), such as extracellular protein regulated protein kinase (ERK), c-Jun N-terminal kinase (JNK) and p38, can activate the NF-κB pathway to produce a series of inflammatory factors.

Nobiletin (2-(3,4-dimethoxyphenyl)-5,6,7,8-tetramethoxychromen-4-one) is one of the major polymethoxyflavones in the peel of citrus fruits, including oranges, mandarins, limes, and lemons [10]. Various pharmacological effects attributed to nobiletin include antioxidant, anti-inflammatory, anti-diabetic, anti-atherogenic and anti-carcinogenic activities [11–15]. The neuroprotective property of nobiletin has been demonstrated in several recent studies. The compound prevents ischemic brain injury by regulating the Akt/cAMP-response element-binding protein (CREP)/Bcl-2 pathway in Sprague–Dawley rats. Nobiletin reduces Aβ-stimulated memory impairment in several AD animal models, including olfactory bulbectomy mice, APP-SL 7–5 Tg mice and 3XTg-AD mice [16–18].

Our previous study demonstrated the ability of nobiletin to non-competitively inhibit BACE1 via hydrogen-bond-mediated interactions with allosteric residues of the enzyme [19]. However, the primary molecular mechanism underlying the neuroprotective effect of nobiletin on Aβ-induced oxidative stress and inflammation has not yet been clearly demonstrated, which prompted the present investigation of the possible effects of nobiletin on PC12 cell models.

2. Materials and Methods

2.1. Cell Culture and $A\beta_{25\text{-}35}$ Stock Solution

PC12 cells were cultured in RPMI1640 medium supplemented with 10% horse serum, 5% fetal bovine serum, and penicillin (100 U/mL) and treated with nobiletin (Sigma-Aldrich, St. Louis, MO, USA) for 24 h at 37 °C in a 5% CO_2 incubator. $A\beta_{25\text{-}35}$ (Sigma-Aldrich) was dissolved in dimethylsulfoxide (DMSO) at an initial concentration of 10 mM and diluted with phosphate buffered saline (PBS). $A\beta_{25\text{-}35}$ solution was incubated at 37 °C for 48 h to permit aggregation before use.

2.2. Cell Viability Analysis

Cell viability was assessed using the established 3-(4,5-dimethylthiazol-2-yl)-2,5-diphenyltetrazolium bromide (MTT) assay. PC12 cells (1×10^5 cells/well in a 96-well plate) were incubated with MTT reagent for 3 h at 37 °C. The medium was removed and the formazan crystals produced by the reduction of MTT were dissolved in DMSO. The absorbance due to the formazan crystals was measured at 570 nm using a model ELX808 spectrophotometer (BioTek, Winooski, VT, USA).

2.3. Intracellular ROS Analysis

PC12 cells were stained with CM-H_2DCFDA in minimum essential medium without serum at 37 °C for 30 min in the dark and resuspended in Hank's balanced salt solution. Cells were placed on glass slides and cultured overnight. Imaging ROS was done by fluorescence spectrophotometry with excitation and emission wavelengths of 485 and 528 nm, respectively, using a model FLX800 spectrometer (BioTek).

2.4. Apoptosis Assay by Hoechst 33342 Staining

PC12 cells were harvested with PBS and fixed in 4% paraformaldehyde for 20 min at 25 °C and then washed with PBS before being exposed to Hoechst 33342 for 15 min in the dark. The apoptotic morphology was observed using fluorescence microscopy (Olympus Optical Co., Tokyo, Japan) at 400× magnification.

2.5. Fluorescence-Activated Cell Sorting (FACS) Analysis

PC12 cells were cultured in 24-well plates (5×10^5 cells/well) and incubated with 50 μM $A\beta_{25\text{-}35}$ for 24 h after pretreatment with various concentrations of nobiletin for 1 h. After incubation, cells were collected and analyzed using the Muse™ Cell Analyzer (Merck Millipore, Darmstadt, Germany). Briefly, FACS analysis following staining with annexin V and 7-aminoactinomycin D (7-AAD) was performed to detect apoptosis, and cell viability was determined using a DNA-binding dye. Cells were fixed in cold 70% ethanol and stained with propidium iodide (PI), a membrane impermeant dye, to analyze the cell cycle.

2.6. Assessment of Levels of Nitric Oxide (NO) and Prostaglandin E2 (PGE_2)

PC12 cells (2×10^6) were seeded on 6-cell plates containing medium and incubated with $A\beta_{25\text{-}35}$ for 24 h in the presence or absence of nobiletin. The formation of NO_2^-, a stable end product that has been extensively used as an indicator of NO accumulation, was assessed using Griess reagent. The media were mixed with Griess reagent at RT for 10 min and the NO levels were analyzed using the aforementioned ELX808 microplate reader at 570 nm.

The supernatants were mixed with primary antibody solution and PGE_2 conjugate for 2 h, followed by washing and addition of stop solution. The absorbance was measured at 450 nm using the ELX808 microplate reader.

2.7. Western Blot Analysis

Proteins (40 μg) were separated by 10% sodium dodecyl sulfate-polyacrylamide gel electrophoresis and transferred to polyvinylidene difluoride (PVDF) membranes. The blots were blocked for 2 h at RT with 5% skim milk in 0.1% Tween 20 in PBS (PBST) and then incubated overnight with specific primary antibodies at 4 °C. The primary antibodies used were against β-actin (1:1000), tumor necrosis factor-alpha (TNF-α; 1:1000), cyclooxygenase-2 (COX-2; 1:1000), inducible NO synthase (iNOS; 1:1000), p-65 (1:1000), phospho (p)-IκB-α (1:1000), p-JNK (1:1000), p-p38 (1:1000), and p-ERK (1:1000). Subsequently, horseradish peroxidase (HRP)-conjugated anti-rabbit IgG secondary antibody or anti-goat IgG was used. Protein bands were visualized using the EZ-capture device (Atto, Tokyo, Japan).

2.8. Statistical Analysis

Statistical analysis was performed using SAS 9.3 software (SAS Institute Inc., Cary, NC, USA). All results are expressed as mean ± SD and are representative of the data obtained from three independent experiments. Statistical comparisons of differences between groups were performed through the Student's t test, considering * $p < 0.05$, ** $p < 0.01$, and *** $p < 0.001$ as being statistically significant.

3. Results and Discussion

3.1. Nobiletin Inhibits Cytotoxicity Evoked by $A\beta_{25\text{-}35}$

To verify the neuroprotective effect of nobiletin, cell viability was assessed using the MTT assay and FACS. As shown in Figure 1a, nobiletin did not adversely affect PC12 cell viability at concentrations of 1 to 25 μM, which were used for further study. Treatment with 50 μM $A\beta_{25\text{-}35}$ for 24 h induced approximately 40% cell death in comparison with the control group ($p < 0.001$; Figure 1b). However, pretreatment with 1, 10, and 25 μM nobiletin significantly increased cell viability up to 78.1% ± 7.4%, 81.3% ± 4.5%, and 82.4% ± 4.7%, respectively. Notably, 10 μM nobiletin exhibited a similar neuroprotective effect to that of resveratrol, a well-known positive control. Consistent with the results of the MTT assay, nobiletin significantly prevented $A\beta_{25\text{-}35}$-induced cell death in FACS analysis (Figure 1c), and this effect was dependent on the dose of nobiletin. These data provided correlative evidence indicating that nobiletin contributes to cell survival in PC12 cells damaged by $A\beta_{25\text{-}35}$.

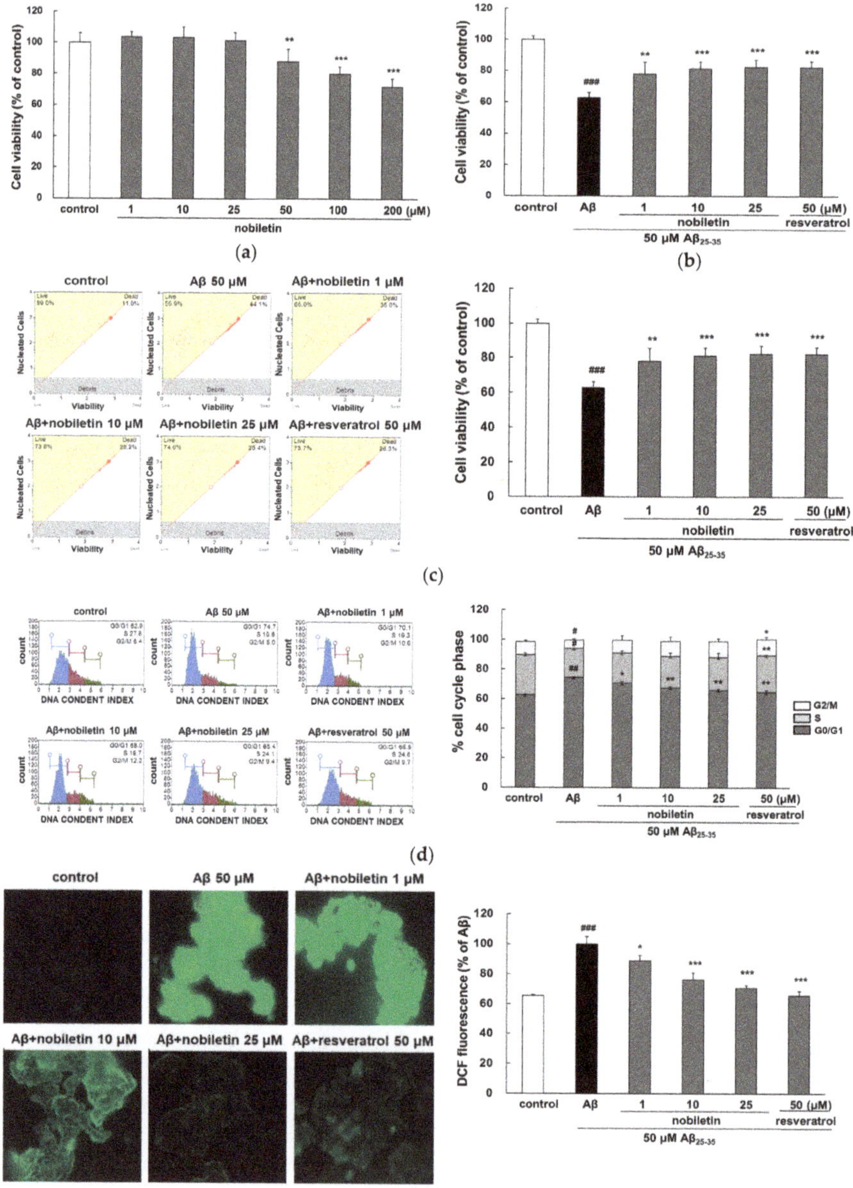

Figure 1. Protective properties of nobiletin against $A\beta_{25\text{-}35}$-mediated cell damage. (**a**) Evaluation of cytotoxicity nobiletin alone in PC12 cells. Cells were pretreated with nobiletin for 1 h followed by exposure to 50 μM of $A\beta_{25–35}$ for 24 h, and cell viability was assessed by (**b**) MTT reduction assay and (**c**) fluorescence-activated cell sorting (FACS) analysis. (**d**) Cell cycle progression was measured by FACS. The percentage of cells in the G_0/G_0, S, and G_0/M phases of the cell cycle was determined using the Muse 1.5 Analysis software. (**e**) Intracellular ROS production was observed by CM-H_2DCFDA fluorescent dye. ### $p<0.001$, ## $p < 0.01$, and # $p < 0.05$ vs. control. *** $p < 0.001$, ** $p < 0.01$, and * $p < 0.05$ vs. $A\beta_{25–35}$.

Cell cycle regulation is a crucial process of cell growth and proliferation in neurons [20]. As shown in Figure 1d, A$\beta_{25\text{-}35}$ significantly induced an increase in cells in the G_0/G_1 phase ($p < 0.01$) and a corresponding decrease in cells in the S phase and G_2/M phase ($p < 0.05$), suggesting that cells had lower rates of growth and tended to be arrested at the G_0/G_1 transition. However, nobiletin restored A$\beta_{25\text{-}35}$-mediated cell cycle dysregulation in a concentration-dependent manner, which may contribute to the enhanced cell viability effect of the compound.

Nobiletin was further evaluated for its antioxidant property in A$\beta_{25\text{-}35}$-injured cells by ROS regulation. As indicated in Figure 1e, fluorescence intensity and large numbers of bright particles in cells were visibly increased by A$\beta_{25\text{-}35}$ exposure, suggesting the presence of intracellular oxidative stress. A$\beta_{25\text{-}35}$ stimulated significant increase in ROS to 100% ± 4.86% ($p < 0.001$). However, pretreatment of nobiletin decreased ROS generation in a dose-dependent manner ($p < 0.05$ and $p < 0.001$). These data were consistent with previous descriptions of the antioxidant property of nobiletin. The compound protected PC12 cells against H_2O_2-triggered damage by scavenging ROS, decreasing malonaldehyde (MDA), and enhancing glutathione (GSH) and superoxide dismutase (SOD) contents [21]. In addition, excessive intracellular ROS stimulates the activation of signal transduction cascades, which disturbs calcium homeostasis and leads to the initiation of apoptosis. Nobiletin prevents mitochondrial calcium overload as well as ROS generation in glutamate-induced cortical neurons [22]. In vivo, the intraperitoneal administration of nobiletin reportedly reduced tau phosphorylation, the index of protein oxidation, and protein carbonyl levels in SAMP8 mice, which were related to the recovery of GSH/glutathione disulfide (GSSG) ratio and increased glutathione peroxidase (GPx) activity [23]. These results demonstrated that the neuroprotective role of nobiletin results, at least in part, from the reduction of oxidative stress.

3.2. Nobiletin Reduces A$\beta_{25\text{–}35}$-Mediated Apoptosis and Caspase-3 Activation

Intracellular oxidative stress plays a central role in the induction of neuronal apoptosis stimulated by Aβ [24]. As shown in Figure 2a, cells treated with A$\beta_{25\text{–}35}$ exhibited uneven morphology of their nuclei as a result of membrane blebbing, chromatin aggregation, and nuclear condensation, whereas the nobiletin pretreated group showed dispersed and weak fluorescence that is typical of live cells. Particularly, nobiletin at 25 μM decreased apoptosis similar to the level of the positive control group. Analysis of morphology alone is not sufficient to distinguish between early and late apoptotic cells. Thus, flow cytometric analysis was employed to quantitatively analyze apoptotic cell death. When exposed to A$\beta_{25\text{-}35}$, early and late apoptosis was significantly increased to 26.39% ± 2.48% and 29.06% ± 2.33%, respectively, compared with the control group ($p < 0.001$, Figure 2b). Nobiletin—at all concentrations—markedly reduced both early and late apoptosis. These results were consistent with the anti-apoptotic activity of nobiletin in endoplasmic reticulum stress-induced PC12 and I/R-exposed Kupffer cells [25].

Caspase-3 is a biomarker of oxidative-stress-stimulated cell death that has also been implicated in the final stage of apoptosis. The caspase-3 was obviously activated in the A$\beta_{25\text{-}35}$-treated group ($p < 0.001$, Figure 2c). However, enhanced caspase-3 activity was dose-dependently decreased by nobiletin. Aβ-triggered apoptotic cell death is related to the reduced anti-apoptotic Bcl-2 protein and the increased pro-apoptotic molecule Bax expression. In a previous study, nobiletin displayed an anti-apoptosis effect by decreasing the ratio of Bcl-2/Bax expression in H_2O_2-stimulated HT22 cells [26].

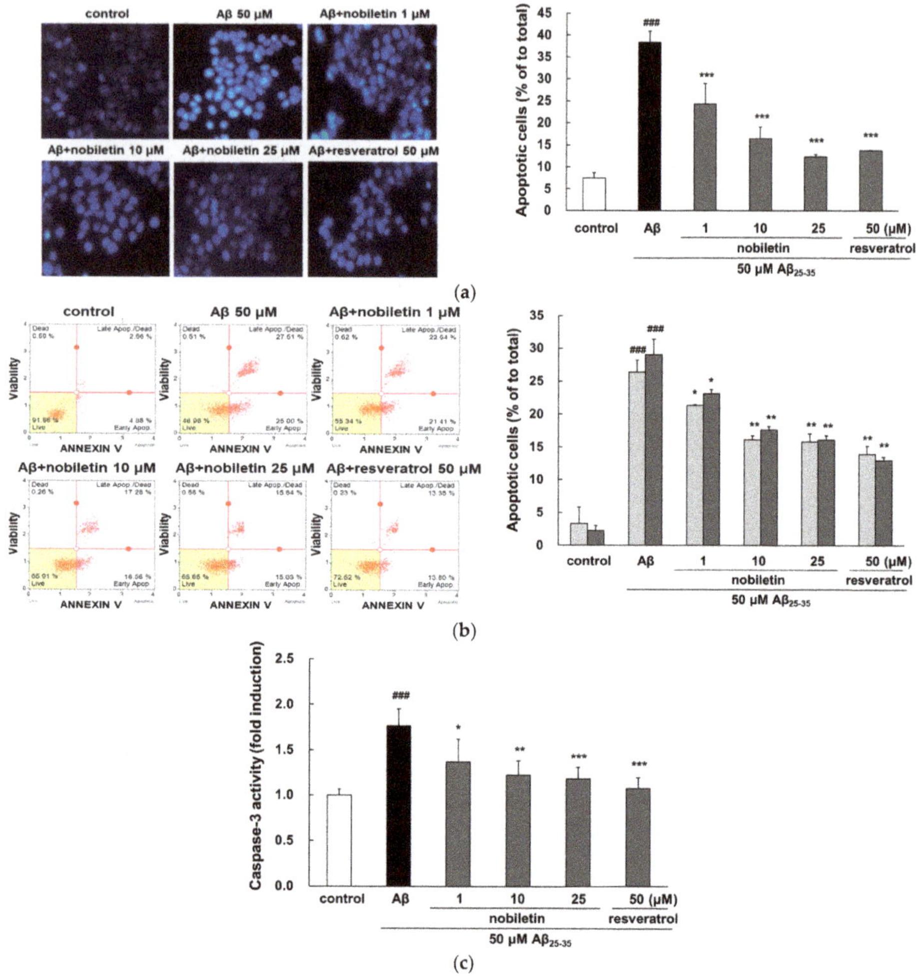

Figure 2. Activity of nobiletin in preventing A$\beta_{25\text{-}35}$-evoked apoptosis and caspase-3. (**a**) Morphological features of apoptotic cells were observed by fluorescence microscopy using Hoechst 33342 staining (magnification ×400). (**b**) Flow cytometric analysis was used to investigate the properties of nobiletin on A$\beta_{25\text{-}35}$-stimulated apoptosis. The cell populations discriminated in each quadrant are live cells in the lower-left quadrant, early apoptotic cells in the lower-right quadrant, late apoptotic cells in the upper-right quadrant, and dead cells in the upper-left quadrant. (**c**) Caspase-3 activity was assessed using the caspase-3 assay kit. $^{\#\#\#}$ $p < 0.001$ vs. control. *** $p < 0.001$, ** $p < 0.01$, and * $p < 0.05$ vs. A$\beta_{25\text{–}35}$.

3.3. Nobiletin Suppresses A$\beta_{25\text{-}35}$-Induced Release of Inflammatory Markers

A$\beta_{25\text{-}35}$ exposure increases the release of NO and PGE_2 by approximately 5-fold compared with the control (Figure 3a,b). Notably, treatment with 10 and 25 μM nobiletin exhibited similar activity as that of 50 μM resveratrol. A$\beta_{25\text{-}35}$ treatment increased the expression level of pro-inflammatory cytokines, such as TNF-α and interleukin (IL)-1β, by 3-fold (Figure 3c), but the expressions were suppressed by 10 and 25 μM nobiletin.

As shown in Figure 3d, the level of iNOS stimulated by A$\beta_{25\text{-}35}$ increased up to 202.5% ± 18.3% compared with the control ($p < 0.001$). However, nobiletin promptly inhibited the expression of iNOS protein. Notably, the highest concentration of nobiletin resulted in almost complete suppression of the enzyme production (118.3% ± 15.7%; $p < 0.001$). In parallel, A$\beta_{25\text{-}35}$-mediated COX-2 expression was also markedly blocked by nobiletin.

Several studies have demonstrated that nobiletin possesses strong anti-inflammatory ability in lipopolysaccharide (LPS)-induced expression of pro-inflammatory cytokines in BV2 microglial cells [27–29]. A recent in vivo study suggested that the oral administration of nobiletin (100 mg/kg/day) for 6 weeks attenuated microglial activation and secretion of pro-inflammatory mediators, leading to the restoration of memory deficits in mice [30].

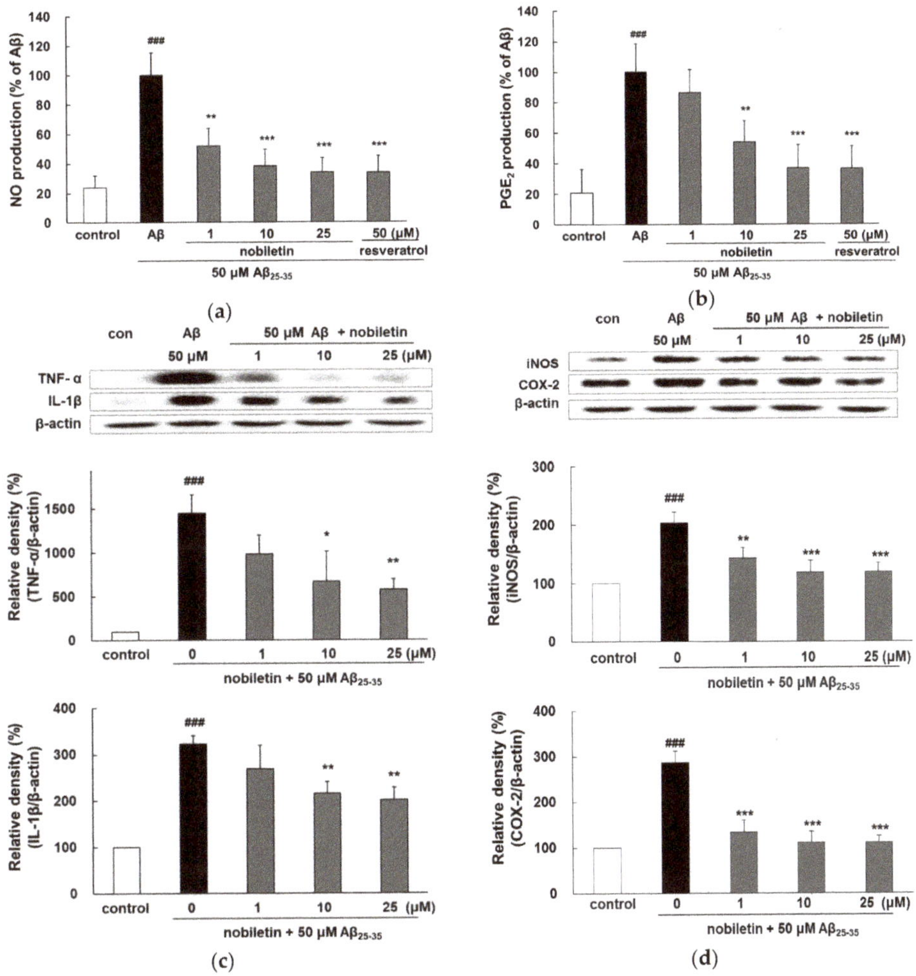

Figure 3. Inhibitory properties of nobiletin on A$\beta_{25\text{-}35}$-mediated expression of (**a**) NO, (**b**) PGE_2, (**c**) TNF-α and IL-β, and (**d**) iNOS and COX-2 in PC12 cells. The cells were pretreated with nobiletin for 1 h and then exposed to A$\beta_{25\text{-}35}$ for 24 h. The culture supernatant was collected to evaluate the NO and PGE_2 formation. Protein expression of TNF- α, IL-β, iNOS, and COX-2 was measured by Western blot analysis. The band intensities were quantified using Image J software and normalized to β-actin. ### $p < 0.001$ vs. control. *** $p < 0.001$, ** $p < 0.01$, and * $p < 0.05$ vs. A$\beta_{25\text{–}35}$.

3.4. Nobiletin Regulates $A\beta_{25\text{-}35}$-Induced NF-κB and MAPK Signaling Pathways

$A\beta_{25\text{-}35}$ obviously elevated the phosphorylation of p65 and IκB-α by 227.4% ± 19.6% and 194.4% ± 10.1%, respectively. In contrast, pretreatment with nobiletin at 10 and 25 μM remarkably repressed p65 expression (Figure 4a). Moreover, in the case of IκB-α, all doses of nobiletin indicated a significant inhibitory effect in respect to $A\beta_{25\text{-}35}$ treatment.

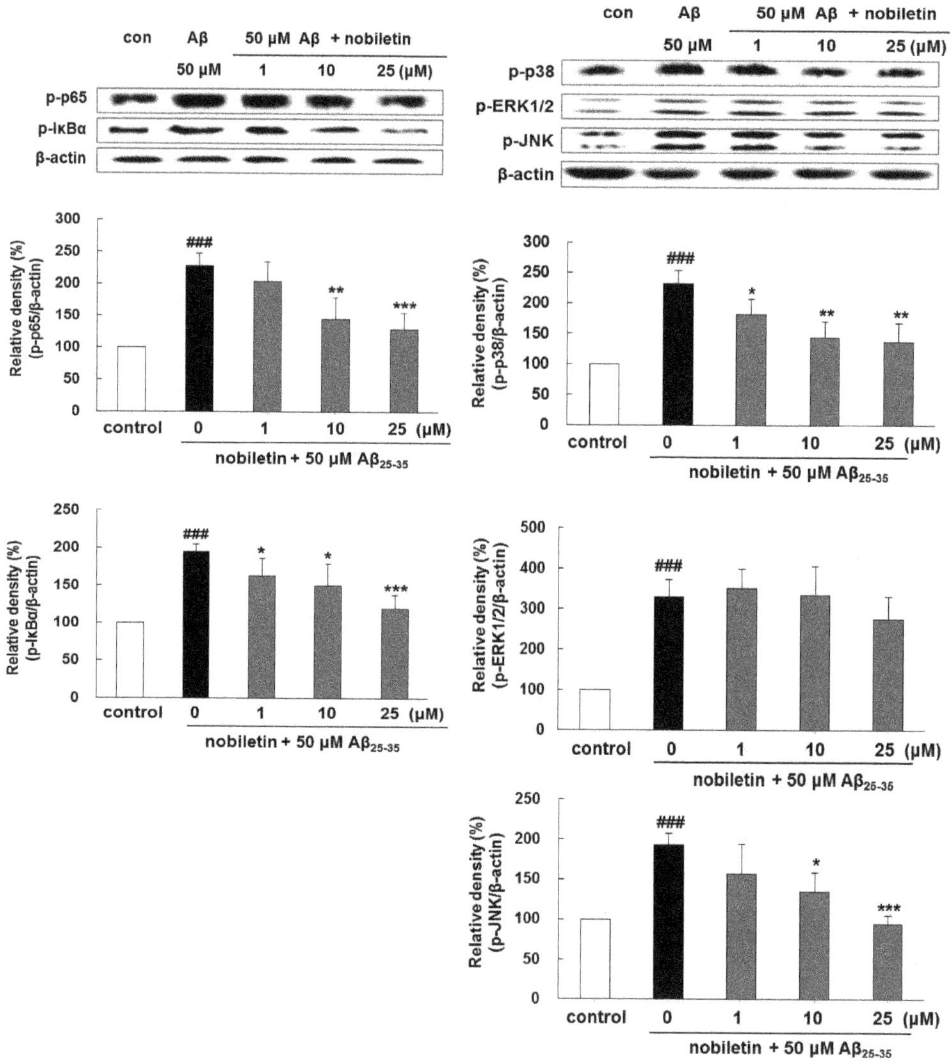

Figure 4. Inhibitory properties of nobiletin in the NF-κB/MAPKs pathway. Phosphorylation of (**a**) p65 and IκB-α, and (**b**) p38, ERK1/2, and JNK was examined by Western blot analysis. The cells were pretreated with nobiletin for 1 h and then exposed to $A\beta_{25\text{-}35}$ for 4 h (p65 and IκB-α) or 1 h (p38, ERK1/2, and JNK). Quantification of band intensities was conducted using Image J software and normalized to β-actin. ### $p < 0.001$ vs. control. *** $p < 0.001$, ** $p < 0.01$, and * $p < 0.05$ vs. $A\beta_{25\text{-}35}$.

As presented in Figure 4b, nobiletin suppressed A$\beta_{25\text{-}35}$-evoked phosphorylation of p38 and JNK. In particular, the level of phosphorylated JNK was markedly suppressed when treated with 25 μM nobiletin (94.4% ± 9.8%, $p < 0.001$), indicating that the phosphorylation of MAPKs was closely related with the inhibitory property of nobiletin on A$\beta_{25\text{-}35}$-evoked activation of p65 and IκB-α.

Previous research has reported that BACE1 promoter transactivation is modulated by the NF-κB signaling pathway, indicating that the suppression of NF-κB leads to inhibition of BACE1 activity [9]. In our previous study, we observed that nobiletin blocked BACE1 activity, suggesting that the compound might modulate BACE1 activation by suppressing the NF-κB signaling pathway [19]. Another study reported that nobiletin inhibited H_2O_2-evoked cell death in HT22 murine hippocampal cells, accompanied by decreased JNK and p38 phosphorylation [26]. In rat primary astrocytes, nobiletin suppressed the overexpression of iNOS and NO by the inhibition of the NF-κB/p38 MAPK pathways.

Several animal studies clearly demonstrated that nobiletin (10–50 mg/kg) improves memory in rats exposed to chronic intracerebroventricular infusion of A$\beta_{1\text{-}40}$ [31]. In addition, daily supplementation of nobiletin (10 mg/kg) for 4 months significantly lowers both A$\beta_{1\text{-}40/42}$ and amyloid plaques in 9-month-old APP-SL 7–8 Tg mice [17]. Recently, Nakajima et al. demonstrated that nobiletin (10 mg/kg) reduces tau phosphorylation in SAMP8 mice [23].

Bioavailability is an essential factor for the development of potential therapeutic agents. In addition, the anti-AD agents must penetrate the blood–brain barrier to attain sufficient concentration for the therapeutic application within the central nervous system. When nobiletin (50 mg/kg) was administered, the content of intact nobiletin was detected in rat brains within 1 h after dosing, suggesting that this compound can rapidly cross the blood–brain barrier and reach the brain. Furthermore, the concentration of nobiletin was 2.4-fold higher in the brain than in plasma. Interestingly, nobiletin was detected for up to 24 h in the brain, whereas in the plasma, it was observed up to 9 h, suggesting that elimination of this compound from the brain was slower in comparison to plasma [32]. Of note, nobiletin has no discernible toxicity in chronic treatments in mice and humans [33].

4. Conclusions

The novel results of the present study indicate that nobiletin, a natural compound in citrus peel, has a neuroprotective effect on A$\beta_{2\ 5\text{-}35}$-induced cytotoxic damage of PC12 cells. The data may provide a preventive and/or therapeutic potential of nobiletin for degenerative disorders of the brain.

Author Contributions: Data curation, S.L.; writing-original draft, K.Y.; supervision, review & editing, M.J.

Funding: This research was funded by Dong-A University.

Conflicts of Interest: All authors declare no conflict of interest.

References

1. Selkoe, D.J. Translating cell biology into therapeutic advances in Alzheimer's disease. *Nature* **1999**, *399*, A23–A31. [CrossRef] [PubMed]
2. Shen, Y.; He, P.; Zhong, Z.; McAllister, C.; Lindholm, K. Distinct destructive signal pathways of neuronal death in Alzheimer's disease. *Trends Mol. Med.* **2006**, *12*, 574–579. [CrossRef] [PubMed]
3. Butterfield, D.A.; Drake, J.; Pocernich, C.; Castegna, A. Evidence of oxidative damage in Alzheimer's disease brain: Central role for amyloid beta-peptide. *Trends Mol. Med.* **2001**, *12*, 548–554. [CrossRef]
4. Garcia-Alloza, M.; Dodwell, S.A.; Meyer-Luehmann, M.; Hyman, B.T.; Bacskai, B.J. Plaque-derived oxidative stress mediates distorted neurite trajectories in the Alzheimer mouse model. *J. Neuropathol. Exp. Neurol.* **2006**, *65*, 1082–1089. [CrossRef]
5. Robinson, R.A.; Lange, M.B.; Sultana, R.; Galvan, V.; Fombonne, J.; Gorostiza, O.; Zhang, J.; Warrier, G.; Cai, J.; Pierce, W.M.; et al. Differential expression and redox proteomics analyses of an Alzheimer disease transgenic mouse model: Effects of the amyloid-β peptide of amyloid precursor protein. *Neuroscience* **2011**, *177*, 207–222. [CrossRef] [PubMed]

6. Butterfield, D.A.; Sultana, R. Methionine-35 of Aβ (1–42): Importance for oxidative stress in Alzheimer disease. *J. Amino Acid.* **2011**, *2011*, 198430. [CrossRef] [PubMed]
7. Wojsiat, J.; Zoltowska, K.M.; Laskowska-Kaszub, K.; Wojda, U. Oxidant/Antioxidant imbalance in alzheimer's disease: Therapeutic and diagnostic prospects. *Oxid. Med. Cell Longev.* **2018**, *2018*, 6435861. [CrossRef]
8. Chen, C.H.; Zhou, W.; Liu, S.; Deng, Y.; Cai, F.; Tone, M.; Tone, Y.; Tong, Y.; Song, W. Increased NF-κB signalling up-regulates BACE1 expression and its therapeutic potential in Alzheimer's disease. *Int. J. Neuropsychopharmacol.* **2012**, *15*, 77–90. [CrossRef]
9. Chami, L.; Buggia-Prévot, V.; Duplan, E.; Del Prete, D.; Chami, M.; Peyron, J.F.; Checler, F. Nuclear factor-κB regulates βAPP and β- and γ-secretases differently at physiological and supraphysiological Aβ concentrations. *J. Biol. Chem.* **2012**, *287*, 24573–24584. [CrossRef]
10. Nogata, Y.; Sakamoto, K.; Shiratsuchi, H.; Ishii, T.; Yano, M.; Ohta, H. Flavonoid composition of fruit tissues of citrus species. *Biosci. Biotechonol. Biochem.* **2006**, *70*, 178–192. [CrossRef]
11. Wang, M.; Meng, D.; Zhang, P.; Wang, X.; Du, G.; Brennan, C.; Li, S.; Ho, C.T.; Zhao, H. Antioxidant protection of nobiletin, 5-dimethylnobiletin, tangeretin, and 5-demethyltangeretin from citrus peel in *Sacch. Cerevisiae*. *J. Agric. Food Chem.* **2018**, *66*, 3155–3160. [CrossRef] [PubMed]
12. Choi, S.Y.; Hwang, J.H.; Ko, H.C.; Park, J.G.; Kim, S.J. Nobiletin from Citrus fruit peel inhibits the DNA-binding activity of NF-kappab and ROS production in LPS-activated RAW 264.7 cells. *J. Ethnopharmacol.* **2007**, *113*, 149–155. [CrossRef] [PubMed]
13. Lee, Y.S.; Cha, B.Y.; Saito, K.; Yamakawa, H.; Choi, S.S.; Yamaguchi, K.; Yonezawa, T.; Teruya, T.; Nagai, K.; Woo, J.T. Nobiletin improves hyperglycemia and insulin resistance in obese diabetic ob/ob mice. *Biochem. Pharmacol.* **2010**, *79*, 1674–1683. [CrossRef] [PubMed]
14. Mulvihill, E.E.; Assini, J.M.; Lee, J.K.; Allister, E.M.; Sutherland, B.G.; Koppes, J.B.; Sawyez, C.G.; Edwards, J.Y.; Telford, D.E.; Charbonneau, A.; et al. Nobiletin attenuates VLDL overproduction, dyslipidemia, and atherosclerosis in mice with diet-induced insulin resistance. *Diabetes* **2011**, *60*, 1446–1457. [CrossRef] [PubMed]
15. Miyamoto, S.; Yasui, Y.; Tanaka, T.; Ohigashi, H.; Murakami, A. Suppressive effects of nobiletin on hyperleptinemia and colitis-related colon carcinogenesis in male ICR mice. *Carcinogenesis* **2008**, *29*, 1057–1063. [CrossRef] [PubMed]
16. Nakajima, A.; Yamakuni, T.; Haraguchi, M.; Omae, N.; Song, S.Y.; Kato, C.; Nakagawasai, O.; Tadano, T.; Yokosuka, A.; Mimaki, Y.; et al. Nobiletin, a citrus flavonoid that improves memory impairment, rescues bulbectomy induced cholinergic neurodegeneration in mice. *J. Pharmacol. Sci.* **2007**, *105*, 122–126. [CrossRef]
17. Onozuka, H.; Nakajima, A.; Matsuzaki, K.; Shin, R.W.; Ogino, K.; Saigusa, D.; Tetsu, N.; Yokosuka, A.; Sashida, Y.; Mimaki, Y.; et al. Nobiletin, a citrus flavonoid, improves memory impairment and Abeta pathology in a transgenic mouse model of Alzheimer's disease. *J. Pharmacol. Exper. Thera.* **2008**, *326*, 739–744. [CrossRef]
18. Nakajima, A.; Aoyama, Y.; Shin, E.J.; Nam, Y.; Kim, H.C.; Nagai, T.; Yokosuka, A.; Mimaki, Y.; Yokoi, T.; Ohizumi, Y.; et al. Nobiletin, a citrus flavonoid, improves cognitive impairment and reduces soluble Abeta levels in a triple transgenic mouse model of Alzheimer's disease (3XTg-AD). *Behav. Brain Res.* **2015**, *289*, 69–77. [CrossRef]
19. Youn, K.; Yu, Y.; Lee, J.; Jeong, W.S.; Ho, C.T.; Jun, M. Polymethoxyflavones: Novel β-secretase (BACE1) inhibitors from Citrus peels. *Nutrients* **2017**, *9*, 973. [CrossRef]
20. Giovanni, A.; Wirtz-Brugger, F.; Keramaris, E.; Slack, R.; Park, D.S. Involvement of cell cycle elements, cycline-dependent kinases, pRb, and E2F x DP, in B-amyloid-induced neuronal death. *J. Biol. Chem.* **1999**, *274*, 19011–19016. [CrossRef]
21. Lu, Y.H.; Su, M.Y.; Huang, H.Y.; Lin-Li; Yuan, C.G. Protective effects of the citrus flavanones to PC12 cells against cytotoxicity induced by hydrogen peroxide. *Neurosci. Lett.* **2010**, *484*, 6–11. [CrossRef] [PubMed]
22. Lee, J.H.; Amarsanaa, K.; Wu, J.; Jeon, S.C.; Cui, Y.; Jung, S.C.; Park, D.B.; Kim, S.J.; Han, S.H.; Kim, H.W.; et al. Nobiletin attenuates neurotoxic mitochondrial calcium overload through K+ influx and ΔΨm across mitochondrial inner membrane. *Korean J. Physiol. Pharmacol.* **2018**, *3*, 311–319. [CrossRef] [PubMed]
23. Nakajima, A.; Aoyama, Y.; Nguyen, T.T.; Shin, E.J.; Kim, H.C.; Yamada, S.; Nakai, T.; Nagai, T.; Yokosuka, A.; Mimaki, Y.; et al. Nobiletin, a citrus flavonoid, ameliorates cognitive impairment, oxidative burden, and hyperphosphorylation of tau in senescence-accelerated mouse. *Behav. Brain Res.* **2013**, *250*, 351–360. [CrossRef] [PubMed]

24. Xian, Y.F.; Lin, Z.X.; Mao, Q.Q.; Ip, S.P.; Su, Z.R.; Lai, X.P. Protective effect of isorhynchophylline against beta-amyloid-induced neurotoxicity in PC12 cells. *Cell Mol. Neurobiol.* **2012**, *32*, 353–360. [CrossRef]
25. Li, Z.R.; Yang, L.; Zhen, J.; Zhao, Y.; Lu, Z.N. Nobiletin protects PC12 cells from ERS-induced apoptosis in OGD/R injury via activation of the PI3K/AKT pathway. *Exp. Ther. Med.* **2018**, *16*, 1470–1476. [CrossRef]
26. Cho, H.W.; Jung, S.Y.; Lee, G.H.; Cho, J.H.; Choi, I.Y. Neuroprotective effect of Citrus unshiu immature peel and nobiletin inhibiting hydrogen peroxide-induced oxidative stress in HT22 murine hippocampal neuronal cells. *Pharmacogn. Mag.* **2015**, *11*, S284–S289. [CrossRef]
27. Cui, Y.; Wu, J.; Jung, S.C.; Park, D.B.; Maeng, Y.H.; Hong, J.Y.; Kim, S.J.; Lee, S.R.; Kim, S.J.; Kim, S.J.; et al. Anti-neuroinflammatory activity of nobiletin on suppression of microglial activation. *Biol. Pharm. Bull.* **2010**, *33*, 1814–1821. [CrossRef]
28. Ho, S.C.; Kuo, C.T. Hesperidin, nobiletin, and tangeretin are collectively responsible for the anti-neuroinflammatory capacity of tangerine peel (Citri reticulatae pericarpium). *Food Chem. Toxicol.* **2014**, *71*, 176–182. [CrossRef]
29. Wang, Y.; Zang, W.; Ji, S.; Cao, J.; Sun, C. Three Polymethoxyflavones purified from Ougan (Citrus reticulata Cv. Suavissima) inhibited LPS-Induced NO elevation in the neuroglia BV-2 cell line via the JAK2/STAT3 pathway. *Nutrients* **2019**, *11*, 791. [CrossRef]
30. Qi, G.; Mi, Y.; Fan, R.; Li, R.; Liu, Z.; Liu, X. Nobiletin Protects against systemic inflammation-stimulated memory impairment via MAPK and NF-κB signaling pathways. *J. Agric. Food Chem.* **2019**, *67*, 5122–5134. [CrossRef]
31. Matsuzaki, K.; Yamakuni, T.; Hashimoto, M.; Haque, A.M.; Shido, O.; Mimaki, Y.; Sashida, Y.; Ohizumi, Y. Nobiletin restoring beta-amyloid-impaired CREB phosphorylation rescues memory deterioration in Alzheimer's disease model rats. *Neurosci. Lett.* **2006**, *400*, 230–234. [CrossRef] [PubMed]
32. Singh, S.P.; Wahajuddin; Tewari, D.; Patel, K.; Jain, G.K. Permeability determination and pharmacokinetic study of nobiletin in rat plasma and brain by validated high-performance liquid chromatography method. *Fitoterapia* **2011**, *82*, 1206–1214. [CrossRef] [PubMed]
33. Lee, Y.S.; Cha, B.Y.; Choi, S.S.; Choi, B.K.; Yonezawa, T.; Teruya, T.; Nagai, K.; Woo, J.T. Nobiletin improves obesity and insulin resistance in high-fat diet-induced obese mice. *J. Nutr. Biochem.* **2013**, *24*, 156–162. [CrossRef] [PubMed]

nutrients

MDPI

Article

Shea Nut Oil Extracts Enhance the Intra-Articular Sodium Hyaluronate Effectiveness on Surgically Induced OA Progression in Rats

Ing-Jung Chen [1,2] and Chih-Shung Wong [1,3,*]

1 Department of Anesthesiology, Cathay General Hospital, Taipei 10630, Taiwan; dr.jungchen@gmail.com
2 Department of Medical Research, Cathay General Hospital, Taipei 10630, Taiwan
3 Graduate Institute of Medical Sciences, National Defense Medical Center, Taipei 11490, Taiwan
* Correspondence: w82556@gmail.com; Tel.: +(886)-2-27082121 (ext. 3510)

Received: 29 February 2020; Accepted: 26 March 2020; Published: 30 March 2020

Abstract: Osteoarthritis (OA) progression is associated with joint pain and stiffness. Intra-articular hyaluronic acid (IAHA) injection in knee OA restores the viscoelasticity of the joint and prevents cartilage damage. Shea nut oil extract (SNO) was shown to provide chondroprotection on surgically-induced OA progression in rats. Here we aim to examine IAHA injection supplemented with SNO diet for a synergetic evaluation on the disease progression in OA rats. We employed an anterior cruciate ligament transection plus medial meniscectomy-induced knee OA rat model with up to 12 weeks of sign/behavior observation (knee width, weight-bearing) and histological assessments of joint damage. We found both IAHA and SNO alone significantly attenuated histological changes of cartilage degeneration and synovial reactions in these knee OA rats. Nonetheless, oral SNO alone mitigated OA pain and inflammation while IAHA alone had no significant impact on the weight-bearing test and knee joint swelling. Moreover, with IAHA-treated rats fed with oral SNO diet, additional anti-inflammatory and anti-nociceptive effects were found, which further enhanced and maintained IAHA protection. Given the differential phenotype of oral SNO vs. IAHA, a regimen of IAHA coupled with SNO supplement provides a long-term effect of IAHA treatment. Taken together, the SNO supplement can be safely used as an adjuvant diet for chronic symptomatic relief of OA coupled with IAHA management.

Keywords: shea nut oil; lipids; triterpenes; osteoarthritis; pain; hyaluronic acid; triterpenes; cartilage

1. Introduction

Osteoarthritis (OA) is a multifactorial joint disease and a common disabling condition affecting the global population [1,2]. The increasing joint pain and stiffness gradually leads to reduced physical function, quality of life, and frequent physician visits [3,4]. Up to 54.4 million adults experienced doctor-diagnosed arthritis during 2013–2015 in United States, and 23.7 million had arthritis-attributable activity limitations. In addition, adults with heart disease, diabetes, and obesity have a higher prevalence of OA (49.3%, 47.1%, and 30.6%, respectively) and consequently arthritis-attributable activity limitations [5]. Therefore, the management of these associated factors are recommended to potentially reduce symptomatic knee and hip OA incidences [6,7].

OA pain is the predominant limiting factor for a patient's activity and life quality, and it leads those individuals affected to seek medical care [8]. The pain tends to be localized to the affected joint and aggravated by joint use while relieved by rest. The ultimate goal of nonsurgical treatment modalities is to reduce the pain and restore function while delaying total knee replacement (TKR), a substantial direct health-care cost in OA patients with end-stage disease. Researchers have urged into the preventive management of OA and development of disease-modifying OA drugs [9].

HA (hyaluronic acid) is an intrinsic component within the knee joint and provides viscoelastic properties to synovial fluid. Increasing HA levels through intra-articular (IA) injection restores the viscoelasticity of the synovial fluid, which aids shock absorption, lubrication, and protection of the joint, along with a good safety profile [10–12]. Moreover, several retrospective studies have shown the potential of IAHA (intra-articular hyaluronic acid) to delay the time for TKR in patients with OA [13–16].

Shea nut oil (SNO), extracted from the African shea tree (Vitellaria paradoxa), contains a high nutritional value with high triterpene and oleic/stearic fatty acid concentration, vitamins, and minerals. Importantly, their high triterpene alcohol and tocopherol content are considered to have anti-inflammatory and anti-oxidant properties [17]. Clinical evidence suggests the bioactive triterpene concentrate has anti-inflammatory effects under daily oral supplementation in OA patients [18]. Additionally, a clinical report demonstrated a decrease in pain and stiffness in patients after daily oral SNO supplement for 16 weeks [19]. We previously reported that preventive oral administration of SNO dose-dependently reduces cartilage degeneration in a rat model of anterior cruciate ligament transection plus medial meniscectomy (ACLT + MMx)-induced OA [20]. It also reduces pain and provides differential cartilage protection in both acute and chronic OA rats [21].

Regarding the complexity of this natural plant oil, the overall protective effect may be derived from a combined mechanism of actions of the different triterpene concentrates (primary α, β-amyrin, lupeol, and butyrospermol), monounsaturated oleic acid, or tocopherol found in this shea nut oil product. For instance, in mouse inflammation models induced by complete Freund's adjuvant and by partial sciatic nerve ligation, daily oral intake of α and β-amyrin showed long-lasting antinociceptive and anti-inflammatory effects via direct activation of cannabinoid receptors and a concomitant inhibition of inflammatory NF-κB, cyclic adenosine monophosphate response element binding (CREB) pathway [22]. Otuki et al.'s report suggested that the antinociceptive properties of mixed amyrins may be involved in the inhibition of protein kinase A and protein kinase C pathways [23]. Others researchers found that lupeol acetate ameliorates collagen-induced arthritis through suppression of inflammatory cytokines and inhibition of bone erosion [24]. Indeed, the potential antioxidant and free radical scavenging effects of amyrin and lupeol have been demonstrated both in vitro and in vivo [25–28]. A recent report showed increased antioxidant activity and suppressed proinflammatory cytokine expressions in obese OA rats fed with SNO, further consolidating our previous findings [29]. Either of the triterpenoids and their potential target mechanisms may play a key regulatory role in our OA model and contribute to the antiarthritic action of SNO.

Although the protective potential of oral SNO supplementation differs from that of IAHA in many aspects (route, dose frequency, and mechanism of action). But the combination of both therapeutic models, like IA corticosteroid/anti-inflammatory drugs combined with IAHA, can lead to significant improvement of the clinical outcome of either agent alone [30]; this however only provides short duration, and is not feasible for persisting injection. Therefore, we sought to determine the role of oral SNO as an adjuvant in combination with IAHA injections and compare the effectiveness in the OA rat model. In this study, we compared the differences in protective potential between SNO and IAHA in OA rats, and the hypothetical synergetic effect of SNO with IAHA on the prevention of OA progression was also examined.

2. Materials and Methods

2.1. ACLT + MMx-Induced OA Animal Model

A rat model of surgically-induced OA was proceeded as described previously [20,21], and all animal care and experimental protocols complied with institutional and international standards (Principles of Laboratory Animal Care, National Institutes of Health) and were approved (Institutional Animal Care and Use Committee [IACUC] no. 107-030) by the IACUC of Cathay General Hospital (Taipei, Taiwan). Adult male Wistar rats were purchased from BioLASCO Taiwan Co., Ltd. (Yilan, Taiwan) and housed in Cathy Medical Research center with free access to the standard diet and water with a 12-h light/dark cycle at a temperature of 22 ± 2 °C and 55% humidity.

All ACLT + MMx surgeries were performed on the right knee of the rats by a single research specialist. Briefly, male Wistar rats (330–350 g) were anesthetized in an induction chamber using 5% isoflurane and then maintained with 2% isoflurane via a custom-made facemask. The right knee joint skin was shaved, and sterilized with povidone-iodine solution. An incision was made in the medial aspect of the joint capsule; the anterior cruciate ligament was transected using a scalpel, and the medial meniscus was removed completely using a tenotomy scissor. Following surgery, the joint was irrigated with normal saline; the joint capsule was sutured with 4–0 Vicryl, and 4–0 monofilament nylon was used for skin closure. Next, the wound area was sterilized, and cefazolin (100 mg/kg/day) was administered intramuscularly for 3 days to prevent infection. For the sham-operated rats, the same procedure was repeated, but neither ACLT nor removal of the medial meniscus was performed.

2.2. Experimental Design

As shown in Figure 1A, ACLT + MMx ($n = 48$) or sham surgery (Sham-OP, $n = 9$) was performed at week 0. Body weights, widths of the knee joints, and weight-bearing symmetry were measured before the surgery as the baseline. After the surgery, ACLT + MMx rats were assigned as the nontreated control (OA-control, $n = 12$), the group treated with SNO (223.2 mg/kg, OA-SNO, $n = 12$, administered oral SNO daily after the surgery, the effective dose was derived from our SNO dose-dependent animal report [20]), the group treated with HA (50 μL per joint/week, OA-HA, $n = 12$, received IA injection of HA weekly at weeks 2–4 and 9–11) and a group with combined treatment, HA plus SNO (OA-SNOHA, IA injection of HA (50 μL per joint/week) at weeks 2–4 and 9–11 plus the daily oral SNO (223.2 mg/kg) beginning from the 2nd week). The SNO concentrate provided by Universal Integrated Corp. (Taipei, Taiwan) was administered by oral gavage with the aid of isoflurane anesthesia, and the HA (Seikagaku Corporation, Ibaraki, Japan) was injected into the OA knee joint with a 25 G-needle syringe. As an injection control, 50 μL of saline solution was injected to OA-control ($n = 6$) at week 2–4 (3 weekly injections), and it showed no statistical difference in both knee width and weight-bearing test from non-injected OA-control rats ($n = 6$) (Figure 1B). The 4th week and 12th week knee joint section confirmed the progressive deterioration of OA joint with a reactive chondrocytes hypertrophy, and increasing cartilage erosion accompanied by chondrocytes loss (Figure 1C).

2.3. Knee Width and Weight-Bearing Test

The width of the knee joint was measured using a steel caliper (resolution 0.01 mm, E-Base Measuring Tools Co., Taipei, Taiwan) biweekly after the surgery, and the width of the contralateral knee served as the naïve control. The data are expressed as the Δ knee width (mm); the value was derived from the OA rats (knee width difference of the operated knee and naïve knee) minus the mean value of the sham-OP rats (knee width difference of the operated knee and naïve knee) and was determined as the actual joint swelling induced by ACLT + MMx.

Hind paw static weight-bearing was measured using an incapacitance tester (Linton Instrumentation, Norfolk, UK) to detect OA-induced changes in postural equilibrium every two weeks. The rats were placed on their hind paws in a box containing an inclined plane (65° from horizontal) that was placed above the apparatus. After a brief accommodation period, the weight that the animals applied to each hind limb was measured independently by the apparatus. Five measurements were taken and averaged for each rat. The data are expressed as the difference between the weight applied to the naïve hind limb and the weight applied to the operated hind limb (Δ Force, g); the change in the weight distribution between the naïve and operated hind limb represents the OA pain of the rats [31,32].

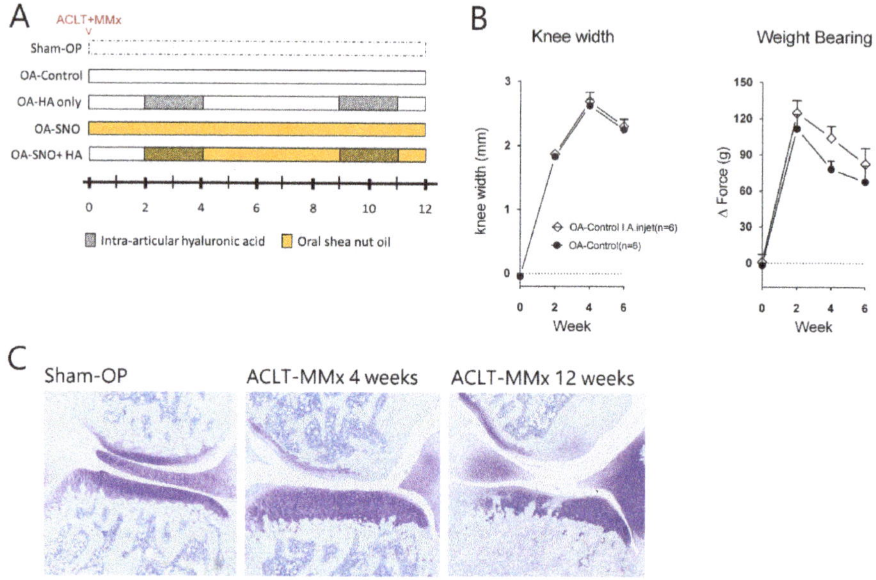

Figure 1. (**A**) The graphic scheme of study. (**B**) The knee width/weight-bearing changes of OA (osteoarthritis) rats with 3 IA (intra-articular) injection (weekly) of 50 μL of saline, compared to non-injected control. (**C**) The representative section of histological change of ACLT + MMx (anterior cruciate ligament transection plus medial meniscectomy)-induced OA from 4 weeks to 12 weeks post-surgery as compared to sham-OP (sham operated) rats.

2.4. Histopathological Examination of Knee Joint

All rats were sacrificed via exsanguination under deep anesthesia on the 12th week post-surgery. The OA knee joints were removed and fixed in 10% formalin for 2 days, followed by a decalcifying solution based on EDTA disodium (12.5%, pH 7.0) for 4 weeks. After decalcification, the joints were embedded in paraffin blocks, and histological coronal sections (5 μm-thick serial section, slides interval: 200 μm) were obtained. Toluidine blue/fast green staining was used to examine morphological changes and the stained sections were digitalized using a Slide Scanner ZEISS Axio Scan Z1 image system (Jena, Germany) and ZEN lite 2.6 (blue edition). The severity of articular cartilage damage on medial tibial plateau was evaluated using the modified Osteoarthritis Research Society International (OARSI) scoring system [33]. The cartilage matrix loss width, tibia cartilage degeneration score, total and significant cartilage degeneration widths, and zonal depth ratio of the lesions and synovial reaction were evaluated.

2.5. Metabolic Profile of Blood Biochemistry Assays

The OA-SNO and OA-control rats fasted for 12 h before the blood sample withdrawal; blood samples were taken from the rat tail vein every 4 weeks post-ACLT + MMx surgery. The blood samples were centrifuged (8000× *g* for 5 min) to separate sera and stored in a −80 °C freezer prior to analysis. The serological levels of uric acid, glucose, total cholesterol (T-CHO), high-density lipoprotein (HDL), and triglyceride (TG) were measured using the FUJI DRI-CHEM 4000i analyzer (FUJIFILM Corporation, Tokyo, Japan) at the Taiwan Mouse Clinic (Academia Sinica, Taipei, Taiwan).

2.6. Statistical Analysis

The data are expressed as the mean ± SEM. All graphical representations and statistical calculations were aided by GraphPad Prism version 6.01 and Microsoft Excel. The Shapiro–Wilk test was used to check the normal distribution of data. Two-way ANOVA, Tukey's multiple comparisons test, Sidak's multiple comparisons test, and Student's t-test were used to analyze the data.

3. Results

3.1. SNO, but Not HA, Attenuates Knee Joint Pain and Inflammatory Swelling in ACLT + MMx-Induced OA Rats

Immediately after ACLT + MMx surgery, rats were assigned to different treatment groups as described in Figure 1A. Following the surgery, in the OA-control group, we found that ACLT + MMx induced a constant and gradually increasing knee width as a result of progressive knee joint inflammation. In contrast, a preventive and long-term oral administration of SNO reduced the OA-induced knee joint swelling at 2 weeks post-treatment and eventually achieved a difference of approximately 27.5% (OA-SNO: 1.89 ± 0.06 vs. OA-control: 2.61 ± 0.05 mm) at week 12. However, such knee reduction was not evident in the HA-treated group (OA-HA: 2.48 ± 0.07 mm vs. OA-control: 2.61 ± 0.05 mm) (Figure 2A). With an oral SNO (daily) added to the HA treatment as an adjuvant supplement since the 2nd week, we started to observe the reduction in knee width from 4th to 12th weeks post-surgery in OA-SNOHA rats, (Figure 2B). At the end, we found a 20.6% reduction in the knee width compared to that of the OA-control group (OA-SNOHA: 2.07 ± 0.10 vs. OA-control: 2.61 ± 0.05 mm).

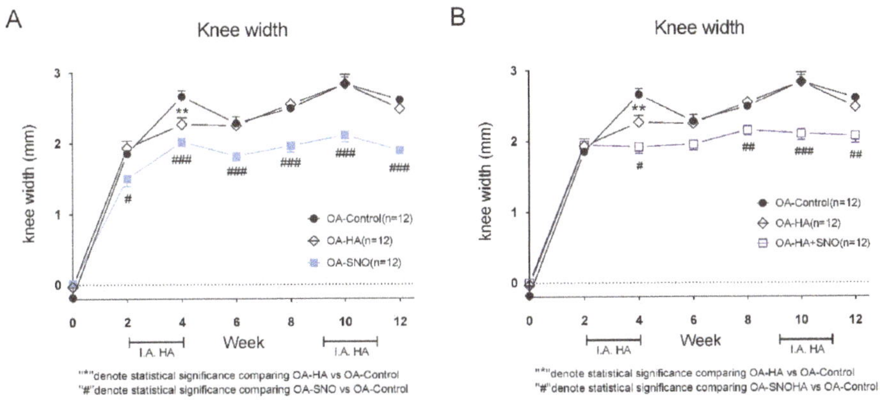

Figure 2. Joint swelling measurement of the OA rats. ACLT + MMx was performed to induce knee OA at week 0, and different treatments were given according to the experimental design. (**A**) Biweekly knee width measurements of the OA-control, OA-HA (hyaluronic acid), and OA-SNO (shea nut oil) rats. (**B**) Biweekly knee width measurements of the OA-control, OA-HA and OA-SNOHA (hyaluronic acid + shea nut oil) rats. The data are presented as the Δ knee width (mm), and the values are expressed as the mean ± SEM. Two-way ANOVA and Sidaks's multiple comparisons test were used to analyze the data. # $p < 0.05$, ##/** $p < 0.01$, ### $p < 0.001$.

We used the weight-bearing test to assess pain behavior during OA progression (Figure 3A). Similar to our previous report, we found that ACLT + MMx-induced OA elicited a constant change in weight-bearing asymmetry compared to that of the sham-OP group, which only presented acute pain in the first few weeks as a result of the surgical procedure. The preventive and daily supplementation of SNO reduced the pain behavior up to 87.9% compared to that of the untreated OA-control group (OA-SNO: 5.22 ± 2.37 g vs. OA-control: 43.21 ± 6.93 g). However, HA alone did not yield any significant reduction in pain during our biweekly weight-bearing measurements; at the end, HA monotherapy yielded only a 25.3% reduction in treated rats compared to OA-control rats (OA-HA: 32.33 ± 5.90 g vs. OA-control: 43.21 ± 6.93 g). When the two treatments were given together since the 2nd week (Figure 3B), the OA-SNOHA rats showed a maximum reduction in pain (71.6%) at the 10th week as compared to O-control (OA-SNOHA: 15.51 ± 6.13 g vs. OA-control: 54.58 ± 7.90 g), and a 75.9% reduction when compared to the rats treated with OA-HA alone (OA-SNOHA: 15.51 ± 6.13 g vs. OA-HA: 64.37 ± 8.67 g).

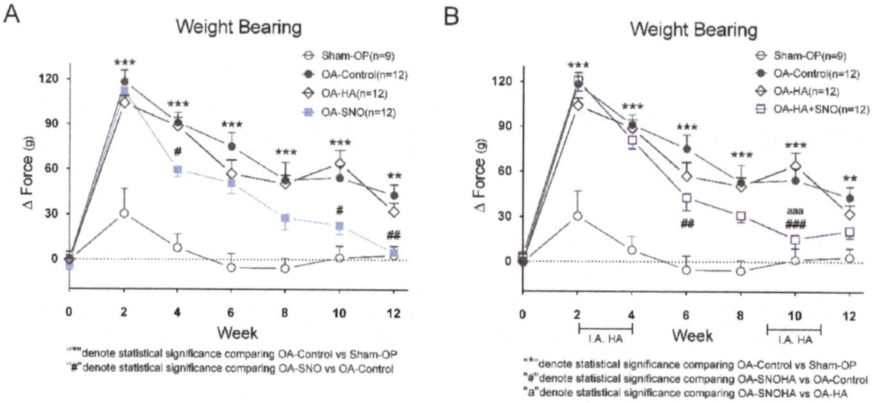

Figure 3. Weight bearing test of the OA rats. ACLT + MMx was performed to induce knee OA at week 0, and different treatments were given according to the experimental design. (**A**) Biweekly weight-bearing measurements of the sham-OP, OA-control, OA-HA, and OA-SNO rats. (**B**) Biweekly weight-bearing measurements of the sham-OP, OA-control, OA-HA, and OA-SNOHA rats. The data are presented as the Δ Force (g) and expressed as the mean ± SEM and two-way ANOVA, and Sidaks's multiple comparisons test were used to analyze the data. # $p < 0.05$, ##/** $p < 0.01$, aaa/###/*** $p < 0.001$.

Next, we evaluated the synovial reaction as a microscopic sign of internal inflammation of the joint. In the sham-operated joint, we found a single layer of synovial lining cells without proliferation of the subsynovial tissue or vascular changes (Figure 4, sham-OP). As a result of the ACLT + MMx, the OA-control joint showed an increased thickness of both synovial lining cells, and the subsynovial tissue contained extensive extracellular matrix. Along with hypervascularity of the subsynovial area, this result suggested a chronic synovial reaction as result of surgically-induced OA (Figure 4, OA-control). Upon IAHA treatment (Figure 4, OA-HA), the histological findings showed a relative reduction in the synovial reaction compared to that of the OA-control joint, which suggested attenuation of OA with the lubricating action of IAHA. Moreover, with the addition of SNO supplementation to IAHA (Figure 4, OA-SNOHA) or oral SNO alone (Figure 4, OA-SNO), the two modalities of treatment offered better anti-inflammatory protection, as shown by the amelioration of the synovial reaction with less synovial proliferation and subsynovial thickness.

3.2. *SNO, HA Alone, and HA Plus SNO, Offered Significant Improvement in Cartilage Integrity in Knee OA Rats*

The cartilage deterioration was evaluated at 12th week post-surgery using the OARSI score system as shown in Table 1 and Figure 5. It is clearly noted that ACLT + MMx (OA-control) caused significant extensive matrix loss and a deformed cartilage surface, while the sham-OP cartilage showed a thin and smooth surface and preserved cartilage integrity. The quantitative data of the OA-HA, OA-SNO, and OA-SNOHA rats, showed a significant attenuation of cartilage matrix loss (specially in surface 0% and mild-depth 50% level) compared to that of the OA-control rats.

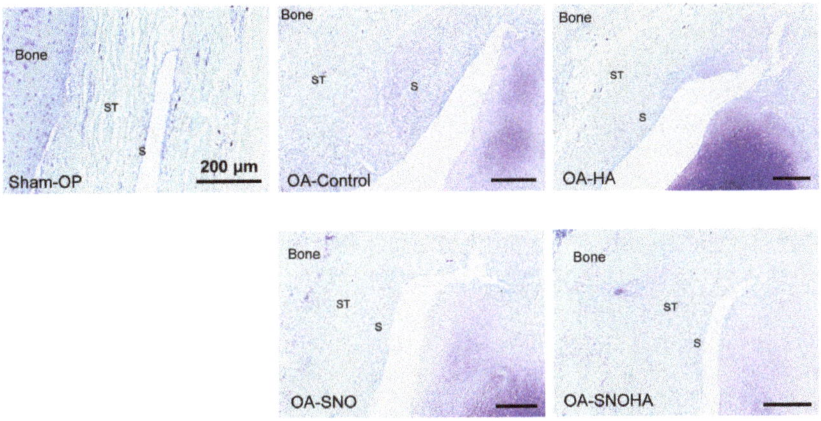

Figure 4. Synovium reaction of the OA joint. Representative sections of the medial femoral condyle joint capsule of the knee joints from the sham-OP, OA-control, OA-HA, OA-SNO, and OA-SNOHA rats at week 12 post-surgery were shown. ST = subsynovial tissue; S = synovial lining cells. Scale bar = 200 µm.

Table 1. Histological scoring of OA knee joint.

	Sham-OP (n = 7)	OA-Control (n = 12)	OA-HA (n = 12)	OA-SNO (n = 12)	OA-SNOHA (n = 12)
Cartilage matrix loss 0% (mm)	0 ***	2.292 ± 0.074	**1.951 ± 0.087** **	**1.864 ± 0.107** **	**1.865 ± 0.060** ***
Cartilage matrix loss 50% (mm)	0 ***	0.792 ± 0.144	**0.386 ± 0.102** *	**0.342 ± 0.103** *	**0.272 ± 0.116** **
Cartilage matrix loss 100% (mm)	0 **	0.372 ± 0.093	0.183 ± 0.076	0.217 ± 0.075	0.160 ± 0.061
Medial Tibia Cartilage Degeneration Score	0 ***	6.208 ± 0.408	**5.000 ± 0.371** *	5.750 ± 0.439	**4.791 ± 0.307** **
Outside zone	0 ***	1.625 ± 0.334	1.458 ± 0.289	1.875 ± 0.326	1.375 ± 0.287
Middle zone	0 ***	1.875 ± 0.163	1.708 ± 0.185	2.042 ± 0.153	1.833 ± 0.130
Inside zone	0 ***	2.708 ± 0.195	**1.833 ± 0.214** **	**1.833 ± 0.177** **	**1.583 ± 0.133** ***
Total cartilage degeneration width (mm)	0 ***	2.475 ± 0.073	**2.213 ± 0.070** *	2.335 ± 0.073	**2.215 ± 0.058** **
Significant cartilage degeneration width (mm)	0 ***	0.947 ± 0.111	**0.612 ± 0.110** *	**0.528 ± 0.127** *	**0.368 ± 0.117** ***
Zonal depth ratio of lesions	0 ***	0.469 ± 0.039	**0.358 ± 0.029** *	**0.331 ± 0.024** **	**0.313 ± 0.020** ***

All operated knee joints were collected at 12 weeks post-surgery and were processed with toluidine/fast green staining for evaluation using OARSI's parameters. The bold text shows the value with statistical significance. Asterisk denote the statistical examination of each group in comparison with OA-control using Student's t-test. * $p < 0.05$, ** $p < 0.01$, *** $p < 0.001$

The cartilage degeneration score showed that ACLT + MMx induced a total lesion score of 6.208 ± 0.408, with the inside zone (2.708 ± 0.195) being the most affected region. The treatment with IAHA injection or combined SNO showed a significantly lower total degeneration score (5.000 ± 0.371 and 4.791 ± 0.307) than the OA-control (6.208 ± 0.408). In addition, we found that those rescues were observed primary in the inside zone, where all treatments (HA, SNO, and SNOHA) showed significant protective effects (1.833 ± 0.214, 1.833 ± 0.177, and 1.583 ± 0.133) compared to OA-control (2.708 ± 0.195).

Next, we quantified the extension (measured in width, mm) of the cartilage degeneration. The parameters are further subclassed as total (any type of degenerative change) or significant (seriously compromised, 50% of chondrocytes are absent or necrotic) and as the zonal depth ratio. Both OA-HA and combined treatment OA-SNOHA led to a significant attenuation of the 3 parameters, while OA-SNO more specifically decreased the significant cartilage degeneration width and zonal depth ratio. In summary, the oral SNO treatment alone significantly improved 5 of the 10 parameters we measured, while both IAHA and SNOHA treatment yielded significant ameliorations of 7 out of 10 parameters.

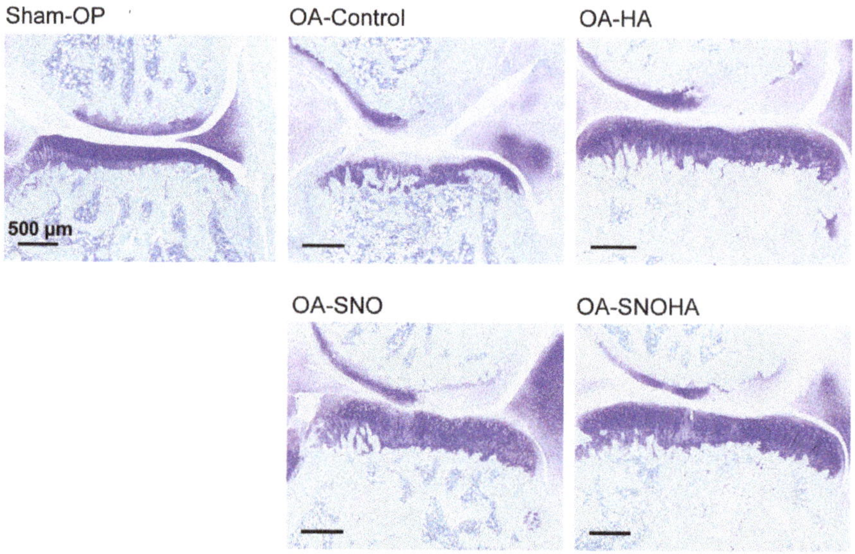

Figure 5. Representative section of medial tibial cartilage from each treatment (OA-HA, OA-SNO, and OA-SNOHA) and control (sham-OP, OA-control) were shown. Scale bar = 500 µm.

3.3. *Metabolic Profiling of OA Rats Receiving Daily Oral Supplementation of SNO*

Firstly, we found no difference of body weight between sham-OP and OA-control, but a significantly, but minor increase of body weight was detected in the OA-HA group at the 12th week (Figure 1A). On the other hand, we found gradual and minor reduction of body weight in OA rats supplemented with oral SNO, including those receiving IAHA at the same time (Figure 1B). We previously demonstrated that long-term oral SNO supplementation in OA rats decreased body weight and blood TG level without altering the blood aspartate transaminase (AST), blood urea nitrogen (BUN), and cholesterol levels [21]. Here, we compared the full metabolic profile at the 4th, 8th, and 12th weeks. Similarly, we found no alteration of uric acid, total cholesterol, and HDL at any of the time-points (Figure 6C–E). Consistent with our previous report, a significant reduction in TG levels was observed as early as the 4th week (118.8 mg/dL vs. 98.08 mg/dL) and remained reduced until the 12th week (126.7 mg/dL vs. 88.25 mg/dL) (Figure 6C–E).

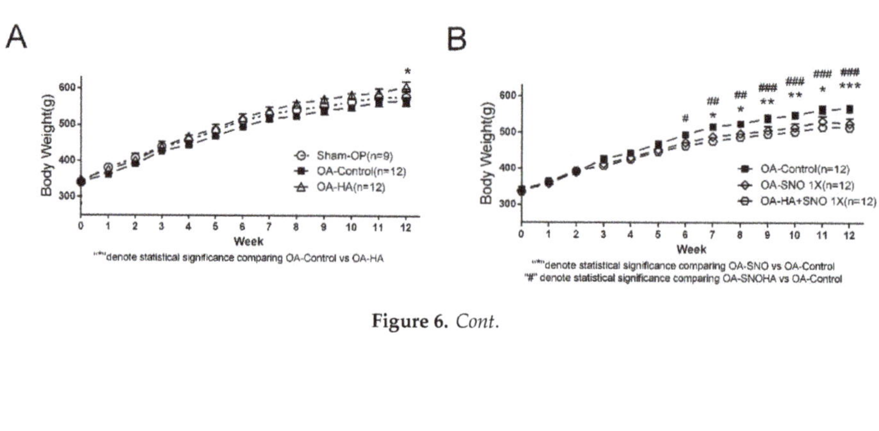

Figure 6. *Cont.*

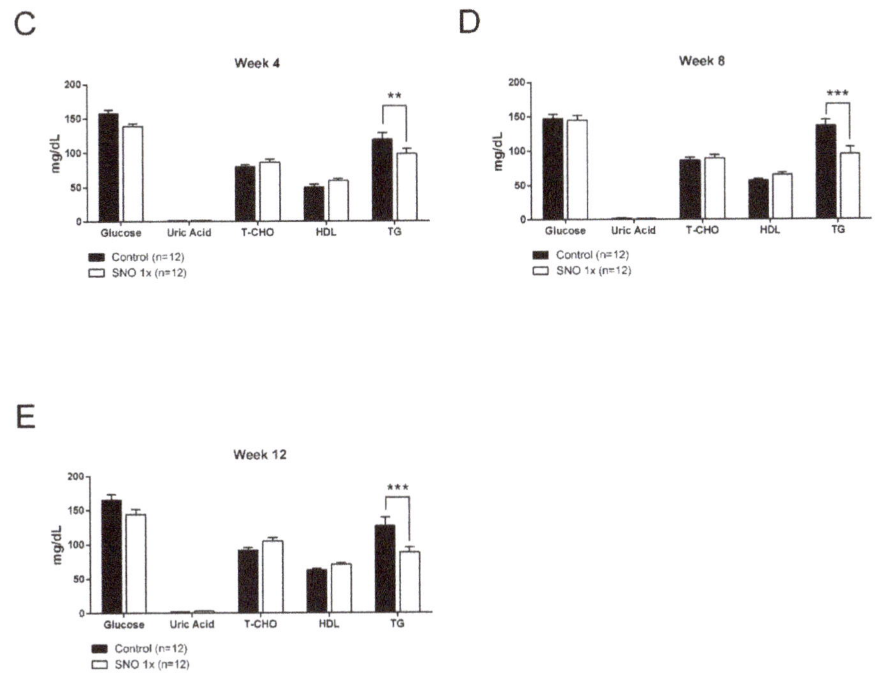

Figure 6. Body weight changes OA-control vs. OA-HA (**A**), or OA-control vs. OA-SNO and SNOHA (**B**). Metabolic profile of OA-control vs. OA-SNO rats (glucose, uric acid, total cholesterol, high-density lipoprotein (HDL), and triglyceride (TG)) at the 4th week (**C**), 8th week (**D**), and 12th week (**E**) after ACLT + MMx surgery. Values are expressed as the mean ± SEM, and two-way ANOVA and Tukey/Sidaks's multiple comparisons test were used to analyze the data. #/* $p < 0.05$, ##/** $p < 0.01$, ##/*** $p < 0.001$.

4. Discussion

In the current study, we found that both daily oral SNO supplementation and 6 weekly doses of IAHA alone are sufficient to attenuate post-traumatic OA-induced cartilage deterioration. In terms of the knee joint swelling and pain assessment, we found a marked difference between the two modalities of treatment, in which the long-term daily oral SNO supplement resolved better reduction in inflammatory signs/symptoms of the knee as well as the synovial reaction in the joint cavity than IAHA alone. The combination of both treatments demonstrated an additive effect, the SNO + IAHA OA rats showed the best histological scores, and at 10 weeks of oral SNO supplement provided additional anti-inflammatory and antinociceptive effects on visco-lubricative IAHA chondro-protection. Moreover, long-term oral SNO supplementation caused no alteration in metabolic profiles, such as serological uric acid, total cholesterol, and HDL levels, compared to those of control OA rats. Surprisingly, there was a minor reduction in fasting blood glucose and significantly reduced TG levels in the SNO-treated rats.

It is worth noting, a reduction of body weight was also observed in SNO and SNOHA treated rats. This similar finding was reported recently in obese rats [29]. The anti-obesity effect of SNO could be the result of reduction of OA pain, or vice versa. Given the fact that the clinical and pathogenic correlation between OA and metabolic disorder has been extensively reviewed [34], thus the management of body weight is strongly recommended by OARSI, American Academy of Orthopaedic Surgeons (AAOS), and the American College of Rheumatology [35–37]. Clinical studies on weight loss and preclinical studies targeting metabolic abnormalities in OA are an area of research interest and have achieved important improvement in OA progression [38]. In rat model of type 2 diabetes mellitus,

Onur et al. demonstrated the metabolic disease itself contributes to the onset and progression of knee osteoarthritis [39]. Mooney et al. showed that surgically-induced OA mice fed a high-fat diet presented not only higher fasting glucose levels and body weights compared to those of lean OA mice, but also had worse OARSI histological scores and less cartilage thickness [40]. Moreover, a recent report also demonstrated that cartilage deterioration was sustained even after the high-fat diet was withdrawn from the OA mice, and the blood glucose and body weight were restored to the levels in normal diet mice [41]. These findings suggested that an increased weight load is not the sole cause of the severity of OA progression; instead, the lipid/glucose metabolic pathways could also jeopardize cartilage integrity and synthesis.

Moreover, an emerging concept of gut–joint axis has associated the gut dysbiosis (perturbation of gut microbiota (GM) biodiversity and function), and the leaky gut syndrome with the joint disease progression [42,43]. In both human and rodent model, an increase in serum level of the pro-inflammatory marker and bacterial metabolites were associated with OA severity [44,45]. This chronic low-grade inflammation as result of dysbiosis explains a new OA phenotype, indicated as the metabolic OA [46]. In fact, long-term diet or prebiotic supplements have shown to shift GM colony with the improvement of cartilage integrity [47]. The diary ingestion of high triterpenes SNO could have potential modulation on the GM diversity, which may associate with both metabolic change and cartilage protection. Future evaluation of GM colonies under long-term oral SNO remains to be explored.

On the other hand, in vitro evidence shows that a mixture of triterpenes (α, β-amyrin) significantly reduced lipid droplet formation via suppression of PPARγ and C/EBPα expression, while enhancing the translocation of glucose transporter GLUT4 onto the plasma membrane of 3T3-L1 cells [48]. Furthermore, reduction in blood glucose, total cholesterol, and TG levels were also observed in streptozotocin-induced diabetic mice treated with an α, β-amyrin mixture [49]. Moreover, triterpene as lupeol was also found to have a hypolipidemic effect (decreased total cholesterol, TG, and phospholipids) in rats fed a high cholesterol diet [50]. The effect of SNO (with a high concentrate of triterpenes) on lipid metabolism may be connected to the molecular mechanism of its chondro-protective effect.

As previously shown, the predominant fatty acids in femoral head cartilage are palmitic (16:0), oleic (18:1), and linoleic (18:2) acids [51]. In the animal model of OA, those fatty acids were significantly reduced in mice after destabilization of the medial meniscus [52]. Nonetheless, oleic acid exposure downregulates the expression of MMP-1 and COX-2 in TNF-α stimulated human chondrocytes culture while linoleic acid increased PGE2 production [53]. These results suggest that local fatty acid concentrations could be results of OA and also contribute to OA progression. The high proportion of oleic acid of SNO could be one of the chondro-protection factors.

The interval and multiple doses of IAHA showed a significant chondro-protective effect in the treated group compared with the OA-control in our ACLT + MMx injured OA rats. Surprisingly, we observed neither an antinociceptive effect nor a reduction in inflammatory signs, as demonstrated by the weight-bearing asymmetry and knee swelling tests. In fact, several animal studies on OA pain reported similar findings on the HA effect. Ikeuchi et al. employed a monoiodoacetate (MIA)-induced OA pain model and found no significant difference in weight-bearing asymmetry in HA-treated rats [54]. Boettger et al. also demonstrated in a rat bradykinin/PGE2 pain model that HA lost its antinociceptive efficacy (shown as weight-bearing asymmetry) from day 7 after injection [55]. Recently, IAHA was found unable to reduce ankle swelling in MIA-induced ankle OA [56], which is similar to the knee width exam in our ACLT + MMx OA model. However, the limited time of follow-up and the small sample size are the two major limitations of this study. An evaluation with an extended follow-up observation will further elucidate the long-term effect of IAHA alone or in combination with oral SNO for the treatment of chronic osteoarthritis.

5. Conclusions

Although pharmacological treatments of OA are rapid and effective for symptomatic relief in regular clinical practice, the long-term use is restricted by the associated adverse effects. IAHA has been conditionally recommended for long-term treatment of knee OA with a favorable safety profile over repeated IA corticosteroid. At and beyond 12 weeks of treatment, it may have the beneficial effects on pain [57]. Nutraceuticals are safe candidates for long-term supplementation to provide persistent effect as treatment adjuvant [58]. In fact, many nutraceutical products have been extensively used for OA pain and their active compounds were identified for potential drug development [59]. In conclusion, the evidence and safety profiles observed in these surgically-induced OA rats suggest that long-term oral SNO supplement can be used as an effective adjuvant for IAHA treatment to enhance the symptomatic relief and delay the disease progression in clinical practice.

Author Contributions: (1) I.-J.C., contributes to formal analysis, investigation, methodology, and writing—original draft. (2) C.-S.W., contributes to conceptualization, funding acquisition, supervision, writing—review & editing and approval of the submitted and final versions. All authors have read and agreed to the published version of the manuscript.

Funding: This study was supported in the form of a contract from Universal Integrated Corp. to evaluate the effect of the product SheaFlex75 on our well-established OA animal model. The Article Processing Charge was funded by Cathy General Hospital.

Acknowledgments: We thank our laboratory assistants Ching-Pai Wang and Tsou-Mei Liu for performing oral gavage, the knee width and weight bearing measurements. Thanks to the Taiwan Animal Consortium for technical service in tissue embedding/sectioning and the biochemical analysis, and the Instrument Center of National Defense Medical Center for providing the slide scanner service.

Conflicts of Interest: The authors declare no conflict of interest. The sponsors had no role in the design, execution, interpretation, or writing of the study.

References

1. Litwic, A.; Edwards, M.H.; Dennison, E.M.; Cooper, C. Epidemiology and Burden of Osteoarthritis. *Br. Med. Bull.* **2013**, *105*, 185–199. [CrossRef] [PubMed]
2. Neogi, T. The Epidemiology and Impact of Pain in Osteoarthritis. *Osteoarthr. Cartil.* **2013**, *21*, 1145–1153. [CrossRef] [PubMed]
3. Dominick, K.L.; Ahern, F.M.; Gold, C.H.; Heller, D.A. Health-Related Quality of Life and Health Service Use among Older Adults with Osteoarthritis. *Arthritis Rheum.* **2004**, *51*, 326–331. [CrossRef] [PubMed]
4. Hunter, D.J.; Riordan, E.A. The Impact of Arthritis on Pain and Quality of Life: An Australian Survey. *Int. J. Rheum. Dis.* **2014**, *17*, 149–155. [CrossRef]
5. Barbour, K.E.; Helmick, C.G.; Boring, M.; Brady, T.J. Vital Signs: Prevalence of Doctor-Diagnosed Arthritis and Arthritis-Attributable Activity Limitation - United States, 2013–2015. *MMWR Morb. Mortal. Wkly. Rep.* **2017**, *66*, 246–253. [CrossRef]
6. Felson, D.T.; Zhang, Y. An Update on the Epidemiology of Knee and Hip Osteoarthritis with a View to Prevention. *Arthritis Rheum.* **1998**, *41*, 1343–1355. [CrossRef]
7. Roos, E.M.; Arden, N.K. Strategies for the Prevention of Knee Osteoarthritis. *Nat. Rev. Rheumatol.* **2016**, *12*, 92–101. [CrossRef]
8. O'Neill, T.W.; Felson, D.T. Mechanisms of Osteoarthritis (Oa) Pain. *Curr. Osteoporos. Rep.* **2018**, *16*, 611–616. [CrossRef]
9. Hunter, D.J.; Schofield, D.; Callander, E. The Individual and Socioeconomic Impact of Osteoarthritis. *Nat. Rev. Rheumatol.* **2014**, *10*, 437–441. [CrossRef]
10. Wehling, P.; Evans, C.; Wehling, J.; Maixner, W. Effectiveness of Intra-Articular Therapies in Osteoarthritis: A Literature Review. *Ther. Adv. Musculoskelet Dis.* **2017**, *9*, 183–196. [CrossRef]
11. Bannuru, R.R.; Osani, M.; Vaysbrot, E.E.; McAlindon, T.E. Comparative Safety Profile of Hyaluronic Acid Products for Knee Osteoarthritis: A Systematic Review and Network Meta-Analysis. *Osteoarthr. Cartil.* **2016**, *24*, 2022–2041. [CrossRef]

12. Busse, P.; Vater, C.; Stiehler, M.; Nowotny, J.; Kasten, P.; Bretschneider, H.; Goodman, S.B.; Gelinsky, M.; Zwingenberger, S. Cytotoxicity of Drugs Injected into Joints in Orthopaedics. *Bone Joint Res.* **2019**, *8*, 41–48. [CrossRef]
13. Altman, R.; Fredericson, M.; Bhattacharyya, S.K.; Bisson, B.; Abbott, T.; Yadalam, S.; Kim, M. Association between Hyaluronic Acid Injections and Time-to-Total Knee Replacement Surgery. *J. Knee Surg.* **2016**, *29*, 564–570. [CrossRef]
14. Delbarre, A.; Amor, B.; Bardoulat, I.; Tetafort, A.; Pelletier-Fleury, N. Do Intra-Articular Hyaluronic Acid Injections Delay Total Knee Replacement in Patients with Osteoarthritis - a Cox Model Analysis. *PLoS ONE* **2017**, *12*, e0187227. [CrossRef]
15. Maheu, E.; Rannou, F.; Reginster, J.Y. Efficacy and Safety of Hyaluronic Acid in the Management of Osteoarthritis: Evidence from Real-Life Setting Trials and Surveys. *Semin. Arthritis Rheum.* **2016**, *45* (Suppl. 4), S28–S33. [CrossRef]
16. Dasa, V.; Lim, S.; Heeckt, P. Real-World Evidence for Safety and Effectiveness of Repeated Courses of Hyaluronic Acid Injections on the Time to Knee Replacement Surgery. *Am. J. Orthop. (Belle Mead NJ)* **2018**, *47*. [CrossRef] [PubMed]
17. Honfo, F.G.; Akissoe, N.; Linnemann, A.R.; Soumanou, M.; Van Boekel, M.A. Nutritional Composition of Shea Products and Chemical Properties of Shea Butter: A Review. *Crit. Rev. Food Sci. Nutr.* **2014**, *54*, 673–686. [CrossRef] [PubMed]
18. Cheras, P.A.; Myers, S.P.; Paul-Brent, P.A.; Outerbridge, K.H.; Nielsen, G.V. Randomized Double-Blind Placebo-Controlled Trial on the Potential Modes of Action of Sheaflex70 in Osteoarthritis. *Phytother. Res.* **2010**, *24*, 1126–1131. [PubMed]
19. Chen, S.P.; Lo, S.F.; Wang, Y.C.; Chou, T.Y.; Chang, K.M.; Chou, L.W. Validating Efficacy of Shea Nut Oil Extract in Knee Osteoarthritis Patients. *Evid. Based Complem. Altern. Med.* **2013**, *2013*, 147163. [CrossRef]
20. Kao, J.H.; Lin, S.H.; Lai, C.F.; Lin, Y.C.; Kong, Z.L.; Wong, C.S. Shea Nut Oil Triterpene Concentrate Attenuates Knee Osteoarthritis Development in Rats: Evidence from Knee Joint Histology. *PLoS ONE* **2016**, *11*, e0162022. [CrossRef]
21. Chen, I.J.; Lin, S.H.; Wong, C.S. Oral Shea Nut Oil Triterpene Concentrate Supplement Ameliorates Pain and Histological Assessment of Articular Cartilage Deterioration in an Aclt Injured Rat Knee Osteoarthritis Model. *PLoS ONE* **2019**, *14*, e0215812. [CrossRef] [PubMed]
22. Da Silva, K.A.; Paszcuk, A.F.; Passos, G.F.; Silva, E.S.; Bento, A.F.; Meotti, F.C.; Calixto, J.B. Activation of Cannabinoid Receptors by the Pentacyclic Triterpene Alpha,Beta-Amyrin Inhibits Inflammatory and Neuropathic Persistent Pain in Mice. *Pain* **2011**, *152*, 1872–1887. [CrossRef] [PubMed]
23. Otuki, M.F.; Ferreira, J.; Lima, F.V.; Meyre-Silva, C.; Malheiros, A.; Muller, L.A.; Cani, G.S.; Santos, A.R.; Yunes, R.A.; Calixto, J.B. Antinociceptive Properties of Mixture of Alpha-Amyrin and Beta-Amyrin Triterpenes: Evidence for Participation of Protein Kinase C and Protein Kinase a Pathways. *J. Pharmacol. Exp. Ther.* **2005**, *313*, 310–318. [CrossRef] [PubMed]
24. Wang, W.H.; Chuang, H.Y.; Chen, C.H.; Chen, W.K.; Hwang, J.J. Lupeol Acetate Ameliorates Collagen-Induced Arthritis and Osteoclastogenesis of Mice through Improvement of Microenvironment. *Biomed. Pharmacother.* **2016**, *79*, 231–240. [CrossRef]
25. Gupta, R.; Sharma, A.K.; Sharma, M.C.; Dobhal, M.P.; Gupta, R.S. Evaluation of Antidiabetic and Antioxidant Potential of Lupeol in Experimental Hyperglycaemia. *Nat. Prod. Res.* **2012**, *26*, 1125–1129. [CrossRef]
26. Singh, D.; Arya, P.V.; Sharma, A.; Dobhal, M.P.; Gupta, R.S. Modulatory Potential of Alpha-Amyrin against Hepatic Oxidative Stress through Antioxidant Status in Wistar Albino Rats. *J. Ethnopharmacol.* **2015**, *161*, 186–193. [CrossRef]
27. Cardoso, B.K.; de Oliveira, H.L.M.; Melo, U.Z.; Cfaa Campo Fernandez, C.M.M.; Goncalves, J.E.; Laverde, A., Jr.; Romagnolo, M.B.; Linde, G.A.; Gazim, Z.C. Antioxidant Activity of Alpha and Beta-Amyrin Isolated from Myrcianthes Pungens Leaves. *Nat. Prod. Res.* **2018**, 1–5. [CrossRef]
28. Parvez, M.K.; Alam, P.; Arbab, A.H.; Al-Dosari, M.S.; Alhowiriny, T.A.; Alqasoumi, S.I. Analysis of Antioxidative and Antiviral Biomarkers Beta-Amyrin, Beta-Sitosterol, Lupeol, Ursolic Acid in Guiera Senegalensis Leaves Extract by Validated Hptlc Methods. *Saudi Pharm. J.* **2018**, *26*, 685–693. [CrossRef]
29. Sudirman, S.; Chen, C.K.; Long, B.T.; Chang, H.W.; Tsou, D.; Kong, Z.L. Vitellaria Paradoxa Nut Triterpene-Rich Extract Ameliorates Symptoms of Inflammation on Post-Traumatic Osteoarthritis in Obese Rats. *J. Pain Res.* **2020**, *2020*, 261–271. [CrossRef]

30. Euppayo, T.; Punyapornwithaya, V.; Chomdej, S.; Ongchai, S.; Nganvongpanit, K. Effects of Hyaluronic Acid Combined with Anti-Inflammatory Drugs Compared with Hyaluronic Acid Alone, in Clinical Trials and Experiments in Osteoarthritis: A Systematic Review and Meta-Analysis. *BMC Musculosk. Disord.* **2017**, *18*, 387. [CrossRef]
31. Bove, S.E.; Calcaterra, S.L.; Brooker, R.M.; Huber, C.M.; Guzman, R.E.; Juneau, P.L.; Schrier, D.J.; Kilgore, K.S. Weight Bearing as a Measure of Disease Progression and Efficacy of Anti-Inflammatory Compounds in a Model of Monosodium Iodoacetate-Induced Osteoarthritis. *Osteoarthr. Cartil.* **2003**, *11*, 821–830. [CrossRef]
32. Fernihough, J.; Gentry, C.; Malcangio, M.; Fox, A.; Rediske, J.; Pellas, T.; Kidd, B.; Bevan, S.; Winter, J. Pain Related Behaviour in Two Models of Osteoarthritis in the Rat Knee. *Pain* **2004**, *112*, 83–93. [CrossRef] [PubMed]
33. Gerwin, N.; Bendele, A.M.; Glasson, S.; Carlson, C.S. The Oarsi Histopathology Initiative - Recommendations for Histological Assessments of Osteoarthritis in the Rat. *Osteoarthr. Cartil.* **2010**, *18* (Suppl. 3), S24–S34. [CrossRef] [PubMed]
34. Mobasheri, A.; Rayman, M.P.; Gualillo, O.; Sellam, J.; van der Kraan, P.; Fearon, U. The Role of Metabolism in the Pathogenesis of Osteoarthritis. *Nat. Rev. Rheumatol.* **2017**, *13*, 302–311. [CrossRef] [PubMed]
35. McAlindon, T.E.; Bannuru, R.R.; Sullivan, M.C.; Arden, N.K.; Berenbaum, F.; Bierma-Zeinstra, S.M.; Hawker, G.A.; Henrotin, Y.; Hunter, D.J.; Kawaguchi, H.; et al. Oarsi Guidelines for the Non-Surgical Management of Knee Osteoarthritis. *Osteoarthr. Cartil.* **2014**, *22*, 363–388. [CrossRef]
36. Hochberg, M.C.; Altman, R.D.; April, K.T.; Benkhalti, M.; Guyatt, G.; McGowan, J.; Towheed, T.; Welch, V.; Wells, G.; Tugwell, P. American College of Rheumatology 2012 Recommendations for the Use of Nonpharmacologic and Pharmacologic Therapies in Osteoarthritis of the Hand, Hip, and Knee. *Arthritis Care Res.* **2012**, *64*, 465–474. [CrossRef]
37. Jevsevar, D.S. Treatment of Osteoarthritis of the Knee: Evidence-Based Guideline, 2nd Edition. *J. Am. Acad. Orthop. Surg.* **2013**, *21*, 571–576.
38. Urban, H.; Little, C.B. The Role of Fat and Inflammation in the Pathogenesis and Management of Osteoarthritis. *Rheumatology* **2018**, *57* (Suppl. 4), iv10–iv21. [CrossRef]
39. Onur, T.; Wu, R.; Metz, L.; Dang, A. Characterisation of Osteoarthritis in a Small Animal Model of Type 2 Diabetes Mellitus. *Bone Joint Res.* **2014**, *3*, 203–211. [CrossRef]
40. Mooney, R.A.; Sampson, E.R.; Lerea, J.; Rosier, R.N.; Zuscik, M.J. High-Fat Diet Accelerates Progression of Osteoarthritis after Meniscal/Ligamentous Injury. *Arthritis Res. Ther.* **2011**, *13*, R198. [CrossRef]
41. Datta, P.; Zhang, Y.; Parousis, A.; Sharma, A.; Rossomacha, E.; Endisha, H.; Wu, B.; Kacprzak, I.; Mahomed, N.N.; Gandhi, R.; et al. High-Fat Diet-Induced Acceleration of Osteoarthritis Is Associated with a Distinct and Sustained Plasma Metabolite Signature. *Sci. Rep.* **2017**, *7*, 8205. [CrossRef] [PubMed]
42. De Sire, A.; de Sire, R.; Petito, V.; Masi, L.; Cisari, C.; Gasbarrini, A.; Scaldaferri, F.; Invernizzi, M. Gut-Joint Axis: The Role of Physical Exercise on Gut Microbiota Modulation in Older People with Osteoarthritis. *Nutrients* **2020**, *12*, 574. [CrossRef]
43. Berthelot, J.M.; Sellam, J.; Maugars, Y.; Berenbaum, F. Cartilage-Gut-Microbiome Axis: A New Paradigm for Novel Therapeutic Opportunities in Osteoarthritis. *RMD Open* **2019**, *5*, e001037. [CrossRef] [PubMed]
44. Huang, Z.Y.; Stabler, T.; Pei, F.X.; Kraus, V.B. Both Systemic and Local Lipopolysaccharide (Lps) Burden Are Associated with Knee Oa Severity and Inflammation. *Osteoarthr. Cartil.* **2016**, *24*, 1769–1775. [CrossRef] [PubMed]
45. Collins, K.H.; Paul, H.A.; Reimer, R.A.; Seerattan, R.A.; Hart, D.A.; Herzog, W. Relationship between Inflammation, the Gut Microbiota, and Metabolic Osteoarthritis Development: Studies in a Rat Model. *Osteoarthr. Cartil.* **2015**, *23*, 1989–1998. [CrossRef] [PubMed]
46. Szychlinska, M.A.; Di Rosa, M.; Castorina, A.; Mobasheri, A.; Musumeci, G. A Correlation between Intestinal Microbiota Dysbiosis and Osteoarthritis. *Heliyon* **2019**, *5*, e01134. [CrossRef]
47. Favazzo, L.J.; Hendesi, H.; Villani, D.A.; Soniwala, S.; Dar, Q.A.; Schott, E.M.; Gill, S.R.; Zuscik, M.J. The Gut Microbiome-Joint Connection: Implications in Osteoarthritis. *Curr. Opin. Rheumatol.* **2020**, *32*, 92–101. [CrossRef]

48. De Melo, K.M.; de Oliveira, F.T.B.; Costa Silva, R.A.; Gomes Quindere, A.L.; Marinho Filho, J.D.B.; Araujo, A.J.; Barros Pereira, E.D.; Carvalho, A.A.; Chaves, M.H.; Rao, V.S.; et al. Alpha, Beta-Amyrin, a Pentacyclic Triterpenoid from Protium Heptaphyllum Suppresses Adipocyte Differentiation Accompanied by Down Regulation of Ppargamma and C/Ebpalpha in 3t3-L1 Cells. *Biomed. Pharmacother.* **2019**, *109*, 1860–1866. [CrossRef]
49. Santos, F.A.; Frota, J.T.; Arruda, B.R.; de Melo, T.S.; da Silva, A.A.; Brito, G.A.; Chaves, M.H.; Rao, V.S. Antihyperglycemic and Hypolipidemic Effects of Alpha, Beta-Amyrin, a Triterpenoid Mixture from Protium Heptaphyllum in Mice. *Lipids Health Dis.* **2012**, *11*, 98. [CrossRef]
50. Sudhahar, V.; Kumar, S.A.; Sudharsan, P.T.; Varalakshmi, P. Protective Effect of Lupeol and Its Ester on Cardiac Abnormalities in Experimental Hypercholesterolemia. *Vasc. Pharmacol.* **2007**, *46*, 412–418. [CrossRef]
51. Lippiello, L.; Walsh, T.; Fienhold, M. The Association of Lipid Abnormalities with Tissue Pathology in Human Osteoarthritic Articular Cartilage. *Metabolism* **1991**, *40*, 571–576. [CrossRef]
52. Wu, C.L.; Kimmerling, K.A.; Little, D.; Guilak, F. Serum and Synovial Fluid Lipidomic Profiles Predict Obesity-Associated Osteoarthritis, Synovitis, and Wound Repair. *Sci. Rep.* **2017**, *7*, 44315. [CrossRef] [PubMed]
53. Bastiaansen-Jenniskens, Y.M.; Siawash, M.; van de Lest, C.H.; Verhaar, J.A.; Kloppenburg, M.; Zuurmond, A.M.; Stojanovic-Susulic, V.; Van Osch, G.J.; Clockaerts, S. Monounsaturated and Saturated, but Not N-6 Polyunsaturated Fatty Acids Decrease Cartilage Destruction under Inflammatory Conditions: A Preliminary Study. *Cartilage* **2013**, *4*, 321–328. [CrossRef] [PubMed]
54. Ikeuchi, M.; Izumi, M.; Aso, K.; Sugimura, N.; Kato, T.; Tani, T. Effects of Intra-Articular Hyaluronic Acid Injection on Immunohistochemical Characterization of Joint Afferents in a Rat Model of Knee Osteoarthritis. *Eur. J. Pain* **2015**, *19*, 334–340. [CrossRef]
55. Boettger, M.K.; Kummel, D.; Harrison, A.; Schaible, H.G. Evaluation of Long-Term Antinociceptive Properties of Stabilized Hyaluronic Acid Preparation (Nasha) in an Animal Model of Repetitive Joint Pain. *Arthritis Res. Ther.* **2011**, *13*, R110. [CrossRef]
56. Jimbo, S.; Terashima, Y.; Teramoto, A.; Takebayashi, T.; Ogon, I.; Watanabe, K.; Sato, T.; Ichise, N.; Tohse, N.; Yamashita, T. Antinociceptive Effects of Hyaluronic Acid on Monoiodoacetate-Induced Ankle Osteoarthritis in Rats. *J. Pain Res.* **2019**, *12*, 191–200. [CrossRef]
57. Bannuru, R.R.; Osani, M.C.; Vaysbrot, E.E.; Arden, N.K.; Bennell, K.; Bierma-Zeinstra, S.M.A.; Kraus, V.B.; Lohmander, L.S.; Abbott, J.H.; Bhandari, M.; et al. Oarsi Guidelines for the Non-Surgical Management of Knee, Hip, and Polyarticular Osteoarthritis. *Osteoarthr. Cartil.* **2019**, *27*, 1578–1589. [CrossRef]
58. Henrotin, Y.; Lambert, C.; Couchourel, D.; Ripoll, C.; Chiotelli, E. Nutraceuticals: Do They Represent a New Era in the Management of Osteoarthritis? - A Narrative Review from the Lessons Taken with Five Products. *Osteoarthr. Cartil.* **2011**, *19*, 1–21. [CrossRef]
59. Wang, A.; Leong, D.J.; Cardoso, L.; Sun, H.B. Nutraceuticals and Osteoarthritis Pain. *Pharmacol. Ther.* **2018**, *187*, 167–179. [CrossRef]

nutrients

MDPI

Article

FlexPro MD®, a Combination of Krill Oil, Astaxanthin and Hyaluronic Acid, Reduces Pain Behavior and Inhibits Inflammatory Response in Monosodium Iodoacetate-Induced Osteoarthritis in Rats

Min Hee Park [1], Jae Chul Jung [2], Stephen Hill [3], Elizabeth Cartwright [3], Margaret H. Dohnalek [3], Min Yu [4], Hee Joon Jun [4], Sang Bae Han [4], Jin Tae Hong [4,*] and Dong Ju Son [4,*]

1 Division of Life and Pharmaceutical Sciences, Ewha Womans University, 52 Ewhayeodae-gil, Sedaemun-gu, Seoul 03760, Korea; 131pcg01@naver.com

2 R&D Center, Novarex Co., Ltd., 60 Gangni 1-gil, Ochang-eup, Cheongwon-gu, Cheongju, Chungbuk 28126, Korea; jcjung@novarex.co.kr

3 US Nutraceuticals, Inc. d/b/a Valensa International, Eustis, FL 32726, USA; S.Hill@valensa.com (S.H.); E.Cartwright@valensa.com (E.C.); M.Dohnalek@valensa.com (M.H.D.)

4 College of Pharmacy and Medical Research Center, Chungbuk National University, 194-21 Osongsaengmyeong 1-ro, Osong-eup, Heungduk-gu, Cheongju, Chungbuk 28160, Korea; yumin0413@chungbuk.ac.kr (M.Y.); hj@chungbuk.ac.kr (H.J.J.); shan@chungbuk.ac.kr (S.B.H.)

* Correspondence: jinthong@chungbuk.ac.kr (J.T.H.); sondj1@chungbuk.ac.kr (D.J.S.); Tel.: +82-43-261-2813 (J.T.H.); +82-43-261-2822 (D.J.S.)

Received: 17 February 2020; Accepted: 26 March 2020; Published: 30 March 2020

Abstract: Osteoarthritis (OA) is a degenerative joint disease and a leading cause of adult disability. Since there is no cure for OA and no effective treatment to slow its progression, current pharmacologic treatments, such as analgesics and non-steroidal anti-inflammatory drugs (NSAIDs), only alleviate symptoms, such as pain and inflammation, but do not inhibit the disease process. Moreover, chronic intake of these drugs may result in severe adverse effects. For these reasons, patients have turned to the use of various complementary and alternative approaches, including diverse dietary supplements and nutraceuticals, in an effort to improve symptoms and manage or slow disease progression. The present study was conducted to evaluate the anti-osteoarthritic effects of FlexPro MD® (a mixture of krill oil, astaxanthin, and hyaluronic acid; FP-MD) in a rat model of OA induced by monosodium iodoacetate (MIA). FP-MD significantly ameliorated joint pain and decreased the severity of articular cartilage destruction in rats that received oral supplementation for 7 days prior to MIA administration and for 21 days thereafter. Furthermore, FP-MD treatment significantly reduced serum levels of the articular cartilage degeneration biomarkers cartilage oligomeric matrix protein (COMP) and crosslinked C-telopeptide of type II collagen (CTX-II), and the pro-inflammatory cytokines tumor necrosis factor alpha (TNF-α), interleukin-1β (IL-1β), and interleukin-6 (IL-6), as well as mRNA expression levels of inflammatory mediators, inducible nitric oxide synthase (iNOS) and cyclooxygenase-2 (COX-2), and matrix-degrading enzymes, matrix metalloproteinase (MMP)-2 and MMP-9, in the knee joint tissue. Our findings suggest that FP-MD is a promising dietary supplement for reducing pain, minimizing cartilage damage, and improving functional status in OA, without the disadvantages of previous dietary supplements and medicinal agents, including multiple adverse effects.

Keywords: FlexPro MD; krill oil; astaxanthin; hyaluronic acid; inflammation; pain; osteoarthritis

1. Introduction

Osteoarthritis (OA), also called degenerative joint disease, is a universal debilitating joint disease, and is the most prevalent type of arthritis characterized by synovial inflammation, the gradual loss of articular cartilage and degenerative changes in other surrounding tissues, including the synovium, menisci, ligaments, and subchondral bone, that are caused by multiple risk factors such as age, weight, excessive joint usage, and metabolic or genetic factors [1,2]. OA is the most common joint disorder in the elderly around the world, which can decrease quality of life due to pain, stiffness, loss of function, and disability [3]. Chronic joint pain is the main symptom of OA and is the primary reason for patients to seek medical and pharmacological treatments. Although structural changes, cartilage degeneration, and subchondral bone remodeling are the main contributors to disease progression and joint pain in patients, inflammatory responses mediated by pro-inflammatory cytokines and inflammatory mediators following joint injury also contribute to these pathological events in OA [4–8].

There is no treatment to prevent the development of OA or cure it once it has set in; thus, the goal of OA treatments currently available in clinical practice is mainly aimed at reducing pain, minimizing cartilage damage, and improving or maintain functional status. Commonly recommended treatments include aerobic exercise, strengthening exercises, and medicinal agents, including non-steroidal anti-inflammatory drugs (NSAIDs), corticosteroids, or acetaminophen, which treat OA primarily by reducing pain and inflammation [9]. However, these medications cannot prevent progressive cartilage degradation or repair the damaged cartilage of OA patients, and long-term use of these drugs can lead to various adverse effects, including renal toxicity, gastrointestinal disturbances, diarrhea, nausea, vomiting, or increased cardiovascular risks [10–12]. In addition, a recent clinical study reported that chronic but not short-term use of NSAIDs was associated with no clinical improvement in pain and a minimal, non-statistically significant improvement in clinical outcomes in persons with stiffness and functional and structural changes due to OA [13]. Despite their questionable efficacy, prescription NSAIDs are widely used in OA treatment due to the lack of more effective medications.

Therefore, the chronic nature of the disease has encouraged researchers and patients to try various complementary and alternative approaches, including dietary supplements, functional health foods, and nutraceuticals, to relieve pain and improve joint function as well as to manage or slow disease progression [14–17]. The most widely used and studied dietary supplements or functional health foods for joint pain include those related to chondroprotection, such as glucosamine, chondroitin sulfate, methylsulfonylmethane (MSM), collagen hydrolysates, and hyaluronic acid [18]. However, their effectiveness remains controversial; moreover, they can cause multiple adverse effects, such as stomach upset, constipation, diarrhea, headache, and rash [19,20]. In addition, natural products and nutraceutical products, including curcumin and turmeric extract, green tea extract, resveratrol, citrus fruit extract, *Boswellia serrata*, omega-3 fatty acids, and many others, have been investigated as potential candidates for joint-health promoting dietary supplements [21]. Interestingly, several studies have hypothesized that gut microbiota modulation through the supplementation of specific dietary ingredients including probiotics and prebiotics are able to modify the onset and the progression of OA [22]. Although previous studies have investigated the gut-joint axis in animal model, clinical studies in humans are lacking and no clear mechanism of action has been determined [23]. Therefore, future perspectives should focus on a more detailed understanding of the effects of microbiota modulation in OA patients. Nevertheless, dietary supplements with proven medical benefits, including managing and slowing disease progression and reducing the symptoms of OA, may offer safer alternatives to currently available pharmacological therapies [24,25].

FlexPro MD® (FP-MD), a novel, patented multi-ingredient dietary supplement formulation consisting of krill oil, natural astaxanthin, and sodium hyaluronate, has been shown to markedly reduce pain in subjects suffering from chronic mild-to-moderate knee joint pain [26]. Moreover, another study demonstrated that FP-MD could effectively inhibit lipopolysaccharide (LPS)-induced mRNA expression of pro-inflammatory cytokines and inflammatory markers by reducing nuclear factor-kappa B (NF-κB) activation, both in RAW264.7 cells and in an LPS-induced arthritis mouse model. In addition, FP-MD effectively suppressed the expression levels of matrix metalloproteinases (MMPs) at the transcriptional level in inflamed knee joint tissues [27]. However, its effects on the structural joint damage and associated joint pain of OA have not been explored. In this study, we investigated the effects of FP-MD on pain, and the structural changes and inflammatory responses known to cause pain, in a rat model of OA induced by monosodium iodoacetate (MIA) injection.

2. Materials and Methods

2.1. Preparation of FP-MD

FlexPro MD® (FP-MD) is a commercially available dietary supplement containing a proprietary combination of Superba® Antarctic krill (*Euphausia superba*) oil (Aker BioMarine Antarctic US LLC; Metuchen, NJ, USA), Zanthin® natural astaxanthin derived from *Haematococcus pluvialis*, Flexonic® sodium hyaluronate (the sodium salt of hyaluronic acid) produced from fermentation by *Streptococcus zooepidemicus* (Valensa International, Inc.; Eustis, FL, USA), and food vehicles (Table 1). A sample of FP-MD was produced as reported previously [28,29], then stored at room temperature until use. The FP-MD sample consisted of 70% krill oil, 7% *Haematococcus pluvialis* extract, and 7% sodium hyaluronate, along with 16% various excipients (beeswax, olive oil, etc.), as determined by high-performance liquid chromatography (HPLC) and gas chromatography (GC), using the methods recommended by the United States Pharmacopeia (USP) and Korean Pharmacopoeia (KP), respectively.

Table 1. Composition of FlexPro MD®.

Ingredient	Amount (mg)	Ratio (%)
Antarctic krill oil	321	70
Haematococcus pluvialis extract (to deliver 2 mg astaxanthin)	25–35	5.5–7.5
Sodium hyaluronate	33	7.1
Excipients	73–83 [1]	15.4–17.4
Total	462	100

[1] adjust the amount depending on the amount of *Haematococcus pluvialis* extract.

2.2. Animals and Ethics Statement

Adult male Sprague-Dawley (SD) rats (aged 7 weeks with body weights ranging 200–214.0 g) were obtained from Orientbio Inc. (Sungnam, Gyeonggi, Korea) and housed in the animal facility at Biotoxtech Co. Ltd. (Cheongju, Chungbuk, Korea), a non-clinical Good Laboratory Practice (GLP)-certified Contract Research Organization (CRO). The rats were acclimated for 7 days and maintained under conventional housing conditions at 23 ± 2 °C with a controlled 12 h light/dark cycle, and were provided filtered tap water and a rodent chow diet (Envigo RMS Inc., Indianapolis, IN, USA) *ad libitum* throughout the experiment. All animal experimental procedures complied with the National Institute of Health Guide for the Care and Use of Laboratory Animals and the Korean National Animal Welfare Law. The experimental animal facility and all protocols involving animals in this study were reviewed and approved by the Institutional Animal Care and Use Committee of Biotoxtech (IACUC approval number 180644).

2.3. Experimental Design and Administration

The experimental design of the study is shown in Supplementary Figure S1. Rats were randomly distributed in six groups of eight rats each, consisting of a normal sham control (Sham) group; an MIA-induced OA control (MIA) group; a positive control (PC) group; and three groups treated with increasing doses of FP-MD (25 mg/kg, 50 mg/kg, or 100 mg/kg). All of the groups received treatments via oral gavage once daily for 7 days before OA induction and then for 21 days thereafter until the end of the experiment. The PC group received Celecoxib administered orally at a dose of 3 mg/kg.

The Sham and MIA groups received equivalent volumes of vehicle (corn oil). The administration volumes were 5 mL/kg body weight for all experimental groups. The animals were closely monitored for individual clinical signs and other complications by a veterinarian. Body weight was measured weekly, and the weight-bearing levels of the two hind limbs were measured at 0, 3, 7, 14, and 21 days post-OA induction.

2.4. MIA-Induced OA Animal Model

Following the 7-day pretreatment period, the animals were anesthetized with 2% isoflurane delivered via a nose cone and then their right hind knees were injected, either with 3 mg MIA (Sigma-Aldrich Inc., St. Louis, MI, USA) in 50 μL saline or an equivalent volume of sterile saline vehicle, through the infrapatellar ligament as previously described [30–32] while contralateral knees remained intact. The choice of MIA dose was supported by previous work demonstrating that 3 mg of MIA not only induced joint degeneration but also produced significant axonal injury to dorsal root ganglion cells, reproducing the neuropathic pain component typically observed at the later stages of OA development [33,34]. Following injection, the animals were allowed to fully recover from the anesthesia and were monitored appropriately before returning them to the cages, and then continued to receive the same oral doses of FP-MD, Celecoxib, or vehicle, respectively, for 3 weeks.

2.5. Weight-Bearing Test (Pain Assessment)

The weight-bearing distributions between the postoperative and normal hind limbs were determined at 0, 3, 7, 14, and 21 days post-OA induction by using an incapacitance meter (Bioseb Co.; Pinellas Park, FL, USA) as previously described [35,36], which independently measures the weight, in grams, that the animal distributes to each hind paw. Generally, normal animals distribute their weight equally on both hind paws, while animals with MIA-induced joint pain tend to favor the normal limb. The rats were placed in a container with their hind paws comfortably resting on two separate sensor plates. When the rats stand, they make natural adjustments to the weight distribution on both rear paws based on the level of joint pain they are experiencing. The equipment was set to average the weight measured by each sensor over a 5-second period, and a total of three measurements were taken and averaged for each rat. The result was expressed as the mean of weight ratio between injured and non-injured limbs.

2.6. Histological Analysis

Histological changes were assessed to determine the effects of FP-MD and Celecoxib on joint cartilage degeneration. After 3 weeks post-OA induction, the rats were euthanized with CO_2, and the affected knee joints were dissected and fixed in 10% formalin for 24 h at 4 °C, decalcified with 5% formic acid for 2 weeks, and then dehydrated in graded acetone and embedded in paraffin. Sections (thickness, 4–5 μm) were stained with hematoxylin and eosin (H&E) and Safranin-O/Fast green to evaluate structural cartilage damage and the proteoglycan loss, respectively, and then observed under Carl Zeiss Axio Imager A2 microscope (Carl Zeiss, Deisenhofen, Germany). All stained slides were histologically evaluated and statistically graded on a scale of 0–13 by double-blind observation, according to the modified Mankin scoring system [37].

2.7. Serum Analysis

Blood samples were collected serially, either from the jugular vein or abdominal vena cava, at 22 days post-OA induction. Blood samples were centrifuged at 3000 rpm for 12 min to separate the serum. The sera were then stored at −70 °C until used for enzyme-linked immunosorbent assay (ELISA) experiments. The serum levels of the pro-inflammatory cytokines tumor necrosis factor alpha (TNF-α), interleukin-1 beta (IL-1β), and interleukin-6 (IL-6), and the cartilage degeneration mediators cartilage oligomeric matrix protein (COMP) and C-telopeptide of type II collagen (CTX-II), were determined using commercial ELISA kits (TNF-α: cat no. BMS622TEM, Invitrogen, MN, USA; IL-1β: cat no.

BMS630; IL-6: cat no. BMS625; COMP: cat no. abx256440, Abbexa Ltd., Cambridge, UK; CTX-II: cat no. E-EL-R2554, Elabscience Biotechnology Inc., Houston, TX, USA) according to manufacturer's protocol, respectively.

2.8. Real-Time PCR Analysis

Total RNA was extracted from articular cartilage tissues by using QIAzol® and purified using the miRNeasy Mini Kit (Qiagen GmbH, Hilden, Germany) according to the manufacturer's protocol. Total RNA was reverse transcripted into complementary DNA (cDNA) using a High Capacity RNA-to-cDNA kit (Applied Biosystems, Foster City, CA, USA), and then subjected to quantitative real-time PCR (qPCR) using QuantiFast® SYBR Green PCR master mix (Qiagen) with custom-designed specific primers using 18S as house-keeping control on a StepOnePlus™ Real-Time PCR System (Applied Biosystems). The primer sequences are listed in Table 2. Relative fold-changes in target gene expression between groups were determined for all targets using the $2^{\Delta\Delta Ct}$ method.

Table 2. List and sequences of real-time PCR primers for mRNA expression.

Gene		Primer Sequence
iNOS	Forward	5′-CTTTACGCCACTAACAGTGGCA-3′
	Reverse	5′-AGTCATGCTTCCCATCGCTC-3′
COX-2	Forward	5′-CCTCGTCCAGATGCTATCTTTG-3′
	Reverse	5′-GAAGGTCGTAGGTTTCCAGTATT-3′
MMP-2	Forward	5′-CACCAAGAACTTCCGACTATCC-3′
	Reverse	5′-TCCAGTACCAGTGTCAGTATCA-3′
MMP-9	Forward	5′-CCCAACCTTTACCAGCTACTC-3′
	Reverse	5′-GTCAGAACCGACCCTACAAAG-3′

2.9. Statistical Analysis

Data were expressed as mean ± standard error of measurement (SEM) for the indicated number of experiments. The statistical significance of the difference between groups was analyzed using independent sample *t*-tests or one-way analysis of variance (ANOVA) followed by Dunnett's post-hoc test. Statistical analysis was performed using SPSS software version 20.0 (IBM Co.; Armonk, NY, USA). For all tests, *p*-values below 0.05 were considered statistically significant.

3. Results

3.1. FP-MD Reduces the OA-Induced Joint Pain in Rats

We initially tested the effect of oral administration of FP-MD on joint pain in an MIA-induced rat model of OA. Since OA is accompanied by pain, the severity of pain is mainly determined by the asymmetric weight-bearing distribution in the hind limbs. Figure 1 demonstrates the latency differences between right (ipsilateral) and left (contralateral) hind-paws in the sham control (Sham) and experimental groups, which were analyzed to evaluate OA-induced pain using a capacitance tester for 3 weeks. The MIA-induced OA control (MIA) group animals showed a quick reduction of weight-bearing distribution in the OA-induced limbs compared to non-OA-induced knees due to pain induced by MIA injection, while the weight distribution did not change in the sham control group. In contrast, FP-MD- and Celecoxib-treated group animals showed decreased ipsilateral latency throughout the experimental period as compared with MIA group. Moreover, rats treated with doses of 50 and 100 mg/kg of FP-MD and the Celecoxib-treated rats were able to balance the right and left hind-paws and practically returned to the normal control condition. These data indicate the significant restoration of hind-limb weight bearing in the FP-MD-treated rats. In addition, we found that FP-MD treatment does not affect regular body weight gain in rats (Supplementary Figure S2).

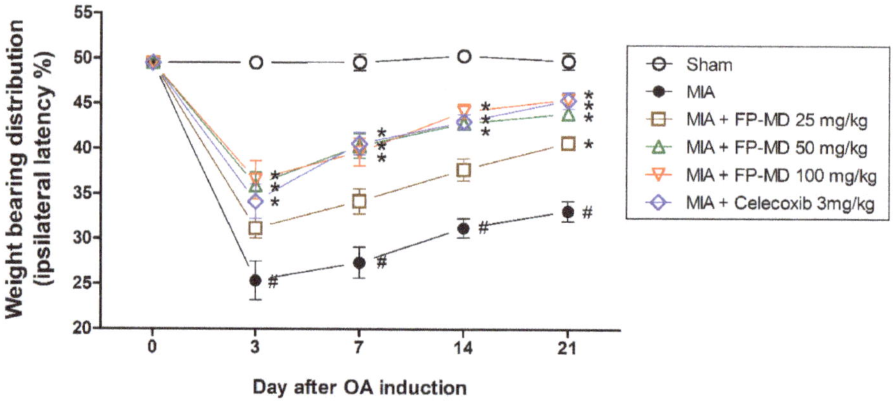

Figure 1. Effects of oral administration of FlexPro MD® (FP-MD) on changes in hind-paws weight-bearing distribution in monosodium iodoacetate (MIA)-induced osteoarthritis (OA) rats. The weight-bearing distribution ratio was measured for 21 days after injection of MIA using an incapacitance tester, compared to that of the vehicle-treated MIA-induced group. Data are expressed as the mean ± S.E.M ($n = 8$). # $p < 0.05$ versus vehicle-treated Sham group and * $p < 0.05$ versus MIA.

3.2. FP-MD Suppressed Articular Cartilage Damage in MIA-Induced OA Rats

As articular cartilage degeneration is the major histopathological feature of OA joints, the effect of FP-MD administration on the histopathological changes and severity of damage in the articular cartilage were evaluated using H&E and Safranin O-fast green staining in the MIA-induced OA rats. As shown in Figure 2A, the sham control group animals exhibited normal articular cartilage structures with smooth articular surfaces, normal chondrocytes with columnar orientation, and intact tide marks and subchondral bone. In contrast, the MIA group showed the severity of surface irregularity and cleft, matrix loss of articular cartilage, degeneration of columnar orientation, degeneration of the tide mark, and the penetration of subchondral bones. We found that oral administration of FP-MD attenuated the structural morphological changes in the articular cartilage, reduced the penetration of the subchondral bones, and reduced the degeneration of the tide marks in comparison to the MIA group. Proteoglycans as one of the major components of the extra cellular matrix (ECM) have various functions in the cartilage. Safranin O-fast green staining revealed that the joints of the MIA-induced OA rats showed joint space narrowing with marked proteoglycan depletion, whereas sham control rats showed the presence of intense proteoglycan in the ECM. Administration with FP-MD and Celecoxib effectively decreased the loss of proteoglycan in the knee joints compared to MIA group. Furthermore, the severity of OA lesions, graded using the modified Mankin's scoring system, and the overall modified Mankin's scores were significantly lower in FP-MD and Celecoxib-treated groups compared with MIA group (Figure 2B).

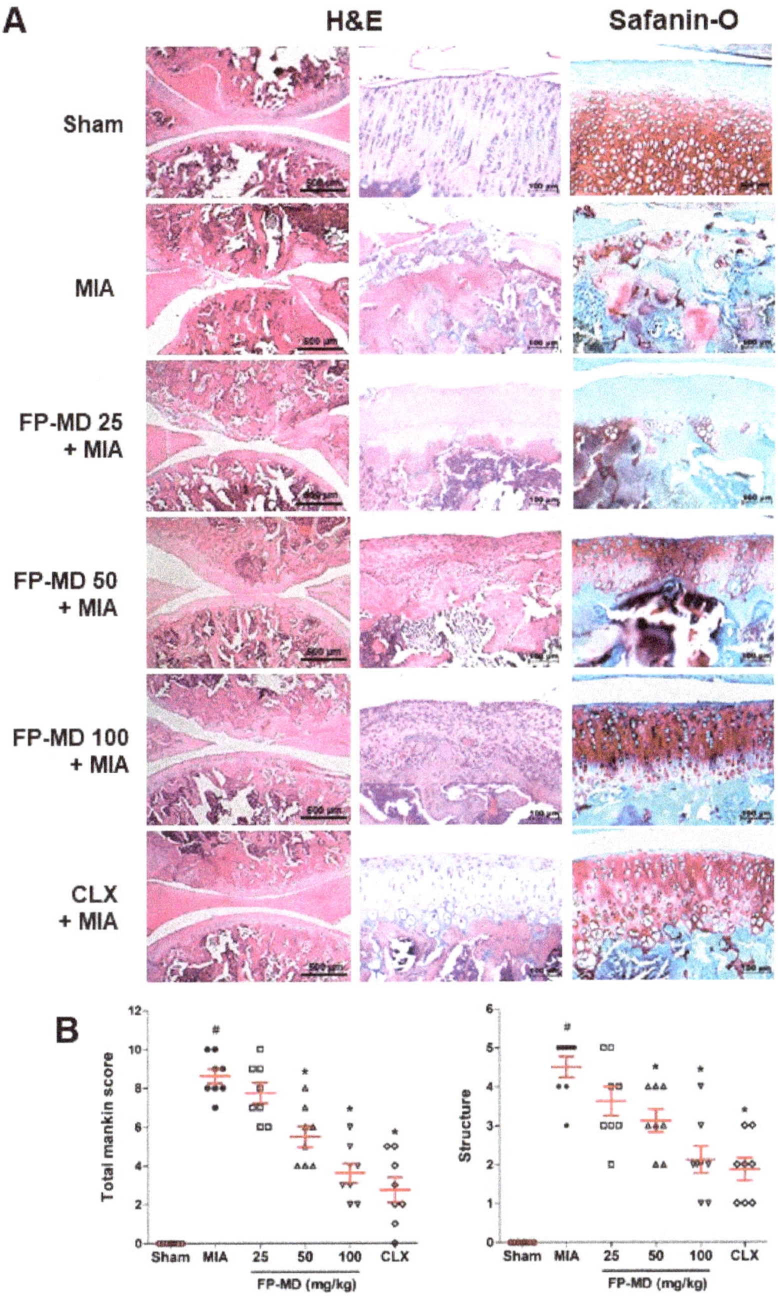

Figure 2. Histological evaluation of joints activity after administration with FlexPro MD® (FP-MD) in monosodium iodoacetate (MIA)-induced osteoarthritis (OA) rats. (**A**) Knee joints of the OA rats were stained with hematoxylin and eosin (H&E) and Safranin O-Fast green. (**B**) The joint lesions were graded on a scale of 0–13 using the modified Mankin's scoring system, giving a combined score for cartilage structure. Data are expressed as the mean ± S.E.M ($n = 8$). # $p < 0.05$ versus vehicle-treated Sham group and * $p < 0.05$ versus MIA. CLX; celecoxib 3 mg/kg-treated.

3.3. FP-MD Suppressed the Pro-Inflammatory Cytokine Levels in MIA-Induced OA Rats

The inflammatory response is an important factor associated with OA pathogenesis, and pro-inflammatory cytokines play a prominent role in the maintenance of tissue injury and chronic inflammation during the progression of OA [38]. We, therefore, examined the effect of FP-MD on the production of inflammatory cytokines associated with OA, such as TNF-α, IL-1β, and IL-6, in MIA-induced OA rats. As shown in Figure 3, the MIA group showed increased serum levels of TNF-α, IL-1β, IL-6, and IFN-γ compared with those in the sham control group. In the FP-MD- and Celecoxib-treated groups, the serum levels of pro-inflammatory cytokine were significantly decreased in a dose-dependent manner when compared to the MIA group. These results suggest that FP-MD protects cartilage in the MIA-induced OA model by modifying these pro-inflammatory cytokines.

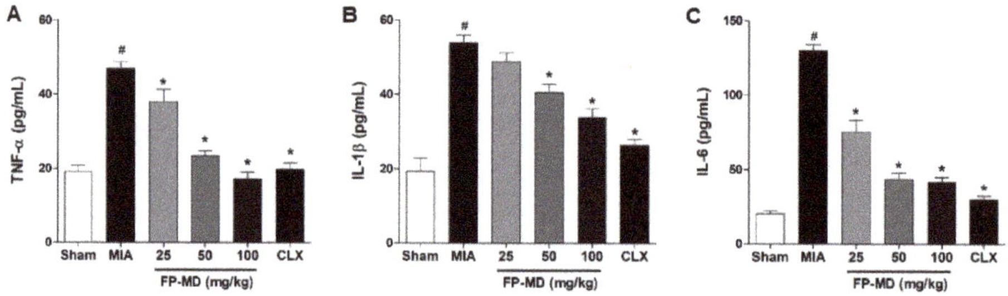

Figure 3. Effects of oral administration of FlexPro MD® (FP-MD) on the level of tumor necrosis factor alpha (TNF-α), interleukin (IL)-1β, and IL-6 in monosodium iodoacetate (MIA)-induced osteoarthritis (OA) rats. The serum concentrations of (**A**) TNF-α, (**B**) IL-1β, and (**C**) IL-6 in FP-MD (25–100 mg/kg) + MIA- or Celecoxib (CLX) (3 mg/kg) + MIA-induced OA rats were compared to those of the vehicle-treated MIA group. Data are expressed as the mean ± S.E.M (n = 8). # $p < 0.05$ versus vehicle-treated Sham group and * $p < 0.05$ versus MIA.

3.4. FP-MD Reduced Biomarkers of Chondrocyte Death in MIA-Induced OA Rats

The effect of FP-MD on the levels of COMP and CTX-II, which are well-established biomarkers for OA diagnosis and progression, which are degradation products of joint tissues, especially the cartilage ECM, during progressive destruction of articular cartilage in OA [39,40], was also investigated. The serum levels of COMP and CTX-II were increased in OAC group by MIA-injection compared with those in the sham control group. In contrast, the serum levels of COMP (Figure 4A) and CTX-II (Figure 4B) were significantly lower in the OA rats treated with FP-MD at doses of 50 and 100 mg/kg, but not 25 mg/kg, than in the OAC group. The MIA-induced overproduction of COMP and CTX-II was also decreased by Celecoxib treatment.

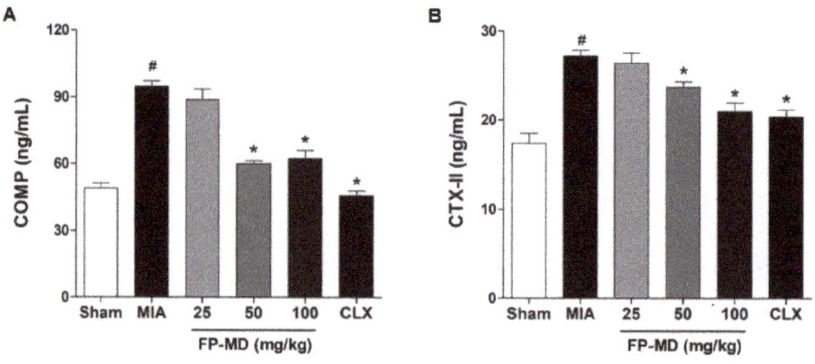

Figure 4. Effects of oral administration of FlexPro MD® (FP-MD) on the production of cartilage oligomeric matrix protein (COMP) and C-telopeptide of type II collagen (CTX-II) in monosodium iodoacetate (MIA)-

induced osteoarthritis (OA) rats. The serum concentrations of (**A**) COMP and (**B**) CTX-II in FP-MD (25–100 mg/kg) + MIA- or Celecoxib (CLX) (3 mg/kg) + MIA-induced OA rats were compared to those of the vehicle-treated MIA group. Data are expressed as the mean ± S.E.M (n = 8). # $p < 0.05$ versus vehicle-treated Sham group and * $p < 0.05$ versus MIA.

3.5. FP-MD Suppressed mRNA Expression of Inflammatory Mediators and Matrix-Degrading Ezzymes in MIA-Induced OA Rats

We next investigated the effect of FP-MD on the mRNA expression levels of inflammatory mediators, inducible nitric oxide synthase (iNOS) and cyclooxygenase-2 (COX-2), and matrix-degrading enzymes, matrix metalloproteinase (MMP)-2 and MMP-9, in the knee joint tissues of MIA-induced OA rats. The mRNA expression of iNOS, COX-2, MMP-2, and MMP-9 increased in OAC group by MIA-injection compared with those in the sham control group (Figure 5). In the FP-MD- and Celecoxib-treated groups, the expression levels of those genes were significantly suppressed. These results suggested that FP-MD inhibited inflammatory responses and articular cartilage damage by inhibiting inflammatory cytokines and MMPs in MIA-induced OA rats.

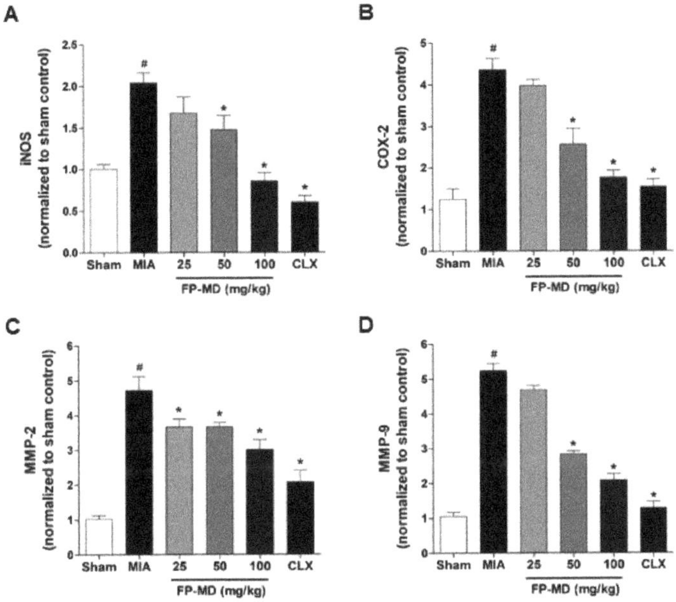

Figure 5. Effects of oral administration of FlexPro MD® (FP-MD) on the expression of inflammatory mediators and metalloproteinases (MMPs) in the knee joint in monosodium iodoacetate (MIA)-induced osteoarthritis (OA) rats. The mRNA expression of (**A**) inducible nitric oxide synthase (iNOS), (**B**) cyclooxygenase-2 (COX-2), (**C**) MMP-2, and (**D**) MMP-9 in FP-MD (25–100 mg/kg) + MIA- or Celecoxib (CLX) (3 mg/kg) + MIA-induced OA rats were compared to those of the vehicle-treated MIA group. Data are expressed as the mean ± S.E.M (n = 8). # $p < 0.05$ versus vehicle-treated Sham group and * $p < 0.05$ vs. MIA.

4. Discussion

In the present study, we evaluated whether or not FP-MD exerts an anti-osteoarthritic effect in the MIA-induced OA rat model, which is one of the well-established animal models for human OA and joint pain. In rats, injection of MIA into joints successfully induces acute inflammation, articular cartilage degradation, and joint pain by the direct interruption of chondrocyte metabolism and the subsequent induction of chondrocyte death, representative of the changes observed in patients with OA [41]. Our data show that oral administration of FP-MD led to a significant reduction in MIA-induced joint pain and a decrease in structural changes, including joint space narrowing and cartilage destruction, which were associated with the reduction of pro-inflammatory cytokines and articular cartilage degeneration biomarkers.

Several clinical studies have demonstrated that daily consumption of krill oil (2 g/day for 30 days) improves the subjective symptoms of knee pain in adults with mild knee pain [42], mitigates the subjective symptoms of OA as assessed by Western Ontario and McMaster Universities Osteoarthritis Index (WOMAC), and reduces C-reactive protein (CRP) levels in patients with rheumatoid arthritis or OA with CRP levels greater than 1.0 mg/dL [43]. In addition, the anti-arthritis, anti-pain, and anti-inflammatory effects of krill oil were also observed in a carrageenan-induced mouse model of inflammatory pain [44,45]. Furthermore, current studies have demonstrated that astaxanthin is a promising anti-inflammatory and anti-pain agent against carrageenan-induced paw edema and pain behavior, as well as neuropathic pain [46–48]. Hyaluronic acid is also known as a useful treatment option for OA that can modify symptoms and relieve joint pain in OA [49–51]. Interestingly, krill oil, astaxanthin, and hyaluronic acid have also been shown to play a role in gut flora modulation, an emerging area of investigation that may also influence the development and progression of osteoarthritis, and symptom management [52–54]. Considering the beneficial therapeutic properties of krill oil, astaxanthin, and hyaluronic acid on arthritis, we have developed a novel multi-ingredient dietary supplement formulation, FP-MD, consisting of krill oil, natural astaxanthin, and proprietary lower molecular weight hyaluronic acid, to address the key factors involved in maintaining joint health. In the present study, we performed in vivo animal studies to further investigate the therapeutic potential of FP-MD for pain relief and chondroprotection in an MIA-induced OA rat model by measuring weight-bearing distribution, histopathological changes, serum levels of pro-inflammatory cytokines, and biomarkers for degradation products of articular cartilage.

Here, we demonstrated that oral administration of FP-MD has pain-relieving effects in MIA-induced OA in rats, consistent with our previous findings in a clinical trial that showed remarkable pain relief in subjects suffering from chronic mild-to-moderate knee joint pain [26]. Moreover, histopathological analysis revealed that the FP-MD-treated group exhibited marked suppression of structural changes, bone resorption, and proteoglycan degradation in the MIA-treated rat knee joint. Taken together, these observations suggest that FP-MD possesses potential pain-relieving activity in OA by reducing articular cartilage damage.

The degradation of articular cartilage, including the degradation of cartilage cells and matrix, is the main pathological characteristic of OA, and matrix degeneration mainly results in the losses of proteoglycans and type II collagen [55]. Several previous studies have demonstrated that the serum levels of some structural molecules and fragments derived from articular cartilage, bone, and the synovium, all of which are affected by OA, are elevated in OA patients and OA animal models, thereby they can be used as diagnostic biomarkers which can potentially predict the increased risk and progression of OA [56,57]. COMP and CTX-II are two biochemical markers that are degradation products of joint tissues, especially the cartilage extracellular matrix, and can potentially predict the destruction of articular cartilage in OA. COMP, a tissue-specific pentametric glycoprotein, is one of the essential components of the extracellular matrix of the cartilage which binds to type II collagen fibers and stabilizes the collagen fiber network of articular cartilage in cooperation with other matrix proteins. However, when the articular cartilage is destroyed during the development of OA and under inflammatory conditions, the levels of COMP is noticeably increased in synovial fluid and serum and is positively correlated with joint damage in knee OA [58]. In addition, levels of CTX-II, a degradation product of type II collagen produced by proteases activated by cartilage injury or degeneration, are also elevated in OA patients as compared with normal individuals and are associated with both the prevalence and progression of OA [59].

Accordingly, the serum levels of these two factors, COMP and CTX-II, have the potential to be prognostic biomarkers for monitoring cartilage degradation in patients with OA. Thus, the serum levels of COMP and CTX-II were analyzed to assess the effects of FP-MD on the degradation of articular cartilage. Consistent with previous studies [60,61], serum levels of both COMP and CTX-II were significantly higher in the OAC group than in the sham control group; however, treatment with FP-MD significantly reduced this increase. These findings suggest that the reduction in COMP and CTX-II serum levels by FP-MD most likely represents suppressed MIA-induced degradation of cartilage, as damaged cartilage is a major contributor to circulating COMP and CTX-II levels.

Increasing evidence has demonstrated that the significantly elevated levels of pro-inflammatory cytokines observed in OA patients play a critical role in the promotion of the catabolic processes in OA, causing cartilage degradation [62,63]. High levels of pro-inflammatory cytokines, such as TNF-α, IL-1β, IL-6, and IL-10, have been found in synovial fluid from OA patients and several experimental animal models of cartilage degradation [64]. Among these cytokines, TNF-α and IL-1β are highly overexpressed in the cartilage as well as in the synovial tissue and are considered the major mediators in the OA pathogenesis [65,66]. These cytokines are known to drive the inflammatory cascade, and their increased production induces catabolic events as they downregulate the synthesis of ECM structural components, including proteoglycan, by inhibiting the anabolic activities of chondrocytes and enhancing MMPs, resulting in the loss of cartilage and increased bone resorption during the development and progression of OA [65,67]. IL-6 has also been reported to play a major role in OA. Although the production of IL-6 by chondrocytes is considerably low in physiological conditions, its production can be stimulated by the number of other cytokines and inflammatory mediators, including TNF-α, IL-1β, IFN-γ, and prostaglandin E_2, resulting in decreased production of type II collagen [68].

Many previous studies have demonstrated that anti-inflammatory agents capable of inhibiting the production of these cytokines might have the potential to control or treat OA [65–67]. Hence, we investigated the anti-inflammatory effects of FP-MD by measuring the serum levels of the pro-inflammatory cytokines TNF-α, IL-1β, and IL-6 in MIA-induced OA rats, and found that oral administration of FP-MD effectively decreased the serum levels of these cytokines. In addition, FP-MD significantly reduced the mRNA expression levels of inflammatory mediators, iNOS and COX-2, and matrix-degrading enzymes, MMP-2 and MMP-9, in the knee joint tissue. Taken together, the present results indicate that FP-MD has the potential to blunt inflammatory responses, and which may subsequently reduce articular cartilage damage.

5. Conclusions

In conclusion, the present study has demonstrated that oral administration of FP-MD effectively attenuates joint pain and the severity of articular cartilage destruction in an MIA-induced OA rat model and that the anti-osteoarthritic effects of FP-MD were associated with the protection of articular cartilage against inflammation-induced degradation thorough the suppression of pro-inflammatory cytokines. Our findings suggest that FP-MD is a promising dietary supplement for reducing pain, minimizing cartilage damage, and improving functional status in OA patients, and could overcome the disadvantages of previous dietary supplements, including glucosamine and chondroitin sulfate, as well as medicinal agents, such as corticosteroids and NSAIDs.

Supplementary Materials: The following are available online at http://www.mdpi.com/2072-6643/12/4/956/s1, Figure S1: Schematic diagram of experimental design and schedule for animal experiments, Figure S2: Effects of oral administration of FP-MD on changes in body weight gain in MIA-induced OA rats.

Author Contributions: M.H.P. participated in data analysis and visualization; J.C.J., conceptualization and review of manuscript; M.Y. and H.J.J. participated in mRNA expression analysis and histological analysis; S.H., E.C., M.H.D., S.B.H., reviewing and editing manuscript; J.T.H. and D.J.S. supervised the overall research, secured funding, and had primary responsibility of the final content. All authors approved the final version of the manuscript.

Funding: This research was supported by a grant from the Ministry of Trade, Industry and Energy (MOTIE) through the Middle Market Enterprises Global Research R&D Project (No. P0000784), the National Research Foundation of Korea (NRF) grant from the Korean government (MRC, 2017R1A5A2015541), and the research grant of the Chungbuk National University in 2019 (2019101027).

Conflicts of Interest: J.C.J. is an employee of Novarex and S.H., E.C. and M.H.D. are employees of US Nutraceuticals, Inc. d/b/a Valensa International.

References

1. Lawrence, R.C.; Helmick, C.G.; Arnett, F.C.; Deyo, R.A.; Felson, D.T.; Giannini, E.H.; Heyse, S.P.; Hirsch, R.; Hochberg, M.C.; Hunder, G.G.; et al. Estimates of the prevalence of arthritis and selected musculoskeletal disorders in the United States. *Arthritis Rheumatol.* **1998**, *41*, 778–799. [CrossRef]
2. Antony, B.; Jones, G.; Jin, X.; Ding, C. Do early life factors affect the development of knee osteoarthritis in later life: A narrative review. *Arthritis Res. Ther.* **2016**, *18*, 202. [CrossRef]
3. Felson, D.T. Clinical practice. Osteoarthritis of the knee. *N. Engl. J. Med.* **2006**, *354*, 841–848. [CrossRef] [PubMed]
4. Van den Bosch, M.H.J. Inflammation in osteoarthritis: Is it time to dampen the alarm(in) in this debilitating disease? *Clin. Exp. Immunol.* **2019**, *195*, 153–166. [CrossRef] [PubMed]
5. Sellam, J.; Berenbaum, F. The role of synovitis in pathophysiology and clinical symptoms of osteoarthritis. *Nat. Rev. Rheumatol.* **2010**, *6*, 625–635. [CrossRef]
6. Kandahari, A.M.; Yang, X.; Dighe, A.S.; Pan, D.; Cui, Q. Recognition of Immune Response for the Early Diagnosis and Treatment of Osteoarthritis. *J. Immunol. Res.* **2015**, *2015*, 192415. [CrossRef]
7. De Lange-Brokaar, B.J.; Ioan-Facsinay, A.; van Osch, G.J.; Zuurmond, A.M.; Schoones, J.; Toes, R.E.; Huizinga, T.W.; Kloppenburg, M. Synovial inflammation, immune cells and their cytokines in osteoarthritis: A review. *Osteoarthr. Cartil.* **2012**, *20*, 1484–1499. [CrossRef]
8. Lieberthal, J.; Sambamurthy, N.; Scanzello, C.R. Inflammation in joint injury and post-traumatic osteoarthritis. *Osteoarthr. Cartil.* **2015**, *23*, 1825–1834. [CrossRef]
9. Zhang, W.; Moskowitz, R.W.; Nuki, G.; Abramson, S.; Altman, R.D.; Arden, N.; Bierma-Zeinstra, S.; Brandt, K.D.; Croft, P.; Doherty, M.; et al. OARSI recommendations for the management of hip and knee osteoarthritis, part I: Critical appraisal of existing treatment guidelines and systematic review of current research evidence. *Osteoarthr. Cartil.* **2007**, *15*, 981–1000. [CrossRef]
10. Wong, R.S.Y. Disease-modifying effects of long-term and continuous use of nonsteroidal anti-inflammatory drugs (NSAIDs) in spondyloarthritis. *Adv. Pharmacol. Sci.* **2019**, *2019*, 5324170. [CrossRef]
11. Le Graverand-Gastineau, M.P. Disease modifying osteoarthritis drugs: Facing development challenges and choosing molecular targets. *Curr. Drug Targets* **2010**, *11*, 528–535. [CrossRef] [PubMed]
12. Crofford, L.J. Use of NSAIDs in treating patients with arthritis. *Arthritis Res. Ther.* **2013**, *15*, S2. [CrossRef] [PubMed]
13. Lapane, K.L.; Yang, S.; Driban, J.B.; Liu, S.H.; Dube, C.E.; McAlindon, T.E.; Eaton, C.B. Effects of prescription nonsteroidal antiinflammatory drugs on symptoms and disease progression among patients with knee osteoarthritis. *Arthritis Rheumatol.* **2015**, *67*, 724–732. [CrossRef]
14. De Silva, V.; El-Metwally, A.; Ernst, E.; Lewith, G.; Macfarlane, G.J.; Arthritis Research UK Working Group on Complementary and Alternative Medicines. Evidence for the efficacy of complementary and alternative medicines in the management of osteoarthritis: A systematic review. *Rheumatology* **2011**, *50*, 911–920. [CrossRef]
15. Lopez, H.L. Nutritional interventions to prevent and treat osteoarthritis. Part I: Focus on fatty acids and macronutrients. *PM R* **2012**, *4*, S145–S154. [CrossRef]
16. Lopez, H.L. Nutritional interventions to prevent and treat osteoarthritis. Part II: Focus on micronutrients and supportive nutraceuticals. *PM R* **2012**, *4*, S155–S168. [CrossRef]
17. Lapane, K.L.; Sands, M.R.; Yang, S.; McAlindon, T.E.; Eaton, C.B. Use of complementary and alternative medicine among patients with radiographic-confirmed knee osteoarthritis. *Osteoarthr. Cartil.* **2012**, *20*, 22–28. [CrossRef]
18. Messina, O.D.; Vidal Wilman, M.; Vidal Neira, L.F. Nutrition, osteoarthritis and cartilage metabolism. *Aging Clin. Exp. Res.* **2019**, *31*, 807–813. [CrossRef]

19. Sawitzke, A.D.; Shi, H.; Finco, M.F.; Dunlop, D.D.; Bingham, C.O., 3rd; Harris, C.L.; Singer, N.G.; Bradley, J.D.; Silver, D.; Jackson, C.G.; et al. The effect of glucosamine and/or chondroitin sulfate on the progression of knee osteoarthritis: A report from the glucosamine/chondroitin arthritis intervention trial. *Arthritis Rheuma.* **2008**, *58*, 3183–3191. [CrossRef]
20. Black, C.; Clar, C.; Henderson, R.; MacEachern, C.; McNamee, P.; Quayyum, Z.; Royle, P.; Thomas, S. The clinical effectiveness of glucosamine and chondroitin supplements in slowing or arresting progression of osteoarthritis of the knee: A systematic review and economic evaluation. *Health Technol. Assess.* **2009**, *13*, 1–148. [CrossRef]
21. Castrogiovanni, P.; Trovato, F.M.; Loreto, C.; Nsir, H.; Szychlinska, M.A.; Musumeci, G. Nutraceutical supplements in the management and prevention of osteoarthritis. *Int. J. Mol. Sci.* **2016**, *17*, 2042. [CrossRef] [PubMed]
22. De Sire, A.; de Sire, R.; Petito, V.; Masi, L.; Cisari, C.; Gasbarrini, A.; Scaldaferri, F.; Invernizzi, M. Gut-Joint axis: The role of physical exercise on gut microbiota modulation in older people with osteoarthritis. *Nutrients* **2020**, *12*, 574. [CrossRef] [PubMed]
23. Favazzo, L.J.; Hendesi, H.; Villani, D.A.; Soniwala, S.; Dar, Q.A.; Schott, E.M.; Gill, S.R.; Zuscik, M.J. The gut microbiome-joint connection: Implications in osteoarthritis. *Curr. Opin. Rheumatol.* **2020**, *32*, 92–101. [CrossRef]
24. Bruyere, O.; Cooper, C.; Al-Daghri, N.M.; Dennison, E.M.; Rizzoli, R.; Reginster, J.Y. Inappropriate claims from non-equivalent medications in osteoarthritis: A position paper endorsed by the European Society for Clinical and Economic Aspects of Osteoporosis, Osteoarthritis and Musculoskeletal Diseases (ESCEO). *Aging Clin. Exp. Res.* **2018**, *30*, 111–117. [CrossRef] [PubMed]
25. Wang, A.; Leong, D.J.; Cardoso, L.; Sun, H.B. Nutraceuticals and osteoarthritis pain. *Pharmacol. Ther.* **2018**, *187*, 167–179. [CrossRef]
26. Study to Evaluate the Safety and Efficacy of a Proprietary "Joint Health" Dietary Supplement in Subjects With Joint and Connective Tissue Pain. Available online: https://ClinicalTrials.gov/show/NCT03209895 (accessed on 13 June 2011).
27. Park, D.R.; Ko, R.; Kwon, S.H.; Min, B.; Yun, S.H.; Kim, M.H.; Minatelli, J.; Hill, S.; Lee, S.Y. FlexPro MD, a mixture of krill oil, astaxanthin, and hyaluronic acid, suppresses lipopolysaccharide-induced inflammatory cytokine production through inhibition of NF-kappaB. *J. Med. Food* **2016**, *19*, 1196–1203. [CrossRef]
28. Minatelli John, A.; Hill, W.S.; Thomas Swati, S.; Rajendran, L.; Moerck Rudi, E. Composition and Method to Alleviate Joint Pain. U.S. Patent 8945608 B2, 3 February 2015.
29. Minatelli John, A.; Hill, W.S.; Moerck Rudi, E. Composition and Method To Alleviate Joint Pain Using Algae Based Oils. U.S. Patent 9238043 B2, 19 January 2016.
30. Fernihough, J.; Gentry, C.; Malcangio, M.; Fox, A.; Rediske, J.; Pellas, T.; Kidd, B.; Bevan, S.; Winter, J. Pain related behaviour in two models of osteoarthritis in the rat knee. *Pain* **2004**, *112*, 83–93. [CrossRef]
31. Guingamp, C.; Gegout-Pottie, P.; Philippe, L.; Terlain, B.; Netter, P.; Gillet, P. Mono-iodoacetate-induced experimental osteoarthritis: A dose-response study of loss of mobility, morphology, and biochemistry. *Arthritis Rheum.* **1997**, *40*, 1670–1679. [CrossRef]
32. Lampropoulou-Adamidou, K.; Lelovas, P.; Karadimas, E.V.; Liakou, C.; Triantafillopoulos, I.K.; Dontas, I.; Papaioannou, N.A. Useful animal models for the research of osteoarthritis. *Eur. J. Orthop. Surg. Traumatol.* **2014**, *24*, 263–271. [CrossRef]
33. Lu, J.; Zhang, T.; Sun, H.; Wang, S.; Liu, M. Protective effects of dioscin against cartilage destruction in a monosodium iodoacetate (MIA)-indcued osteoarthritis rat model. *Biomed. Pharmacother.* **2018**, *108*, 1029–1038. [CrossRef]
34. Sim, B.Y.; Choi, H.J.; Kim, M.G.; Jeong, D.G.; Lee, D.G.; Yoon, J.M.; Kang, D.J.; Park, S.; Ji, J.G.; Joo, I.H.; et al. Effects of ID-CBT5101 in preventing and alleviating osteoarthritis symptoms in a monosodium iodoacetate-induced rat model. *J. Microbiol. Biotechnol.* **2018**, *28*, 1199–1208. [CrossRef] [PubMed]
35. Bove, S.E.; Calcaterra, S.L.; Brooker, R.M.; Huber, C.M.; Guzman, R.E.; Juneau, P.L.; Schrier, D.J.; Kilgore, K.S. Weight bearing as a measure of disease progression and efficacy of anti-inflammatory compounds in a model of monosodium iodoacetate-induced osteoarthritis. *Osteoarthr. Cartil.* **2003**, *11*, 821–830. [CrossRef]
36. Pitcher, T.; Sousa-Valente, J.; Malcangio, M. The monoiodoacetate model of osteoarthritis pain in the mouse. *J. Vis. Exp.* **2016**. [CrossRef]

37. Mankin, H.J.; Dorfman, H.; Lippiello, L.; Zarins, A. Biochemical and metabolic abnormalities in articular cartilage from osteo-arthritic human hips. II. Correlation of morphology with biochemical and metabolic data. *J. Bone Jt. Surg. Am. Vol.* **1971**, *53*, 523–537. [CrossRef]
38. Chen, D.; Shen, J.; Zhao, W.; Wang, T.; Han, L.; Hamilton, J.L.; Im, H.J. Osteoarthritis: Toward a comprehensive understanding of pathological mechanism. *Bone Res.* **2017**, *5*, 16044. [CrossRef] [PubMed]
39. Charni-Ben Tabassi, N.; Desmarais, S.; Bay-Jensen, A.C.; Delaisse, J.M.; Percival, M.D.; Garnero, P. The type II collagen fragments Helix-II and CTX-II reveal different enzymatic pathways of human cartilage collagen degradation. *Osteoarthr. Cartil.* **2008**, *16*, 1183–1191. [CrossRef] [PubMed]
40. Verma, P.; Dalal, K. Serum cartilage oligomeric matrix protein (COMP) in knee osteoarthritis: A novel diagnostic and prognostic biomarker. *J. Orthop. Res. Off. Publ. Orthop. Res. Soc.* **2013**, *31*, 999–1006. [CrossRef]
41. Marker, C.L.; Pomonis, J.D. The monosodium iodoacetate model of osteoarthritis pain in the rat. *Methods Mol. Biol.* **2012**, *851*, 239–248. [CrossRef]
42. Suzuki, Y.; Fukushima, M.; Sakuraba, K.; Sawaki, K.; Sekigawa, K. Krill oil improves mild knee joint pain: A randomized control trial. *PLoS ONE* **2016**, *11*, e0162769. [CrossRef]
43. Deutsch, L. Evaluation of the effect of Neptune Krill Oil on chronic inflammation and arthritic symptoms. *J. Am. Coll. Nutr.* **2007**, *26*, 39–48. [CrossRef]
44. Zadeh-Ardabili, P.M.; Rad, S.K. Anti-pain and anti-inflammation like effects of Neptune krill oil and fish oil against carrageenan induced inflammation in mice models: Current statues and pilot study. *Biotechnol. Rep.* **2019**, *22*, e00341. [CrossRef] [PubMed]
45. Ierna, M.; Kerr, A.; Scales, H.; Berge, K.; Griinari, M. Supplementation of diet with krill oil protects against experimental rheumatoid arthritis. *BMC Musculoskelet. Disord.* **2010**, *11*, 136. [CrossRef] [PubMed]
46. Kuedo, Z.; Sangsuriyawong, A.; Klaypradit, W.; Tipmanee, V.; Chonpathompikunlert, P. Effects of astaxanthin from litopenaeus vannamei on carrageenan-induced edema and pain behavior in mice. *Molecules* **2016**, *21*, 382. [CrossRef] [PubMed]
47. Sharma, K.; Sharma, D.; Sharma, M.; Sharma, N.; Bidve, P.; Prajapati, N.; Kalia, K.; Tiwari, V. Astaxanthin ameliorates behavioral and biochemical alterations in in-vitro and in-vivo model of neuropathic pain. *Neurosci. Lett.* **2018**, *674*, 162–170. [CrossRef] [PubMed]
48. Fakhri, S.; Dargahi, L.; Abbaszadeh, F.; Jorjani, M. Astaxanthin attenuates neuroinflammation contributed to the neuropathic pain and motor dysfunction following compression spinal cord injury. *Brain Res. Bull.* **2018**, *143*, 217–224. [CrossRef] [PubMed]
49. Altman, R.D.; Hochberg, M.C.; Moskowitz, R.W. Recommendations for the medical management of osteoarthritis of the hip and knee: 2000 update. *Arthritis Rheum.* **2000**, *43*, 1905–1915. [CrossRef]
50. Guidolin, D.D.; Ronchetti, I.P.; Lini, E.; Guerra, D.; Frizziero, L. Morphological analysis of articular cartilage biopsies from a randomized, clinical study comparing the effects of 500–730 kDa sodium hyaluronate (Hyalgan) and methylprednisolone acetate on primary osteoarthritis of the knee. *Osteoarthr. Cartil.* **2001**, *9*, 371–381. [CrossRef]
51. Moreland, L.W. Intra-articular hyaluronan (hyaluronic acid) and hylans for the treatment of osteoarthritis: Mechanisms of action. *Arthritis Res. Ther.* **2003**, *5*, 54–67. [CrossRef]
52. Lu, C.; Sun, T.; Li, Y.; Zhang, D.; Zhou, J.; Su, X. Modulation of the gut microbiota by krill oil in mice fed a high-sugar high-fat diet. *Front. Microbiol.* **2017**, *8*, 905. [CrossRef]
53. Liu, H.; Liu, M.; Fu, X.; Zhang, Z.; Zhu, L.; Zheng, X.; Liu, J. Astaxanthin prevents alcoholic fatty liver disease by modulating mouse gut microbiota. *Nutrients* **2018**, *10*, 1298. [CrossRef]
54. Lee, Y.; Sugihara, K.; Gillilland, M.G., 3rd; Jon, S.; Kamada, N.; Moon, J.J. Hyaluronic acid-bilirubin nanomedicine for targeted modulation of dysregulated intestinal barrier, microbiome and immune responses in colitis. *Nat. Mater.* **2020**, *19*, 118–126. [CrossRef] [PubMed]
55. Dunn, S.; Kolomytkin, O.V.; Marino, A.A. Pathophysiology of osteoarthritis: Evidence against the viscoelastic theory. *Pathobiol. J. Immunopathol. Mol. Cell. Biol.* **2009**, *76*, 322–328. [CrossRef] [PubMed]
56. Van Spil, W.E.; DeGroot, J.; Lems, W.F.; Oostveen, J.C.; Lafeber, F.P. Serum and urinary biochemical markers for knee and hip-osteoarthritis: A systematic review applying the consensus BIPED criteria. *Osteoarthr. Cartil.* **2010**, *18*, 605–612. [CrossRef] [PubMed]

57. Bai, B.; Li, Y. Combined detection of serum CTX-II and COMP concentrations in osteoarthritis model rabbits: An effective technique for early diagnosis and estimation of disease severity. *J. Orthop. Surg. Res.* **2016**, *11*, 149. [CrossRef]
58. Saberi Hosnijeh, F.; Siebuhr, A.S.; Uitterlinden, A.G.; Oei, E.H.; Hofman, A.; Karsdal, M.A.; Bierma-Zeinstra, S.M.; Bay-Jensen, A.C.; van Meurs, J.B. Association between biomarkers of tissue inflammation and progression of osteoarthritis: Evidence from the Rotterdam study cohort. *Arthr. Res. Ther.* **2016**, *18*, 81. [CrossRef] [PubMed]
59. Jansen, N.W.; Roosendaal, G.; Lundin, B.; Heijnen, L.; Mauser-Bunschoten, E.; Bijlsma, J.W.; Theobald, M.; Lafeber, F.P. The combination of the biomarkers urinary C-terminal telopeptide of type II collagen, serum cartilage oligomeric matrix protein, and serum chondroitin sulfate 846 reflects cartilage damage in hemophilic arthropathy. *Arthritis Rheum.* **2009**, *60*, 290–298. [CrossRef]
60. Jeong, J.W.; Kim, J.; Choi, E.O.; Kwon, D.H.; Kong, G.M.; Choi, I.W.; Kim, B.H.; Kim, G.Y.; Lee, K.W.; Kim, K.Y.; et al. Schisandrae Fructus ethanol extract ameliorates inflammatory responses and articular cartilage damage in monosodium iodoacetate-induced osteoarthritis in rats. *EXCLI J.* **2017**, *16*, 265–277. [CrossRef]
61. Lee, H.J.; Lee, S.M.; Moon, Y.G.; Jung, Y.S.; Lee, J.H.; Venkatarame Gowda Saralamma, V.; Kim, Y.S.; Pak, J.E.; Lee, H.J.; Kim, G.S.; et al. Membrane-free Stem cell components inhibit interleukin-1alpha-stimulated inflammation and cartilage degradation in vitro and in vivo: A rat model of osteoarthritis. *Int. J. Mol. Sci.* **2019**, *20*, 4869. [CrossRef]
62. Kapoor, M.; Martel-Pelletier, J.; Lajeunesse, D.; Pelletier, J.P.; Fahmi, H. Role of proinflammatory cytokines in the pathophysiology of osteoarthritis. *Nat. Rev. Rheumatol.* **2011**, *7*, 33–42. [CrossRef]
63. Rahmati, M.; Mobasheri, A.; Mozafari, M. Inflammatory mediators in osteoarthritis: A critical review of the state-of-the-art, current prospects, and future challenges. *Bone* **2016**, *85*, 81–90. [CrossRef]
64. Goldring, M.B.; Otero, M. Inflammation in osteoarthritis. *Curr. Opin. Rheumatol.* **2011**, *23*, 471–478. [CrossRef] [PubMed]
65. Jotanovic, Z.; Mihelic, R.; Sestan, B.; Dembic, Z. Role of interleukin-1 inhibitors in osteoarthritis: An evidence-based review. *Drugs Aging* **2012**, *29*, 343–358. [CrossRef] [PubMed]
66. Mabey, T.; Honsawek, S. Cytokines as biochemical markers for knee osteoarthritis. *World J. Orthop.* **2015**, *6*, 95–105. [CrossRef] [PubMed]
67. Kellesarian, S.V.; Al-Kheraif, A.A.; Vohra, F.; Ghanem, A.; Malmstrom, H.; Romanos, G.E.; Javed, F. Cytokine profile in the synovial fluid of patients with temporomandibular joint disorders: A systematic review. *Cytokine* **2016**, *77*, 98–106. [CrossRef]
68. Poree, B.; Kypriotou, M.; Chadjichristos, C.; Beauchef, G.; Renard, E.; Legendre, F.; Melin, M.; Gueret, S.; Hartmann, D.J.; Mallein-Gerin, F.; et al. Interleukin-6 (IL-6) and/or soluble IL-6 receptor down-regulation of human type II collagen gene expression in articular chondrocytes requires a decrease of Sp1.Sp3 ratio and of the binding activity of both factors to the COL2A1 promoter. *J. Biol. Chem.* **2008**, *283*, 4850–4865. [CrossRef]

nutrients

MDPI

Communication

Omega-3 Polyunsaturated Fatty Acids Inhibit the Function of Human URAT1, a Renal Urate Re-absorber

Hiroki Saito [1,2,†], Yu Toyoda [2,†], Tappei Takada [2,*,†], Hiroshi Hirata [1], Ami Ota-Kontani [1], Hiroshi Miyata [2], Naoyuki Kobayashi [1], Youichi Tsuchiya [1] and Hiroshi Suzuki [2]

1 Frontier Laboratories for Value Creation, Sapporo Holdings Ltd., 10 Okatome, Yaizu, Shizuoka 425-0013, Japan; hiroki.saito@sapporoholdings.co.jp (H.S.); hiroshi.hirata@sapporoholdings.co.jp (H.H.); ami.ota@sapporoholdings.co.jp (A.O.-K.); naoyuki.kobayashi@sapporoholdings.co.jp (N.K.); yoichi.tsuchiya@sapporoholdings.co.jp (Y.T.)

2 Department of Pharmacy, The University of Tokyo Hospital, 7-3-1 Hongo, Bunkyo-ku, Tokyo 113-8655, Japan; ytoyoda-tky@umin.ac.jp (Y.T.); hmiyata-tky@umin.ac.jp (H.M.); suzukihi-tky@umin.ac.jp (H.S.)

* Correspondence: tappei-tky@umin.ac.jp

† These authors contributed equally to this work.

Received: 2 May 2020; Accepted: 27 May 2020; Published: 29 May 2020

Abstract: The beneficial effects of fatty acids (FAs) on human health have attracted widespread interest. However, little is known about the impact of FAs on the handling of urate, the end-product of human purine metabolism, in the body. Increased serum urate levels occur in hyperuricemia, a disease that can lead to gout. In humans, urate filtered by the glomerulus of the kidney is majorly re-absorbed from primary urine into the blood via the urate transporter 1 (URAT1)-mediated pathway. URAT1 inhibition, thus, contributes to decreasing serum urate concentration by increasing net renal urate excretion. Here, we investigated the URAT1-inhibitory effects of 25 FAs that are commonly contained in foods or produced in the body. For this purpose, we conducted an in vitro transport assay using cells transiently expressing URAT1. Our results showed that unsaturated FAs, especially long-chain unsaturated FAs, inhibited URAT1 more strongly than saturated FAs. Among the tested unsaturated FAs, eicosapentaenoic acid, α-linolenic acid, and docosahexaenoic acid exhibited substantial URAT1-inhibitory activities, with half maximal inhibitory concentration values of 6.0, 14.2, and 15.2 μM, respectively. Although further studies are required to investigate whether the ω-3 polyunsaturated FAs can be employed as uricosuric agents, our findings further confirm FAs as nutritionally important substances influencing human health.

Keywords: Docosahexaenoic acid; eicosapentaenoic acid; functional food; gout; human health; hyperuricemia; PUFA; SLC22A12; transporter; uric acid; uricosuric activity

1. Introduction

Fatty acids (FAs) are physiologically important as energy sources and membrane constituents; moreover, FAs have diverse biological activities that modulate numerous cell/tissue properties in living organisms [1]. Accumulating evidence suggests that via such actions, dietary FAs can influence human health, well-being, and the risk of disease development. For instance, intake of polyunsaturated fatty acids (PUFAs) of the ω-3 family, such as eicosapentaenoic acid (EPA) and docosahexaenoic acid (DHA), has been considered to reduce the risk of chronic diseases including cardiovascular diseases; however, other effects of FAs consumption on human health remain controversial and need further investigation [2–5]. In addition to cardiovascular diseases, it is currently acknowledged that FAs influence a range of other diseases, such as metabolic and inflammatory diseases, as well as cancer [1]. However, little is known about the association between FAs and hyperuricemia, a common urate-related disease, especially regarding the effects of FAs on urate-handling machineries in the body.

Hyperuricemia, which is characterized by elevated serum uric acid (SUA) levels, is a lifestyle-related disease with high prevalence [6]. As hyperuricemia is a risk factor for gout, a very common form of inflammatory arthritis, SUA management at appropriate levels is becoming increasingly important in daily life [7,8]. Due to the lack of functional uricase (urate-degrading enzyme) in humans [9], uric acid is the end-product of human purine metabolism; urate excretion from the body is, therefore, necessary for the maintenance of uric acid homeostasis. The kidney is responsible for elimination of approximately two-thirds of urate [10]. However, only 3%–10% of the urate filtered by the glomerulus of the kidney is secreted to the urine [11] because most of it is re-absorbed from the primary urine into the blood in the proximal tubule by the urate transporter 1 (URAT1, also known as SLC22A12)-mediated pathway [12]. Thus, inhibition of this urate re-absorption pathway contributes to SUA lowering via the increase of net renal urate excretion.

URAT1 is a physiologically important renal urate re-absorber expressed on the brush border membrane of proximal tubular cells. Among the already identified urate re-absorbers expressed on the apical side of the renal cells, URAT1 is most strongly associated with SUA levels in humans [7], as supported by the fact that *URAT1* is the causative gene for renal hypouricemia type 1 [12], an inherited disorder characterized by impaired urate re-absorption in the kidney that results in extremely low SUA levels (SUA ≤ 2 mg/dL; normal range: 3.0–7.0 mg/dL). With hyperuricemia patients, this transporter is also the pharmacological target of uricosuric agents, which promote the excretion of urate, such as benzbromarone [12], lesinurad [13], and dotinurad [14]. In this context, daily consumption of nutrients with URAT1-inhibitory activity may have a beneficial effect on SUA management in subjects with high SUA levels. Actually, food ingredients that inhibit URAT1 function have attracted great interest; we and other groups identified some such natural ingredients from fruit flavonoids [15], coumarins [16], and wood pigments [17]. Nevertheless, despite the nutritional significance of FAs, their effects on URAT1 activity remain to be elucidated.

In the present study, we examined the URAT1-inhibitory effects of 25 FAs using an in vitro transport assay with mammalian cells transiently expressing URAT1. The cell-based assay revealed that unsaturated FAs inhibited URAT1 more strongly than saturated FAs.

2. Materials and Methods

2.1. Materials

Critical materials and resources used in this study are summarized in Table 1. All other chemicals used were commercially available and of analytical grade. The FAs were re-dissolved with dimethyl sulfoxide (DMSO; Nacalai Tesque, Kyoto, Japan) after the solvents were gently evaporated with nitrogen gas on the heat block at 50 °C. All experiments were conducted with the same lot of each vector plasmid for URAT1 (URAT1 wild-type in pEGFP-C1) or mock (pEGFP-C1), which were derived from our previous study [15].

2.2. Cell Culture

Human embryonic kidney 293-derived 293A cells were maintained in Dulbecco's modified Eagle's medium (Nacalai Tesque, Kyoto, Japan) supplemented with 10% fetal bovine serum (Biowest, Nuaillé, France), 1% penicillin–streptomycin (Nacalai Tesque, Kyoto, Japan), 2 mM L-glutamine (Nacalai Tesque, Kyoto, Japan), and 1 × non-essential amino acid (Life Technologies, Carlsbad, CA, USA) at 37 °C in a humidified atmosphere of 5% (v/v) CO_2 in air.

URAT1-expressing or mock plasmids were transfected into 293A cells using polyethylenimine "MAX" (PEI-MAX) (Polysciences, Warrington, PA, USA) as described previously [18], with some modifications. In brief, before transfection, 293A cells were seeded onto 12-well cell culture plates at a concentration of 0.92×10^5 cells/cm^2. Then, 24 h after seeding, each plasmid vector was transiently transfected into the cells using PEI-MAX (1 µg of plasmid/5 µL of PEI-MAX/well). The medium was replaced with fresh medium 24 h after transfection.

Table 1. Key resources.

Reagent or Resource	Source	Identifier
Antibodies		
Rabbit polyclonal anti-EGFP	Life Technologies	Cat# A11122; RRID: AB_221569; 1:1,000 dilution [1]
Rabbit polyclonal anti-α-tubulin	Abcam	Cat# ab15246; RRID: AB_301787; 1:1,000 dilution [1]
Donkey anti-rabbit IgG-horseradish peroxidase (HRP)-conjugate	GE Healthcare	Cat# NA934V; RRID: AB_772206; 1:3,000 dilution [1]
Chemicals		
[8-^{14}C]-Uric acid (53 mCi/mmol)	American Radiolabeled Chemicals	Cat# ARC0513
Arachidonic acid	Cayman Chemical	Cat# 90010; CAS: 506-32-1; Purity: ≥98%
Benzbromarone	FUJIFILM Wako Pure Chemical	Cat# 028-1585; CAS: 3562-84-3; Purity: >98%
Butyric acid	SIGMA-ALDRICH	Cat# B103500-5ML; CAS: 107-92-6; Purity: ≥99%
Decanoic acid	FUJIFILM Wako Pure Chemical	Cat# 033-01073; CAS: 334-48-5; Purity: ≥98%
Dimethyl sulfoxide	Nacalai Tesque	Cat# 13445-74; CAS: 67-68-5
Docosadienoic acid	Cayman Chemical	Cat# 20749; CAS: 17735-98-7; Purity: ≥98%
Docosahexaenoic acid	Cayman Chemical	Cat# 90310; CAS: 6217-54-5; Purity: ≥98%
Docosatetraenoic acid	Cayman Chemical	Cat# 90300; CAS: 28874-58-0; Purity: ≥98%
Dodecanoic acid	SIGMA-ALDRICH	Cat# L556-25G; CAS: 143-07-7; Purity: ≥98%
Eicosadienoic acid	Cayman Chemical	Cat# 90330; CAS: 2091-39-6; Purity: ≥98%
Eicosapentaenoic acid	Cayman Chemical	Cat# 90110; CAS: 10417-94-4; Purity: ≥98%
Eicosatrienoic acid	Cayman Chemical	Cat# 90192; CAS: 20590-32-3; Purity: ≥98%
Henicosapentaenoic acid	Cayman Chemical	Cat# 10670; CAS: 24257-10-1; Purity: ≥95%
Hexanoic acid	FUJIFILM Wako Pure Chemical	Cat# 081-06292; CAS: 142-62-1; Purity: ≥99%
Linoleic acid	Cayman Chemical	Cat# 90150; CAS: 60-33-3; Purity: ≥98%
Myristic acid	FUJIFILM Wako Pure Chemical	Cat# 130-03432; CAS: 544-63-8; Purity: ≥98%
Octanoic acid	SIGMA-ALDRICH	Cat# C2875-10ML; CAS: 124-07-2; Purity: ≥99%
Oleic acid	Cayman Chemical	Cat# 90260; CAS: 112-80-1; Purity: ≥98%
Palmitic acid	Cayman Chemical	Cat# 10006627; CAS: 57-10-3; Purity: ≥98%
Palmitoleic acid	Cayman Chemical	Cat# 10009871; CAS: 373-49-9; Purity: ≥99%
Polyethelenimine "MAX"	Polysciences	Cat# 24765; CAS: 49553-93-7
Stearic acid	SIGMA-ALDRICH	Cat# S4751-1G; CAS: 57-11-4; Purity: ≥98.5%
α-Eleostearic acid	Cayman Chemical	Cat# 10008349; CAS: 506-23-0; Purity: ≥95%
α-Linolenic acid	Cayman Chemical	Cat# 90210; CAS: 463-40-1; Purity: ≥98%
γ-Linolenic acid	Cayman Chemical	Cat# 90220; CAS: 506-26-3; Purity: ≥98%
ω-3 Eicosatetraenoic acid	Larodan Fine Chemicals	Cat# 10-2024; CAS: 24880-40-8; Purity: ≥98%
ω-3 Docosapentaenoic acid	Cayman Chemical	Cat# 90165; CAS: 24880-45-3; Purity: ≥98%
Critical Commercial Assays		
Pierce™ BCA Protein Assay Reagent A & B	Thermo Fisher Scientific	Cat# 23223, Cat# 23224
PureLink™ HiPure Plasmid Filter Midiprep Kit	Thermo Fisher Scientific	Cat# K210015
Recombinant DNA		
The complete URAT1 cDNA	Miyata et al., 2016 [18]	NCBI Reference Sequence: NM_144585.3
Experimental Models: Cell Lines		
293A	Invitrogen	R70507
Software and Algorithms		
Excel 2019	Microsoft	https://products.office.com/ja-jp/home
Statcel4 add-in software	OMS Publishing	http://www.oms-publ.co.jp/

[1] All antibodies were used at indicated dilutions in Tris-buffered saline containing 0.05% Tween 20 and 1% bovine serum albumin for 1 h at room temperature.

2.3. Preparation of Protein Lysates and Immunoblotting

Whole-cell lysates were prepared with cell lysis buffer A containing 50 mM Tris/HCl (pH 7.4), 1 mM dithiothreitol, 1% (*w/v*) Triton X-100, and protease inhibitor cOmplete, EDTA free (Roche, Basel, Switzerland), and were treated with peptide *N*-glycosidase F (PNGase F) (New England Biolabs, Ipswich, MA, USA) as described previously [19]. Protein concentration was determined using the

Pierce™ BCA Protein Assay Kit (Thermo Fisher Scientific, Kanagawa, Japan) with BSA as a standard, according to the manufacturer's protocol.

Whole-cell lysate samples were separated by SDS-PAGE and transferred to an Immobilon-P PVDF membrane (Millipore, Bedford, MA, USA) by electroblotting at 15 V for 60 min, as described previously [15]. Blots were probed with appropriate antibodies (Table 1), and the signals were visualized by chemiluminescence and detected using a multi-imaging Analyzer Fusion Solo 4™ system (Vilber Lourmat, Eberhardzell, Germany).

2.4. Confocal Microscopy

For confocal laser scanning microscopic observation, 48 h after the transfection, 293A cells were fixed with 4% paraformaldehyde for 15 min at room temperature, and further processed according to previous studies [15,20]. In brief, the cells were treated with a fluorescent wheat germ agglutinin conjugate (WGA, Alexa Fluor® 594 conjugate; Thermo Fisher Scientific) to visualize plasma membranes, followed by nuclear staining using TO-PRO-3 Iodide (Molecular Probes, Eugene, OR, USA). Then, the cells were mounted in VECTASHIELD Mounting Medium (Vector Laboratories, Burlingame, CA, USA). To analyze the localization of EGFP-fused URAT1 protein, fluorescence was observed using the FV10i Confocal Laser Scanning Microscope (Olympus, Tokyo, Japan).

2.5. Urate Uptake Assay Using URAT1-Expressing 293A Cells

The urate uptake assay using URAT1-expressing 293A cells was conducted according to our previous studies [15,18] with minor modifications. In brief, 48 h after plasmid transfection, cells were washed twice with Cl^--free transport buffer (Buffer T2: 125 mM Na-gluconate, 4.8 mM K-gluconate, 1.2 mM KH_2PO_4, 1.2 mM $MgSO_4$, 1.3 mM Ca-gluconate, 25 mM HEPES, 5.6 mM D-glucose, and pH 7.4) and pre-incubated in Buffer T2 for 15 min at 37 °C. The buffer was then exchanged with pre-warmed fresh Buffer T2 containing 5 μM [8-^{14}C]-urate with or without test compound at the indicated concentrations (0, 0.1, 0.3, 1, 3, 10, 30, 100, or 300 μM), and the cells were further incubated for 20 sec; 1% DMSO was used as a vehicle control. The cells were subsequently washed five times with ice-cold Buffer T2 and then lysed with 500 μL of 0.2 M NaOH on ice with gentle shaking for 1 h. The lysates were neutralized with 100 μL of 1 M HCl. Then, the radioactivity in the lysate was measured using a liquid scintillator (Tri-Carb 3110TR; PerkinElmer, Waltham, MA, USA). The protein concentration was determined using the Pierce™ BCA Protein Assay Kit. The urate transport activity was calculated as the incorporated clearance (μL/mg protein/min): (incorporated level of urate [disintegrations per minute (DPM)/mg protein/min]/urate level in the incubation mixture [DPM/μL]). URAT1-dependent urate transport activity was calculated by subtracting the urate transport activity of mock cells from that of URAT1-expressing cells.

Urate uptake was measured in the presence of several concentrations of each test compound to address their half maximal inhibitory concentration (IC_{50}) values. URAT1-mediated transport activities were then expressed as a percentage of control (100%). Based on the calculated values, fitting curves were obtained according to the following formula using the least-squares method with Excel 2019 (Microsoft, Redmond, WA, USA):

$$\text{Predicted value } [\%] = 100 - \left(\frac{E_{max} \times C^n}{EC_{50}^n + C^n}\right) \quad (1)$$

where E_{max} is the maximum effect, EC_{50} is the half maximal effective concentration, C is the concentration of the test compound, and n is the sigmoid-fit factor. Finally, based on the results, the IC_{50} was calculated as described previously [15].

2.6. Quantification and Statistical Analysis

All statistical analyses were performed using Excel 2019 with Statcel4 add-in software (OMS publishing, Saitama, Japan). Different statistical tests were used for different experiments as described in the figure legends, which include the number of biological replicates (n). Briefly, when analyzing multiple groups, the similarity of variance between groups was compared using Bartlett's test. When passing the test for homogeneity of variance, a parametric Tukey–Kramer multiple-comparison test for all pairwise comparisons was used. To investigate the inhibitory effect of each FA on URAT1 function (vs. vehicle control indicated as 100%) in the screening stage, one-sample t-test (one-sided) was conducted. Statistical significance was defined as $p < 0.05$ or 0.01.

Each experiment was designed to use samples required to obtain informative results and sufficient material for subsequent studies. No specific statistical test was used to pre-determine the sample sizes empirically determined in the current study. All experiments were monitored in a non-blinded fashion.

3. Results

3.1. URAT1-Mediated Urate Uptake in 293A Cells

Prior to screening the inhibitory effects of 25 FAs on URAT1, we verified our cell-based assay system—an in vitro urate transport assay with mammalian cells transiently expressing URAT1 (Figure 1). Expression of EGFP-tagged URAT1 (EGFP-URAT1) as a matured N-linked glycoprotein (Figure 1a) and its plasma membrane localization (Figure 1b) in 293A cells were confirmed 48 h after plasmid transfection by immunoblotting and confocal microscopy, respectively. Next, we successfully detected URAT1-mediated urate uptake into URAT1-expressing cells, which showed a much stronger transport activity compared to mock cells, representing background urate uptake, indicating that the assay was suitable for the screening (Figure 1c). As expected, URAT1-mediated urate uptake was almost completely inhibited by benzbromarone (30 μM), a URAT1 inhibitor employed as a uricosuric drug. These results were consistent with our previous study [15]. A schematic illustration of this urate transport assay is shown in Figure 1d.

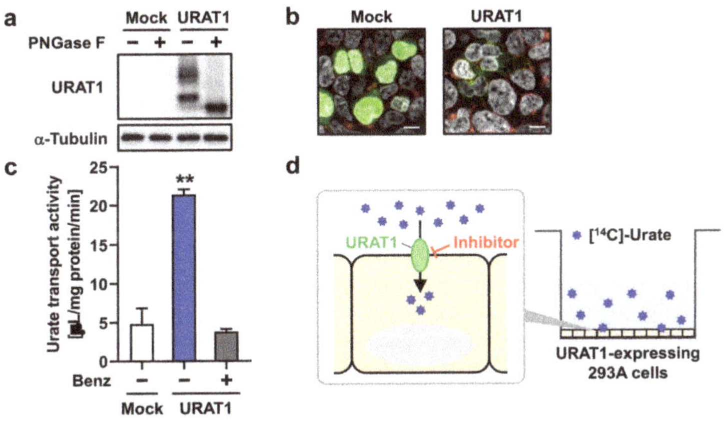

Figure 1. Cell-based urate transport assay with 293A cells transiently expressing URAT1. (**a**) Immunoblot detection of URAT1 protein in whole-cell lysates prepared 48 h after the transfection. α-Tubulin, a loading control. (**b**) Intracellular localization of URAT1. Confocal microscopy images were obtained 48 h after the transfection. Nuclei were stained with TO-PRO-3 iodide (gray); plasma membrane was labeled with Alexa Fluor® 594-conjugated wheat germ agglutinin (red). Bars, 10 μm. (**c**) Urate transport activities. Urate uptake into cells treated with or without 30 μM of benzbromarone (Benz) was measured. Data are expressed as the mean ± SD; $n = 3$. **, $p < 0.01$ (Tukey–Kramer multiple-comparison test). (**d**) Schematic illustration of URAT1-mediated urate transport examined using 293A cells transiently expressing URAT1.

3.2. *Unsaturated Fatty Acids Are Stronger Inhibitors of URAT1 Activity Than Saturated Fatty Acids*

Next, we examined the inhibitory effects of 25 FAs—8 saturated (Figure A1) and 17 unsaturated (Figure A2) FAs—at a concentration of 100 μM on URAT1 function (Figure 2). Despite some exceptions, in this study, almost all of the unsaturated FAs showed a stronger inhibitory effect on URAT1 than saturated FAs. Among the eight saturated FAs, relatively short FAs with chain lengths ranging from C4 to C8 had little effect on URAT1-mediated urate transport; instead, the others (C10 to C18) mildly inhibited URAT1 at the screening concentration. This result suggested that the length of FA could have a substantial effect on the URAT1-inhibitory activity of FAs. Among the 17 unsaturated FAs, 9 inhibited URAT1 activity by over 50%. We therefore focused on these candidates.

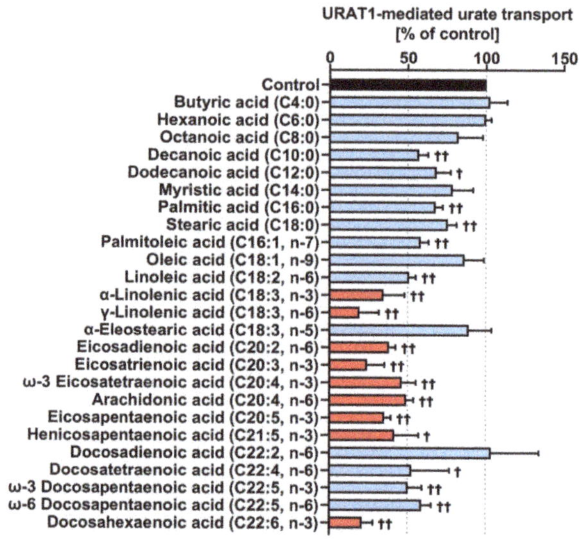

Figure 2. Inhibitory effects of each fatty acid on URAT1-mediated urate transport. The effects of each fatty acid (100 μM) on URAT1-mediated urate transport were investigated with the urate uptake assay. Control, vehicle (non-fatty acid treated) control. Data are expressed as the mean ± SD; $n = 3$–4. †, $p < 0.05$; ††, $p < 0.01$ vs. control (one-sample t-test). Red bars mean that the tested fatty acids inhibited URAT1-mediated urate transporter activity by over 50% compared to control.

3.3. *ω-3 Fatty Acids Are the Most Effective URAT1 Inhibitors*

Further investigation of the dose-dependent inhibitory effects of the nine unsaturated FAs on URAT1 determined the IC_{50} values that are illustrated in Figure 3. Based on the IC_{50} values, EPA was the strongest URAT1 inhibitor among the nine unsaturated FAs examined. Furthermore, EPA inhibited URAT1 activity more strongly than the other unsaturated FAs at low concentrations (≤1 μM) (Figure 3g). Second to EPA, its biosynthetic precursor α-linolenic acid (ALA) (Figure 3a) as well as its product DHA (Figure 3i) strongly inhibited URAT1, while ω-3 docosapentaenoic acid (DPA) showed an IC_{50} of > 100 μM (Figures 2 and A3). Additionally, contrary to ALA, linoleic acid (LA) had a high IC_{50} (133 μM); however, LA could inhibit URAT1 at low concentrations (≤1 μM) (Figure A3). Interestingly, considering the biosynthetic pathways of the above-described FAs distinguished by their structural feature (ω-3 or ω-6 family) (Figure A4), ω-3 FAs seem to inhibit URAT1 more effectively than ω-6 FAs. Given that elevated intake of ω-6 FAs may reportedly promote inflammation, while ω-3 FAs help reduce it [21], ω-3 FAs will be preferable to ω-6 FAs for the prevention of hyperuricemia/gout.

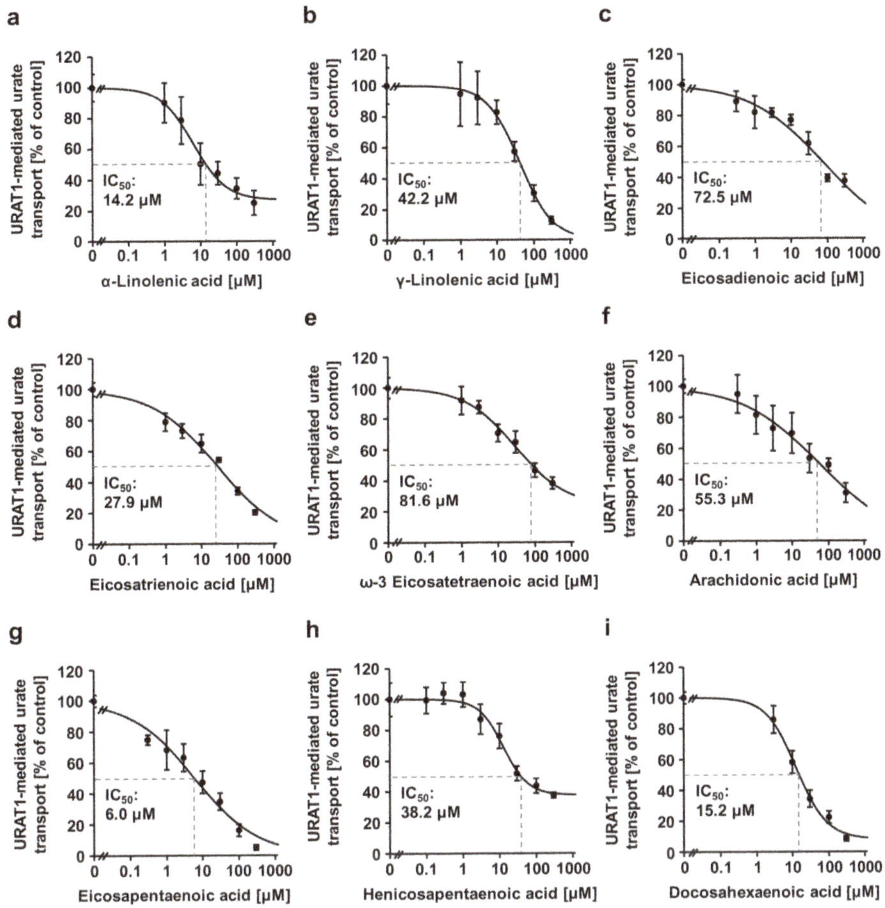

Figure 3. Concentration-dependent inhibition of URAT1-mediated urate transport by unsaturated fatty acids. The effects of each unsaturated fatty acid (0, 0.1, 0.3, 1, 3, 10, 30, 100, or 300 μM) on URAT1-mediated urate transport were investigated with the urate uptake assay. (**a**) α-Linolenic acid (ALA); (**b**) γ-linolenic acid; (**c**) eicosadienoic acid; (**d**) eicosatrienoic acid; (**e**) ω-3 eicosatetraenoic acid; (**f**) arachidonic acid (ω-6 eicosatetraenoic acid); (**g**) eicosapentaenoic acid (EPA); (**h**) henicosapentaenoic acid; (**i**) docosahexaenoic acid (DHA). Data are expressed as the mean ± SD; $n = 4$.

4. Discussion

In this study, to the best of our knowledge, we revealed for the first time the inhibitory effects of FAs on URAT1-dependent urate transport in vitro. While further studies are highly warranted to address the pathophysiological impact of our findings in terms of the possible effect on SUA and renal urate excretion in hyperuricemia model animals as well as in humans, the obtained results will open a new avenue for FAs as nutritionally important substances influencing human health.

Our findings may extend the potentially beneficial effects of PUFAs of the ω-3 family on reducing the risk of hyperuricemia/gout. In fact, recent studies have shown that a dietary intake of certain ω-3 PUFAs, such as EPA and DHA can, at least partly, reduce a wide number of inflammation-related biological reactions [1,4,5,22]. With respect to gout, a small case–control study on patients with this inflammatory condition showed an association between high ω-3 FA levels in the blood and lower frequency of gout attacks [23]. Additionally, ω-3 PUFA-rich fish consumption was reportedly associated with lower risk of recurrent gout attacks in a case-crossover study [24]. Moreover, the anti-inflammatory effects of ω-3 PUFAs were supported by the results obtained in animal models of acute inflammation induced by monosodium urate crystals [25,26]. Thus, in addition to the anti-inflammatory effects previously observed, it should be confirmed in future studies whether ω-3 PUFAs can exhibit uricosuric effects in hyperuricemia models or not.

The potential effects of the daily consumption of ω-3 PUFA-enriched foods on SUA will also be of interest. While the available information is limited and currently inconclusive, a randomized controlled trial in young healthy subjects showed that daily intake of fish oil (2 g; majorly consisting of DHA and EPA) resulted in a significant decrease of SUA after 4 and 8 weeks of supplementation [27]. A similar significant decline in SUA was also observed in healthy elderly men consuming daily supplement pills characterized by ω-3 FAs such as DHA and EPA for three months, although the change was not extensive [28]. Based on these pieces of evidence, increasing the daily intake of ω-3 FAs via eating pattern changes, such as appropriate choice of aliments and cooking oils, might be beneficial to health in terms of SUA management. On the other hand, the behavior of ω-3 FAs in the body has been hardly investigated in those clinical investigations, which warrants further studies focusing on the beneficial effects of dietary and/or endogenously produced ω-3 FAs on the renal urate handling in the body. Additionally, in such cases, not only the URAT1-inhibitory activity but also the disposition of target FAs should be considered.

There were some limitations to our study. First, the present study could not reveal how the FAs inhibited URAT1 function. Addressing this issue in the future will provide a deeper insight into the latent mechanistic features of URAT1. As an antiporter, URAT1 mediates urate transport in exchange for monocarboxylates such as lactate [12], which suggests that URAT1 must have at least two substrate recognition sites in its protein structure. Considering that FAs are carboxylic acids with a long aliphatic chain, they might affect the recognition and/or subsequent membrane transport of the counterpart substrates by URAT1 rather than the recognition of urate. Second, we could not exclude the possibility that the FAs affected the plasma membrane properties, which might result in the indirect decrease of URAT1 function. Nonetheless, given that the interaction of free FAs with cellular membranes occurs within minutes [21] and usually requires biochemical conversion of FAs into phospholipids, the experimental period we used in this study, 20 s incubation for urate uptake, was so short that the tested FAs must have had a negligible effect on the plasma membranes during the assay. Finally, the effects of FAs on other physiologically important urate transporters—GLUT9/SLC2A9 [29,30], OAT10/SLC22A13 [31], and ABCG2/BCRP [32–34]—remain to be elucidated. Since such urate transporters, including URAT1, coordinately regulate the behavior of urate in the human body, comprehensive understanding of the latent interaction between FAs and these transporters should be addressed in the future. Among them, GLUT9 expressed on the basal membrane of proximal tubular cells is involved in the urate transport from the cells to the blood as a counterpart of URAT1 [7]; *GLUT9* is the causative gene for renal hypouricemia type 2 [30]. Given these pieces of information, GLUT9 has the highest priority in the future investigation.

5. Conclusions

In conclusion, we herein found that FAs, especially ω-3 PUFAs such as EPA and DHA, could inhibit URAT1. To gain insight into the potential SUA-lowering effects of certain FAs, further investigations including human studies are required.

Author Contributions: Conceptualization, H.S. (Hiroki Saito), Y.T. (Yu Toyoda), and T.T.; methodology, H.S. (Hiroki Saito), Y.T. (Yu Toyoda) and H.H.; validation, H.S. (Hiroki Saito), Y.T. (Yu Toyoda) and T.T.; formal analysis, H.S. (Hiroki Saito) and Y.T. (Yu Toyoda); investigation, H.S. (Hiroki Saito), Y.T. (Yu Toyoda), H.H., A.O.-K. and H.M.; data curation, H.S. (Hiroki Saito), Y.T. (Yu Toyoda), and T.T.; writing—original draft preparation, H.S. (Hiroki Saito) and Y.T. (Yu Toyoda); writing—review and editing, Y.T. (Yu Toyoda) and T.T.; visualization, H.S. (Hiroki Saito) and Y.T. (Yu Toyoda); supervision, Y.T. (Youichi Tsuchiya) and H.S. (Hiroshi Suzuki); project administration, T.T. and N.K.; funding acquisition, T.T. All authors have read and agreed to the published version of the manuscript.

Funding: This study was supported by JSPS KAKENHI (grant numbers 16H01808 and 20H00568 to T.T.); T.T. has received research grants from Gout and uric acid foundation of Japan, MSD Life Science Foundation, Public Interest Incorporated Foundation, and Takeda Science Foundation. The funders had no role in the study design, data collection and analysis, decision to publish, or preparation of the manuscript.

Acknowledgments: The authors thank Syuichi Segawa for his continuous encouragement and support for this study. Also, we would like to thank Editage (www.editage.jp) for English language editing.

Conflicts of Interest: H. Saito, H.H., A.O.-K., N.K., and Y.Tsuchiya were the employees of Sapporo Holdings Ltd. H.Saito, Y.Toyoda, T.T., H.H., A.O-K., and H. Suzuki have a patent pending related to the work reported in this article. The remaining authors declare no competing interests.

Appendix A

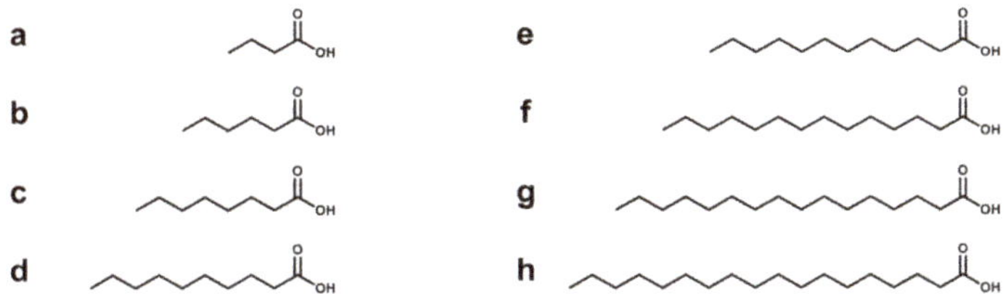

Figure A1. Chemical structures of saturated fatty acids tested in this study. (**a**) Butyric acid; (**b**) hexanoic acid; (**c**) octanoic acid; (**d**) decanoic acid; (**e**) dodecanoic acid; (**f**) myristic acid; (**g**) palmitic acid; (**h**) stearic acid.

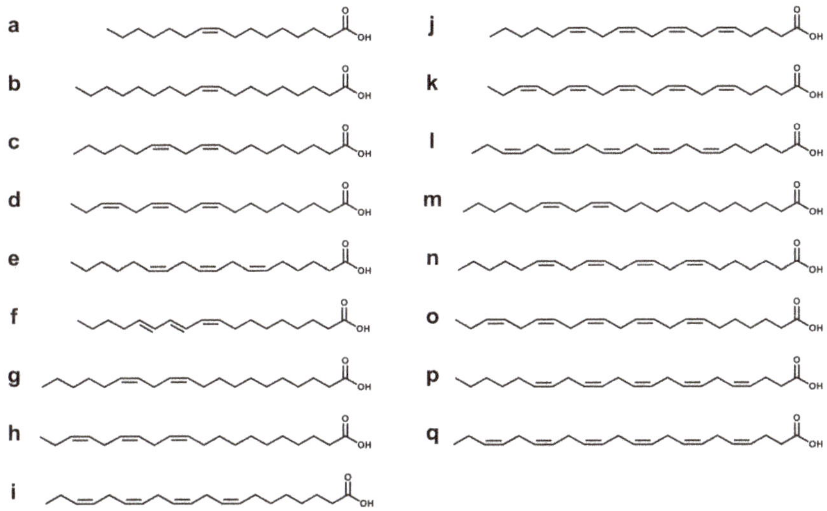

Figure A2. Chemical structures of unsaturated fatty acids tested in this study. (**a**) Palmitoleic acid; (**b**) oleic acid; (**c**) linoleic acid; (**d**) α-linolenic acid (ALA); (**e**) γ-linolenic acid; (**f**) α-eleostearic acid; (**g**) eicosadienoic acid; (**h**) eicosatrienoic acid; (**i**) ω-3 eicosatetraenoic acid; (**j**) arachidonic acid (ω-6 eicosatetraenoic acid); (**k**) eicosapentaenoic acid (EPA); (**l**) henicosapentaenoic acid; (**m**) docosadienoic acid; (**n**) docosatetraenoic acid; (**o**) ω-3 docosapentaenoic acid; (**p**) ω-6 docosapentaenoic acid; (**q**) docosahexaenoic acid (DHA).

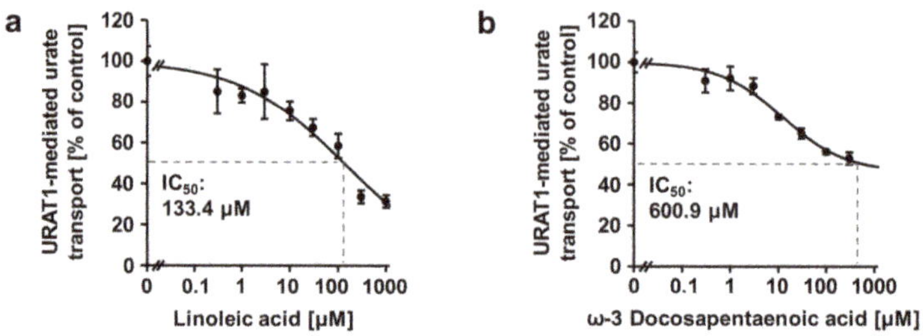

Figure A3. Concentration-dependent inhibition of URAT1-mediated urate transport by each fatty acid. Data are expressed as the mean ± SD; $n = 4$. (**a**) Linoleic acid; (**b**) ω-3 docosapentaenoic acid.

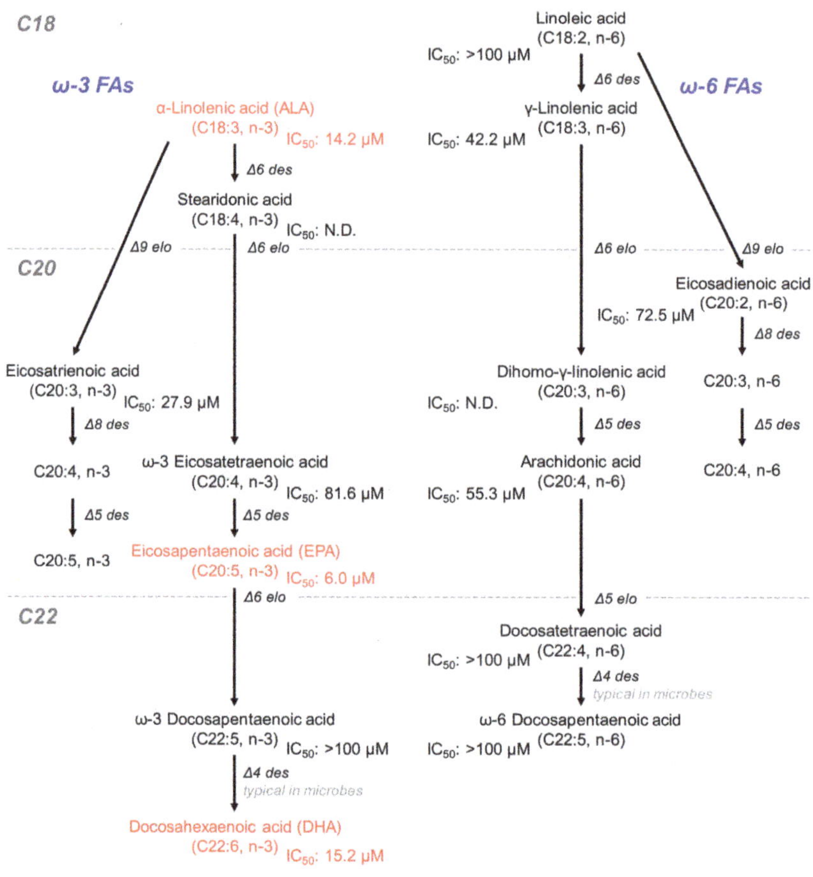

Figure A4. Biosynthetic route of fatty acids and half maximal inhibitory concentration (IC_{50}) values of each tested fatty acid against the urate transport mediated by URAT1. The IC_{50} values are from Figures 2, 3 and A3. Three polyunsaturated fatty acids with substantial URAT1-inhibitory activities are indicated in red. N.D., not determined in this study; FAs, fatty acids; *des*, desaturase; *elo*, elongase. The metabolic pathway is adapted from a previous report [4], with some modifications.

References

1. Calder, P.C. Functional Roles of Fatty Acids and Their Effects on Human Health. *JPEN J. Parenter. Enteral. Nutr.* **2015**, *39*, 18S–32S. [CrossRef] [PubMed]
2. Nguyen, Q.V.; Malau-Aduli, B.S.; Cavalieri, J.; Malau-Aduli, A.E.O.; Nichols, P.D. Enhancing Omega-3 Long-Chain Polyunsaturated Fatty Acid Content of Dairy-Derived Foods for Human Consumption. *Nutrients* **2019**, *11*, 743. [CrossRef] [PubMed]
3. Shahidi, F.; Ambigaipalan, P. Omega-3 Polyunsaturated Fatty Acids and Their Health Benefits. *Annu. Rev. Food Sci. Technol.* **2018**, *9*, 345–381. [PubMed]
4. Zarate, R.; El Jaber-Vazdekis, N.; Tejera, N.; Perez, J.A.; Rodriguez, C. Significance of long chain polyunsaturated fatty acids in human health. *Clin. Transl. Med.* **2017**, *6*, 25.
5. Bowen, K.J.; Harris, W.S.; Kris-Etherton, P.M. Omega-3 Fatty Acids and Cardiovascular Disease: Are There Benefits? *Curr. Treat. Options Cardiovasc. Med.* **2016**, *18*, 69.
6. Dalbeth, N.; Merriman, T.R.; Stamp, L.K. Gout. *Lancet* **2016**, *388*, 2039–2052. [CrossRef]
7. Dalbeth, N.; Choi, H.K.; Joosten, L.A.B.; Khanna, P.P.; Matsuo, H.; Perez-Ruiz, F.; Stamp, L.K. Gout. *Nat. Rev. Dis. Primers.* **2019**, *5*, 69. [CrossRef]
8. Major, T.J.; Dalbeth, N.; Stahl, E.A.; Merriman, T.R. An update on the genetics of hyperuricaemia and gout. *Nat. Rev. Rheumatol.* **2018**, *14*, 341–353.
9. Wu, X.W.; Muzny, D.M.; Lee, C.C.; Caskey, C.T. Two independent mutational events in the loss of urate oxidase during hominoid evolution. *J. Mol. Evol.* **1992**, *34*, 78–84. [CrossRef]
10. Bobulescu, I.A.; Moe, O.W. Renal transport of uric acid: evolving concepts and uncertainties. *Adv. Chronic Kidney Dis.* **2012**, *19*, 358–371. [CrossRef]
11. Hyndman, D.; Liu, S.; Miner, J.N. Urate Handling in the Human Body. *Curr. Rheumatol. Rep.* **2016**, *18*, 34. [CrossRef] [PubMed]
12. Enomoto, A.; Kimura, H.; Chairoungdua, A.; Shigeta, Y.; Jutabha, P.; Cha, S.H.; Hosoyamada, M.; Takeda, M.; Sekine, T.; Igarashi, T.; et al. Molecular identification of a renal urate anion exchanger that regulates blood urate levels. *Nature* **2002**, *417*, 447–452. [CrossRef] [PubMed]
13. Miner, J.N.; Tan, P.K.; Hyndman, D.; Liu, S.; Iverson, C.; Nanavati, P.; Hagerty, D.T.; Manhard, K.; Shen, Z.; Girardet, J.-L.; et al. Lesinurad, a novel, oral compound for gout, acts to decrease serum uric acid through inhibition of urate transporters in the kidney. *Arthritis. Res. Ther.* **2016**, *18*, 214. [CrossRef] [PubMed]
14. Taniguchi, T.; Ashizawa, N.; Matsumoto, K.; Saito, R.; Motoki, K.; Sakai, M.; Chikamatsu, N.; Hagihara, C.; Hashiba, M.; Iwanaga, T. Pharmacological Evaluation of Dotinurad, a Selective Urate Reabsorption Inhibitor. *J. Pharmacol. Exp. Ther.* **2019**, *371*, 162–170. [CrossRef]
15. Toyoda, Y.; Takada, T.; Saito, H.; Hirata, H.; Ota-Kontani, A.; Kobayashi, N.; Tsuchiya, Y.; Suzuki, H. Inhibitory effect of Citrus flavonoids on the in vitro transport activity of human urate transporter 1 (URAT1/SLC22A12), a renal re-absorber of urate. *N.P.J. Sci. Food.* **2020**, *4*, 3. [CrossRef]
16. Tashiro, Y.; Sakai, R.; Hirose-Sugiura, T.; Kato, Y.; Matsuo, H.; Takada, T.; Suzuki, H.; Makino, T. Effects of Osthol Isolated from Cnidium monnieri Fruit on Urate Transporter 1. *Molecules* **2018**, *23*, 2837. [CrossRef]
17. Yu, Z.; Fong, W.P.; Cheng, C.H. Morin (3,5,7,2',4'-pentahydroxyflavone) exhibits potent inhibitory actions on urate transport by the human urate anion transporter (hURAT1) expressed in human embryonic kidney cells. *Drug. Metab. Dispos.* **2007**, *35*, 981–986. [CrossRef]
18. Miyata, H.; Takada, T.; Toyoda, Y.; Matsuo, H.; Ichida, K.; Suzuki, H. Identification of Febuxostat as a New Strong ABCG2 Inhibitor: Potential Applications and Risks in Clinical Situations. *Front. Pharmacol.* **2016**, *7*, 518. [CrossRef]
19. Toyoda, Y.; Sakurai, A.; Mitani, Y.; Nakashima, M.; Yoshiura, K.-I.; Nakagawa, H.; Sakai, Y.; Ota, I.; Lezhava, A.; Hayashizaki, Y.; et al. Earwax, osmidrosis, and breast cancer: why does one SNP (538G>A) in the human ABC transporter ABCC11 gene determine earwax type? *FASEB J.* **2009**, *23*, 2001–2013. [CrossRef]
20. Toyoda, Y.; Mancikova, A.; Krylov, V.; Morimoto, K.; Pavelcová, K.; Bohatá, J.; Pavelka, K.; Suzuki, H.; Matsuo, H.; Takada , T.; et al. Functional Characterization of Clinically-Relevant Rare Variants in ABCG2 Identified in a Gout and Hyperuricemia Cohort. *Cells* **2019**, *8*, 363. [CrossRef]
21. Ibarguren, M.; Lopez, D.J.; Escriba, P.V. The effect of natural and synthetic fatty acids on membrane structure, microdomain organization, cellular functions and human health. *Biochim. Biophys. Acta.* **2014**, *1838*, 1518–1528. [CrossRef]

22. Calder, P.C. Omega-3 polyunsaturated fatty acids and inflammatory processes: nutrition or pharmacology? *Br. J. Clin. Pharmacol.* **2013**, *75*, 645–662. [CrossRef] [PubMed]
23. Abhishek, A.; Valdes, A.M.; Doherty, M. Low omega-3 fatty acid levels associate with frequent gout attacks: a case control study. *Ann. Rheum. Dis.* **2016**, *75*, 784–785. [CrossRef] [PubMed]
24. Zhang, M.; Zhang, Y.; Terkeltaub, R.; Chen, C.; Neogi, T. Effect of Dietary and Supplemental Omega-3 Polyunsaturated Fatty Acids on Risk of Recurrent Gout Flares. *Arthritis Rheumatol.* **2019**, *71*, 1580–1586. [CrossRef] [PubMed]
25. Iverson, C.; Bacong, A.; Liu, S.; Baumgartner, S.; Lundstrom, T.; Oscarsson, J.; Miner, J.N. Omega-3-carboxylic acids provide efficacious anti-inflammatory activity in models of crystal-mediated inflammation. *Sci. Rep.* **2018**, *8*, 1217. [CrossRef] [PubMed]
26. Tate, G.A.; Mandell, B.F.; Karmali, R.A.; Laposata, M.; Baker, D.G.; Schumacher, H.R., Jr.; Zurier, R.B. Suppression of monosodium urate crystal-induced acute inflammation by diets enriched with gamma-linolenic acid and eicosapentaenoic acid. *Arthritis Rheum.* **1988**, *31*, 1543–1551. [CrossRef] [PubMed]
27. Huang, T.; Li, K.; Asimi, S.; Chen, Q.; Li, D. Effect of vitamin B-12 and n-3 polyunsaturated fatty acids on plasma homocysteine, ferritin, C-reaction protein, and other cardiovascular risk factors: a randomized controlled trial. *Asia. Pac. J. Clin. Nutr.* **2015**, *24*, 403–411.
28. Stea, T.H.; Stolevik, S.B.; Berntsen, S.; Ezzathkah Bastani, N.; Paulsen, G.; Lohne Seiler, H.; Hetlelid, K.J.; Blomhoff, R.; Mansoor, M.A. Effect of Omega-3 and Vitamins E + C Supplements on the Concentration of Serum B-Vitamins and Plasma Redox Aminothiol Antioxidant Status in Elderly Men after Strength Training for Three Months. *Ann. Nutr. Metab.* **2016**, *68*, 145–155. [CrossRef]
29. Vitart, V.; Rudan, I.; Hayward, C.; Gray, N.K.; Floyd, J.; Knott, S.A.; Kolcic, I.; Polasek, O.; Graessler, J.; Wilson, J.F.; et al. SLC2A9 is a newly identified urate transporter influencing serum urate concentration, urate excretion and gout. *Nat. Genet.* **2008**, *40*, 437–442. [CrossRef]
30. Matsuo, H.; Chiba, T.; Nagamori, S.; Nakayama, A.; Domoto, H.; Phetdee, K.; Wiriyasermkul, P.; Kikuchi, Y.; Oda, T.; Nishiyama, J.; et al. Mutations in glucose transporter 9 gene SLC2A9 cause renal hypouricemia. *Am. J. Hum. Genet.* **2008**, *83*, 744–751. [CrossRef]
31. Higashino, T.; Morimoto, K.; Nakaoka, H.; Toyoda, Y.; Kawamura, Y.; Shimizu, S.; Nakamura, T.; Hosomichi, K.; Nakayama, A.; Ooyama, K.; et al. Dysfunctional missense variant of OAT10/SLC22A13 decreases gout risk and serum uric acid levels. *Ann Rheum Dis* **2020**, *79*, 164–166. [CrossRef] [PubMed]
32. Ichida, K.; Matsuo, H.; Takada, T.; Nakayama, A.; Murakami, K.; Shimizu, T.; Yamanashi, Y.; Kasuga, H.; Nakashima, H.; Nakamura, T.; et al. Decreased extra-renal urate excretion is a common cause of hyperuricemia. *Nat. Commun.* **2012**, *3*, 764. [PubMed]
33. Matsuo, H.; Takada, T.; Ichida, K.; Nakamura, T.; Nakayama, A.; Ikebuchi, Y.; Ito, K.; Kusanagi, Y.; Chiba, T.; Tadokoro, S.; et al. Common defects of ABCG2, a high-capacity urate exporter, cause gout: A function-based genetic analysis in a Japanese population. *Sci. Transl. Med.* **2009**, *1*, 5ra11. [CrossRef] [PubMed]
34. Woodward, O.M.; Kottgen, A.; Coresh, J.; Boerwinkle, E.; Guggino, W.B.; Kottgen, M. Identification of a urate transporter, ABCG2, with a common functional polymorphism causing gout. *Proc. Natl. Acad. Sci. USA* **2009**, *106*, 10338–10342. [PubMed]

nutrients MDPI

Article

Soy Isoflavone Genistein Inhibits an Axillary Osmidrosis Risk Factor ABCC11: In Vitro Screening and Fractional Approach for ABCC11-Inhibitory Activities in Plant Extracts and Dietary Flavonoids

Hiroki Saito [1,2,†], Yu Toyoda [2,*,†], Hiroshi Hirata [1], Ami Ota-Kontani [1], Youichi Tsuchiya [1], Tappei Takada [2] and Hiroshi Suzuki [2]

1 Frontier Laboratories for Value Creation, Sapporo Holdings Ltd., 10 Okatome, Yaizu, Shizuoka 425-0013, Japan; hiroki.saito@sapporoholdings.co.jp (H.S.); hiroshi.hirata@sapporoholdings.co.jp (H.H.); ami.ota@sapporoholdings.co.jp (A.O.-K.); yoichi.tsuchiya@sapporoholdings.co.jp (Y.T.)

2 Department of Pharmacy, The University of Tokyo Hospital, 7-3-1 Hongo, Bunkyo-ku, Tokyo 113-8655, Japan; tappei-tky@umin.ac.jp (T.T.); suzukihi-tky@umin.ac.jp (H.S.)

* Correspondence: ytoyoda-tky@umin.ac.jp

† These authors contributed equally to this work.

Received: 2 July 2020; Accepted: 12 August 2020; Published: 14 August 2020

Abstract: Axillary osmidrosis (AO) is a common chronic skin condition characterized by unpleasant body odors emanating from the armpits, and its aetiology is not fully understood. AO can seriously impair the psychosocial well-being of the affected individuals; however, no causal therapy has been established for it other than surgical treatment. Recent studies have revealed that human ATP-binding cassette transporter C11 (ABCC11) is an AO risk factor when it is expressed in the axillary apocrine glands—the sources of the offensive odors. Hence, identifying safe ways to inhibit ABCC11 may offer a breakthrough in treating AO. We herein screened for ABCC11-inhibitory activities in 34 natural products derived from plants cultivated for human consumption using an in vitro assay system to measure the ABCC11-mediated transport of radiolabeled dehydroepiandrosterone sulfate (DHEA-S—an ABCC11 substrate). The water extract of soybean (*Glycine max*) was found to exhibit the strongest transport inhibition. From this extract, via a fractionation approach, we successfully isolated and identified genistein, a soy isoflavone, as a novel ABCC11 inhibitor with a half-maximal inhibitory concentration value of 61.5 μM. Furthermore, we examined the effects of other dietary flavonoids on the ABCC11-mediated DHEA-S transport to uncover the effects of these phytochemicals on ABCC11 function. While further human studies are needed, our findings here about the natural compounds will help develop a non-surgical therapy for AO.

Keywords: axillary osmidrosis treatment; bioactivity investigation of food extract; body odor; food ingredient; functional food; *Glycine max*; health promotion; MRP8; phytochemicals; transporter

1. Introduction

Offensive or strong body odors can be a source of social embarrassment. Axillary osmidrosis (AO) is a chronic skin condition characterized by such body odors and excessive sweating from the armpits [1]. In Asian countries such as Japan and China where fewer people have strong body odor, AO is perceived even more negatively [2]. However, except for surgical treatments, no causal therapy has been established for AO.

The inhibition of human ATP-binding cassette transporter C11 (ABCC11, also known as MRP8)—a risk factor of AO—may induce physiological changes related to body odors [1,3–5]. ABCC11 is one of the ABC proteins that transport various molecules across cellular membranes in an ATP-dependent manner [6,7]. A non-synonymous single nucleotide polymorphism c.538G>A (p.Gly180Arg) in the *ABCC11* gene, which codes a functionally null variant with a high allele frequency in East Asians [8], has been found to be a determinant of AO risk [1,3–5]. Considering the facts that (1) genetically *ABCC11*-deficient subjects carry little AO risk and (2) the ABCC11 wild-type (WT) is expressed in human axillary apocrine glands that produce a variety of odor precursors [9], the inhibition of ABCC11 may lead to ways to prevent and treat AO. However, no medication is currently approved for AO treatment by ABCC11 inhibition. Hence, the exploration and identification of biologically safe ABCC11 inhibitors is an important issue.

In this study, we examined the ABCC11-inhibitory activities of 34 dietary plant products using an in vitro transport assay system. By screening the plant extracts and a subsequent fractional approach, genistein, a well-recognized soy isoflavone, was identified as a novel ABCC11 inhibitor with a half-maximal inhibitory concentration (IC_{50}) of 61.5 μM. Moreover, since little is known about food ingredients with the potential to inhibit ABCC11, we further investigated the effects of other dietary flavonoids on the ABCC11 function.

2. Materials and Methods

2.1. Materials

The key materials and resources used in this study are summarized in Table 1. All other chemicals used were commercially available and of analytical grade. The full-length human ABCC11 WT (NCBI accession: NM_033151) open reading frame in pcDNA3.1/hygro(−) plasmid [4] and recombinant adenoviruses for the expression of the human ABCC11 WT [10] were constructed in our previous studies; the plasmid/adenovirus vectors and the corresponding control vectors were prepared as a new experimental lot in this study. The plant materials (Table A1) were purchased from local supermarkets in Shizuoka, Japan, between July 2016 and July 2017.

Table 1. Key resources.

Reagent or Resource	Source	Identifier
Antibodies		
Rat monoclonal anti-MRP8 (ABCC11) antibody	Abcam, Cambridge, MA, USA	Cat# ab91452 [M8I-74]; RRID: AB_2049125
Rabbit polyclonal anti-Na^+/K^+-ATPase α antibody	Santa Cruz Biotechnology, Santa Cruz, CA, USA	Cat# sc-28800; RRID: AB_2290063
Goat anti-rat IgG–horseradish peroxidase (HRP) conjugate	GE Healthcare, Buckinghamshire, UK	Cat# NA935V; RRID: AB_772207
Donkey anti-rabbit IgG–HRP conjugate	GE Healthcare, Buckinghamshire, UK	Cat# NA934V; RRID: AB_772206
Chemicals		
Clear-sol II	Nacalai Tesque, Kyoto, Japan	Cat# 09136-83
Dehydroepiandrosterone sulfate, sodium salt, [1,2,6,7-^{3}H(N)]	PerkinElmer, Waltham, MA, USA	Cat# NET860; 60.0 Ci/mmol
Dimethyl Sulfoxide	Nacalai Tesque, Kyoto, Japan	Cat# 13445-74; CAS: 67-68-5
Methanol	Nacalai Tesque, Kyoto, Japan	Cat# 21929-23; CAS: 67-56-1
3-Hydroxyflavone	Tokyo Chemical Industry, Tokyo, Japan	Cat# H0379; CAS: 577-85-5; Purity: ≥98%
Apigenin	FUJIFILM Wako Pure Chemical, Osaka, Japan	Cat# 016-18911; CAS: 520-36-5; Purity: ≥95%
Cardamonin	R&D systems, Minneapolis, MN, USA	Cat# 2509/10; CAS: 19309-14-9; Purity: ≥98%
Daidzein	FUJIFILM Wako Pure Chemical, Osaka, Japan	Cat# 043-28071; CAS: 486-66-8; Purity: ≥98%
Daidzein 7-β-D-glucuronide 4′-sulfate disodium salt	Toronto Research Chemicals, North York, ON, Canada	Cat# D103525; CAS: 1041134-19-3; Purity: N/A

Table 1. *Cont.*

Reagent or Resource	Source	Identifier
Chemicals		
Dihydromyricetin	EXTRASYNTHESE, Genay, France	Cat# 1351-10mg; CAS: 27200-12-0; Purity: ≥95%
Fisetin	LKT Labs, Minneapolis, MN, USA	Cat# F3473; CAS: 528-48-3; Purity: ≥97%
Galangin	ChromaDex, Irvine, CA, USA	Cat# ASB-00007030-010; CAS: 548-83-4; Purity: N/A
Genistein	FUJIFILM Wako Pure Chemical, Osaka, Japan	Cat# 073-05531; CAS: 446-72-0; Purity: ≥98%
Genistein 7-β-D-glucuronide 4′-sulfate disodium salt	Toronto Research Chemicals, North York, ON, Canada	Cat# G349980; CAS: 176045-29-7; Purity: N/A
Glycitein	FUJIFILM Wako Pure Chemical, Osaka, Japan	Cat# 070-04701; CAS: 40957-83-3; Purity: ≥98%
Gossypetin	ChromaDex, Irvine, CA, USA	Cat# ASB-00007390-010; CAS: 489-35-0; Purity: N/A
Hesperetin	FUJIFILM Wako Pure Chemical, Osaka, Japan	Cat# 320-93841; CAS: 520-33-2; Purity: ≥96%
Isoliquiritigenin	Tokyo Chemical Industry, Tokyo, Japan	Cat# I0822; CAS: 961-29-5; Purity: ≥97%
Kaempferol	FUJIFILM Wako Pure Chemical, Osaka, Japan	Cat# 110-00451; CAS: 520-18-3; Purity: ≥95%
Luteolin	Cayman Chemical, Ann Arbor, MI, USA	Cat# 10004161; CAS: 491-70-3; Purity: ≥98%
Morin	Combi-Blocks, San Diego, CA, USA	Cat# QC-0527; CAS: 480-16-0; Purity: ≥98%
Myricetin	FUJIFILM Wako Pure Chemical, Osaka, Japan	Cat# 137-16791; CAS: 529-44-2; Purity: ≥98%
Naringenin	Tokyo Chemical Industry, Tokyo, Japan	Cat# N0072-5g; CAS: 67604-48-2; Purity: ≥93%
Naringenin chalcone	ChromaDex, Irvine, CA, USA	Cat# ASB-00014207-005; CAS: 73692-50-9; Purity: N/A
Nobiletin	FUJIFILM Wako Pure Chemical, Osaka, Japan	Cat# 149-09341; CAS: 478-01-3; Purity: N/A
Phloretin	FUJIFILM Wako Pure Chemical, Osaka, Japan	Cat# 160-17781; CAS: 60-82-2; Purity: ≥98%
Quercetagetin	ChromaDex, Irvine, CA, USA	Cat# ASB-00017020-005; CAS: 90-18-6; Purity: N/A
Quercetin	ChromaDex, Irvine, CA, USA	Cat# ASB-00017030-010; CAS: 117-39-5: Purity: ≥97%
(*S*)-Equol	Cayman Chemical, Ann Arbor, MI, USA	Cat# 10010173; CAS: 531-95-3; Purity: ≥98%
Taxifolin	EXTRASYNTHESE, Genay, France	Cat# 1036; CAS: 17654-26-1; Purity: N/A
Xanthohumol	TOKIWA PHYTOCHEMICAL, Chiba, Japan	Cat# P2217; CAS: 569-83-5; Purity: ≥98%
(+)-Catechin	FUJIFILM Wako Pure Chemical, Osaka, Japan	Cat# 038-23461; CAS: 154-23-4; Purity: ≥99%
(−)-Catechin gallate	Nagara Science, Gifu, Japan	Cat# NH021302; CAS: 130405-40-2; Purity: ≥98%
(−)-Epicatechin	FUJIFILM Wako Pure Chemical, Osaka, Japan	Cat# 059-06751; CAS: 490-46-0; Purity: ≥98%
(−)-Epicatechin gallate	FUJIFILM Wako Pure Chemical, Osaka, Japan	Cat# 052-06741; CAS: 1257-08-5; Purity: ≥98%
(−)-Epigallocatechin	FUJIFILM Wako Pure Chemical, Osaka, Japan	Cat# 059-08951; CAS: 970-74-1; Purity: ≥99%
(−)-Epigallocatechin gallate	FUJIFILM Wako Pure Chemical, Osaka, Japan	Cat# 056-08961; CAS: 989-51-5; Purity: ≥99%
(+)-Gallocatechin	FUJIFILM Wako Pure Chemical, Osaka, Japan	Cat# 075-06331; CAS: 970-73-0; Purity: ≥99%
(−)-Gallocatechin gallate	Nagara Science, Gifu, Japan	Cat# NH021402; CAS: 4233-96-9; Purity: ≥98%

Table 1. *Cont.*

Reagent or Resource	Source	Identifier
Adenoviruses		
ABCC11-expressing adenovirus	Toyoda et al. 2017 [9]	N/A
EGFP-expressing adenovirus	Toyoda et al. 2017 [9]	N/A
Recombinant DNA		
The complete human ABCC11 cDNA in pcDNA3.1/hyg(−)	Toyoda et al. 2009 [4]	NCBI Reference Sequence: NM_033151
Experimental Models: Cell Lines		
293A	Invitrogen, Waltham, MA, USA	R70507

N/A, not available.

2.2. Preparation of Plant Extracts

After the fruits were cleaned, the peels and pulps were carefully separated. The fresh and dried materials (summarized in Table A1) were finely chopped with a knife and ground using a mill (Crush Millser IFM-C20G; Iwatani, Tokyo, Japan), respectively. In the subsequent extraction step, approximately 50 g of the preprocessed plant material were well liquidized in 100 mL of distilled water using a juicer (Crush Millser IFM-C20G; Iwatani) and stirred for 30 min at room temperature. The suspension was centrifuged at 12,000× *g* at 4 °C for 10 min to remove the debris. The supernatant was collected and passed through ordinary filter paper. The filtrate was dialyzed against distilled water (500 mL) at 4 °C overnight with a dialysis membrane with a molecular weight cut-off of 14,000 (Spectrum Chemical Mfg, New Brunswick, NY, USA). The distilled water containing the small molecules that passed the dialysis membrane was lyophilized using FDU-2000 (EYELA, Tokyo, Japan). The freeze-dried extracts were stored at −20 °C, dissolved in ultrapure water at 10 mg/mL (10,000 ppm), and subjected to sonication as appropriate before use. Then, 5 μL of the solution were mixed with 20 μL of a transport buffer (10 mM Tris/HCl, 250 mM sucrose, and 10 mM $MgCl_2$, and pH 7.4); 1 μL of this clear liquid was used for a vesicle transport assay (total 20 μL/sample), as described below.

2.3. Cell Culture

Human embryonic kidney 293 (HEK293)-derived 293A cells were maintained in Dulbecco's Modified Eagle's Medium (Nacalai Tesque, Kyoto, Japan) supplemented with 10% fetal bovine serum (Biowest, Nuaillé, France), 1% penicillin-streptomycin (Nacalai Tesque), 2 mM L-glutamine (Nacalai Tesque), and 1 × non-essential amino acid (Life Technologies, Tokyo, Japan) at 37 °C in a humidified atmosphere of 5% CO_2 in air (*v*/*v*), following our previous study [11]. To obtain ABCC11-expressing 293A cells for the plasma membrane vesicles, we performed plasmid transfection using polyethylenimine MAX (1 mg/mL in Milli-Q water, pH 7.0; Polysciences, Warrington, PA, USA) [12] or adenovirus infection [9], as described previously.

2.4. Preparation of ABCC11-Expressing Plasma Membrane Vesicles

Plasma membrane vesicles were prepared from ABCC11-expressing 293A cells or control cells, as described previously [12], and then rapidly frozen in liquid N_2 and stored at −80 °C until use. Unless otherwise indicated, the plasma membrane vesicles used in the present study were derived from 293A cells 48 h after the plasmid transfection. The protein concentration of the plasma membrane vesicles was quantified using a BCA Protein Assay Kit (Pierce, Rockford, IL, USA) with bovine serum albumin as a standard according to the manufacturer's protocol.

2.5. Immunoblotting

The expression of ABCC11 protein in plasma membrane vesicles was examined by immunoblotting, as described previously [4,9] with minor modifications. Briefly, the prepared samples were electrophoretically separated on poly-acrylamide gels and transferred to a Hybond® ECL™ nitrocellulose membrane

(GE Healthcare, Buckinghamshire, UK) by electroblotting at 15 V for 70 min. After blocking by Tris-buffered saline containing 0.05% Tween 20 and 5% skim milk (TBST-skim milk) at 4 °C overnight, blots on the membrane were probed with a rat monoclonal anti-ABCC11 antibody (M8I-74; Abcam, Cambridge, MA, USA; diluted 200 fold) and a rabbit polyclonal anti-Na^+/K^+-ATPase α antibody (sc-28800; Santa Cruz Biotechnology, Santa Cruz, CA, USA; diluted 1000 fold), followed by incubation with a goat anti-rat immunoglobulin G (IgG)–horseradish peroxidase (HRP) conjugated antibody (NA935V; GE Healthcare; diluted 2000 fold) and a donkey anti-rabbit IgG–HRP conjugated antibody (NA934V; GE Healthcare; diluted 3000 fold), respectively. All antibodies were used in TBST-skim milk. HRP-dependent luminescence was developed using the ECL™ Prime Western Blotting Detection Reagent (GE Healthcare) and detected using a multi-imaging Analyzer Fusion Solo 4™ system (Vilber Lourmat, Eberhardzell, Germany).

2.6. Vesicle Transport Assay

The inhibitory effects of the various target extracts and compounds on the ABCC11 function were examined using the in vitro vesicle transport assay, a well-established method to quantitatively evaluate ABC transporter function [13]. For this purpose, the ATP-dependent transport of [1,2,6,7-^{3}H(N)]-dehydroepiandrosterone sulfate (DHEA-S) (PerkinElmer, Waltham, MA, USA), which is an ABCC11 substrate [7], into the ABCC11-expressing and control plasma membrane vesicles was quantified following our previous study [9] with some minor modifications in the rapid filtration technique, as described below.

In brief, the plasma membrane vesicles (0.25 mg/mL or indicated concentrations) were incubated with [1,2,6,7-^{3}H(N)]-DHEA-S (100 nM or indicated concentrations) in a reaction mixture (total 20 μL: 10 mM Tris/HCl, 250 mM sucrose, 10 mM $MgCl_2$, 10 mM creatine phosphate, 1 mg/mL creatine phosphokinase, 50 mM ATP or AMP as a substitute of ATP, and pH 7.4) for 5 min at 37 °C, either without (i.e., with only vehicle control) or with the individual target fractions/authentic chemicals at the indicated concentrations. As the vehicle control, 1% water was used for plant extracts; 1% methanol (Nacalai Tesque) or 1% dimethyl sulfoxide (DMSO; Nacalai Tesque) was used for the individual target fractions, as described below. Since stock solutions of authentic chemicals were prepared with DMSO at 10 mM, 1% DMSO was employed as the vehicle control for them. After incubation, the reaction mixture was mixed with 980 μL of an ice-cold stop buffer (2 mM EDTA, 0.25 M sucrose, 0.1 M NaCl, 10 mM Tris-HCl, and pH 7.4) and rapidly filtered on a membrane filter (MF-Millipore Membrane (HAWP02500; Millipore, Tokyo, Japan) for extract screening or Whatman™ Grade GF/F Glass Microfiber Filter Paper (GE Healthcare) for the other experiments). After washing with 5 mL of the ice-cold stop buffer three times, the plasma membrane vesicles trapped on the membrane filter were dissolved in Clear-sol II (Nacalai Tesque). Then, the radioactivity incorporated into the plasma membrane vesicles was measured with a liquid scintillator (Tri-Carb 3110TR; PerkinElmer).

The transport activity in each group was calculated as the incorporated clearance (μL/mg protein/min = incorporated level of DHEA-S (disintegrations per minute (DPM)/mg protein/min)/ DHEA-S level in the incubation mixture (DPM/μL)). ATP-dependent DHEA-S transport was calculated by the difference in transport activity with and without ATP. Similarly, ABCC11-mediated DHEA-S transport activity was calculated by subtracting the ATP-dependent DHEA-S transport activity of control plasma membrane vesicles from that of ABCC11-expressing ones. Unless otherwise indicated, effects of the target fractions/compounds on the ATP-dependent DHEA-S transport activity were also examined for the control plasma membrane vesicles.

2.7. Fractionation of Soybean (Glycine max) Extract

Medium-pressure liquid chromatography (MPLC) was conducted using a dual channel automated flash chromatography system (EPCLC-W-Prep 2XY; YAMAZEN, Osaka, Japan), as described below. All the eluates were evaporated to dryness and then stored at −20 °C. They were reconstituted in an appropriate solvent before use in the vesicle transport assay for the evaluation of ABCC11-inhibitory activities and/or chemical characterization by mass spectrometry (MS) analysis.

The water extract of dry soybeans was separated into 12 fractions (Fr.#1-12) by MPLC on an octadecyl-silica (ODS) column (DispoPackAT ODS-25; particle size 25 μm, column size 120 g, i.d. 40 × 188 mm; YMC, Kyoto, Japan). The separation was performed in the linear gradient elution mode with solvent A (0.2% formic acid in water) and solvent B (0.2% formic acid in acetonitrile) (solvent A:solvent B (*v*/*v*): 0–5 min 95:5; 5–25 min 95:5 to 0:100; and 25–35 min 0:100) at a flow rate of 40 mL/min, with UV monitoring at 265 nm using an equipped UV detector. Each fraction was reconstituted (10 mg/mL) in an appropriate solvent (i.e., water for Fr.#1 and Fr.#2, 50% methanol for Fr.#3-11, and methanol for Fr.#12) before use.

Among the 12 fractions, Fr.#11 (the target fraction reconstituted in 50% methanol) was further subjected to MPLC over an ODS column (RediSep ODS GOLD; 5.5 g media, 20–40 μm spherical; Teledyne Isco, Lincoln, NE, USA) in the stepwise elution mode using a mixture of the same A and B solvents (solvent A:solvent B (*v*/*v*): 0–2 min 80:20; 2–9 min 50:50; and 9–17 min 0:100) at a flow rate of 15 mL/min with UV monitoring at 254 nm. This gives three subfractions (Fr.#11-1 to Fr.#11-3) plus a dominant peak eluted from 3.0 to 5.2 min. The dominant peak was collected and then further separated in the same column with a linear gradient of 10–50% of solvent B in solvent A to give three more subfractions (Fr.#11-4 to Fr.#11-6).

Finally, to further separate ABCC11-inhibitory ingredients, Fr.#11-5—the most active subfraction among Fr.#11-1 to Fr.#11-6 in terms of ABCC11 inhibition—was purified by a recycling preparative HPLC system (LaboACE LC-5060; Japan Analytical Industry, Tokyo, Japan) equipped with a gel permeation column (JAIGEL-GS310; i.d. 20 × 500 mm; Japan Analytical Industry), using methanol as a mobile phase at 5 mL/min and with refractive index monitoring and UV monitoring at 254 nm. In brief, Fr.#11-5 was separated by the recycling mode for 120 min. Then, Fr.#11-5-1 and Fr.#11-5-2 were collected from 123 to 126 min and from 160 to 176 min, respectively. All the wastes were collected and further processed as Fr.#11-5-3. Additionally, all the subfractions were evaporated to dryness and then stored at −20 °C. They were reconstituted in DMSO (2 mg/mL) before use.

2.8. Chemical Characterizations

For the qualitative determination of the isolated compounds, chromatographic separations, and subsequent MS (or MS/MS) analyses were carried out with an LC-quadrupole time-of-flight (Q-TOF)-MS/MS system consisting of an HPLC instrument (Agilent 1100 Series equipped with a diode array and multiple wavelength detector (DAD) (G1316A); Agilent Technologies, Santa Clara, CA, USA) coupled with an Agilent 6510 Q-TOF (Agilent Technologies). The chromatographic conditions and MS setting were drawn from our previous study [14] with some minor modifications. Briefly, the separation was performed on a Zorbax Eclipse Plus C18 column (2.1 × 100 mm; Agilent Technologies) maintained at 40 °C under gradient mobile conditions with a mixture of solvent C (0.1% formic acid in water) and solvent D (acetonitrile) (solvent C:solvent D (*v*/*v*): 0–8 min 95:5 to 5:95, and 8–12 min 5:95) with a flow rate of 0.5 mL/min. The detection range of the DAD was set from 190 to 400 nm, and the MS detection system operated in the positive ionization mode at an MS scan range of *m*/*z* 100–1700. Peak analysis was performed using the Agilent MassHunter Workstation software (version B.03.01; Agilent Technologies).

2.9. Calculation of the Half-Maximal Inhibitory Concentration Values

To calculate the IC_{50} value of genistein against DHEA-S transport by ABCC11, the DHEA-S transport activities were measured in the presence of genistein at several concentrations. The ABCC11-mediated DHEA-S transport activities were expressed as a percentage of the control (100%). Based on the calculated values, fitting was carried out with the following formula using the least-squares methods in Excel 2019 (Microsoft, Redmond, WA, USA), as described previously [15]:

$$\text{Predicted value } [\%] = 100 - \left(\mathrm{E_{max}} \times \mathrm{C^n} / \mathrm{EC_{50}^n} + \mathrm{C^n}\right) \tag{1}$$

where E_{max} is the maximum effect, EC_{50} is the half maximal effective concentration, C is the concentration of the test compound, and n is the sigmoid-fit factor. IC_{50} was calculated based on these results.

2.10. Statistical Analysis

All statistical analyses were performed using Excel 2019 with the Statcel4 add-in software (OMS publishing, Saitama, Japan). Various statistical tests were used for different experiments, as described in the figure legends. Briefly, when analyzing multiple groups, the similarity of variance between groups was compared using Bartlett's test. When passing the test for homogeneity of variance, a parametric Tukey–Kramer multiple-comparison test or a Dunnett's test for comparisons with a control group was used. To investigate the inhibitory effect of each dietary food ingredient on ABCC11 function (vs. vehicle control as 100%), one-sample *t*-test (one-sided) was conducted. Statistical significance was defined in terms of $p < 0.05$ or 0.01. The sample sizes were empirically determined to ensure informative results and sufficient material for subsequent studies, and no specific statistical test was used in deciding them. All experiments were monitored in a non-blinded fashion.

2.11. Availability of Data and Material

Data supporting the results of this study are included in this published article and its appendix or are available from the corresponding author on reasonable request.

3. Results

3.1. Confirmation of ABCC11-Mediated Transport Activity

Prior to screening the ABCC11-inhibitory activities of natural products, we verified the transport assay system used in the present study. Immunoblotting with the anti-ABCC11 antibody confirmed the expression of ABCC11 protein as a matured *N*-linked glycoprotein in the plasma membrane vesicles prepared from the ABCC11-expressing cells (Figure 1a). No detectable expression of ABCC11 was observed in the control vesicles. We then measured the ATP-dependent DHEA-S transport into the ABCC11-expressing plasma membrane vesicles (Figure 1b). The DHEA-S transport activities of the ABCC11 vesicles were remarkably higher than those of the mock vesicles, which was enough for the quantitative evaluation of ABCC11-mediated DHEA-S transport activity in subsequent processes.

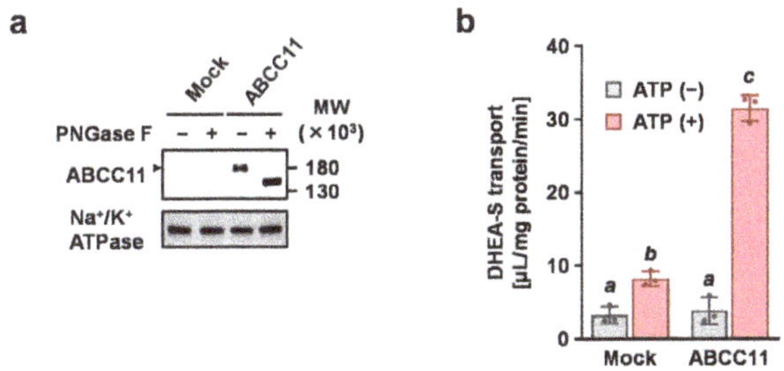

Figure 1. Expression and function of ABCC11. (**a**) Immunoblot detection of ABCC11 protein in the plasma membrane vesicles using an anti-ABCC11 antibody. Mock means plasma membrane vesicles that were prepared from control cells transfected with an empty pcDNA3.1/hyg(−) vector. Arrowhead: matured ABCC11 as an *N*-linked glycosylated protein. Na^+/K^+-ATPase (a plasma membrane protein) was used for a loading control. (**b**) [1,2,6,7-^{3}H(N)]-dehydroepiandrosterone sulfate (DHEA-S) transport activities. Plasma membrane vesicles were incubated with or without ATP for 5 min. In this assay, all incubation mixtures contained 1% dimethyl sulfoxide (DMSO). Data are expressed as the mean ± SD; $n = 3$. Statistical analyses for significant differences were performed using Bartlett's test, followed by a parametric Tukey–Kramer multiple-comparison test. Different letters indicate significant differences between groups ($p < 0.05$).

3.2. Screening the ABCC11-Inhibitory Activities of Plant Extracts

For the ABCC11-inhibitory properties of natural products, we focused on plants commonly found in the human diet including citruses, tea leaves, soybeans, and miso, a traditional grain-based fermented food in Japan [16]. Each sample was extracted with water and then dialyzed, and the resulting outer layer was lyophilized and reconstituted in water at 10 mg/mL. The 34 obtained concentrates (final concentration at 100 ppm) were used for screening the ABCC11-inhibitory activity (Figure 2). Since the extract of soybean (*Glycine max*) showed the highest inhibitory activities (approximately 70% inhibition) and soybean is a common crop consumed globally, we further explored the ingredients therein responsible for the ABCC11-inhibitory activity.

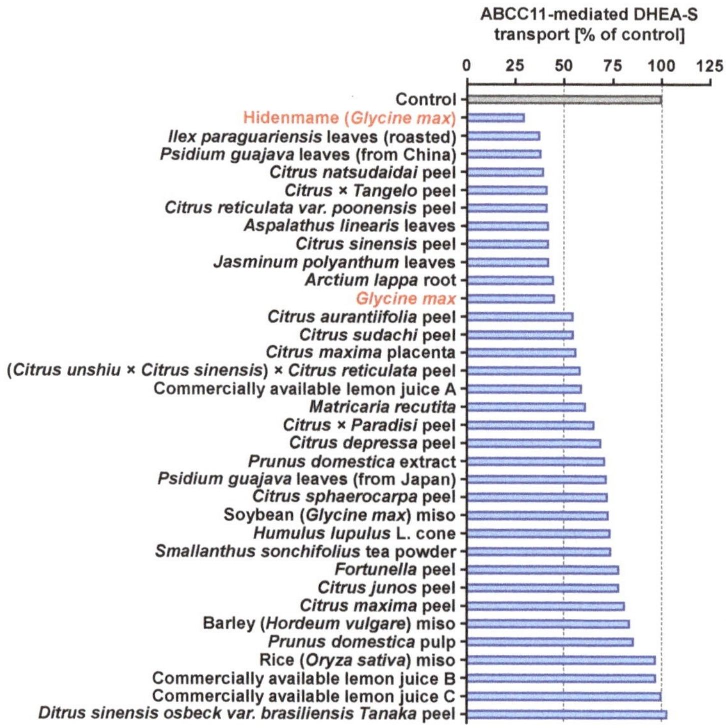

Figure 2. Screening of inhibitory effects of various plant extracts on the transport activity of ABCC11. Inhibitory effect of each plant extract on the ABCC11-mediated [1,2,6,7-^{3}H(N)]-DHEA-S transport activity was investigated by the vesicle transport assay. Plasma membrane vesicles (0.375 mg/mL in the reaction mixture) were incubated with the extract (100 ppm) in the presence of 50 μM [1,2,6,7-^{3}H(N)]-DHEA-S for 5 min; 1% water was used for the vehicle control. Data are expressed as % of vehicle control, and they represent averages of two independent experiments.

3.3. Fractionation and Isolation of Glycine Max (Soybean) Extract by Chromatographic Separations

To determine the ABCC11-inhibitory ingredients in the water extract of soybeans, further fractionation was conducted with liquid chromatographic separations in a total of three steps (Figure 3). First, the water extract was separated with a preparative MPLC system to yield 12 fractions (Fr.#1-12) (Figure 4a). The ABCC11-inhibitory activities of these 12 fractions were measured at 100 ppm (Figure 4b). Fr.#1-9 showed no significant effect, whereas Fr.#10-12 significantly inhibited the ABCC11-mediated DHEA-S transport. Secondly, since Fr.#11 exhibited the highest activity, we next further separated it with a similar preparative MPLC to give a total of six subfractions (Fr.#11-1 to Fr.#11-6), as described in Materials and Methods (Section 2.7). Monitoring the MPLC effluent at 254 nm showed that, among the six subfractions, the main compounds were collected in Fr.#11-5, which had the highest ABCC11-inhibitory activity (approximately 22% of inhibition at 50 ppm) among the six subfractions.

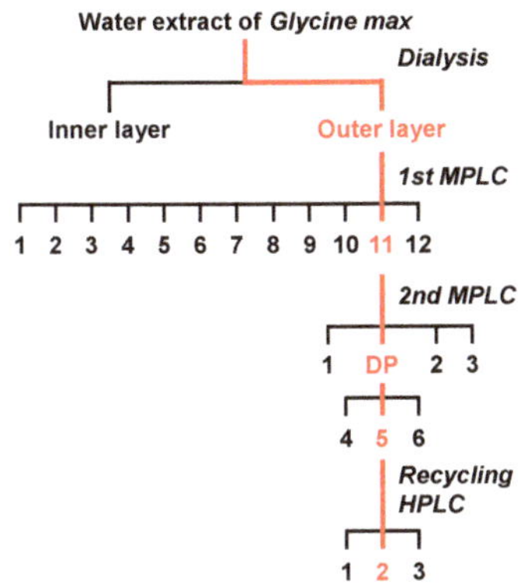

Figure 3. Separation scheme used to fractionate ABCC11 inhibitors in the soybean extract. In each separation step, the fraction with the highest ABCC11-inhibitory activity is colored in red. DP: dominant peak (details are described in Materials and Methods).

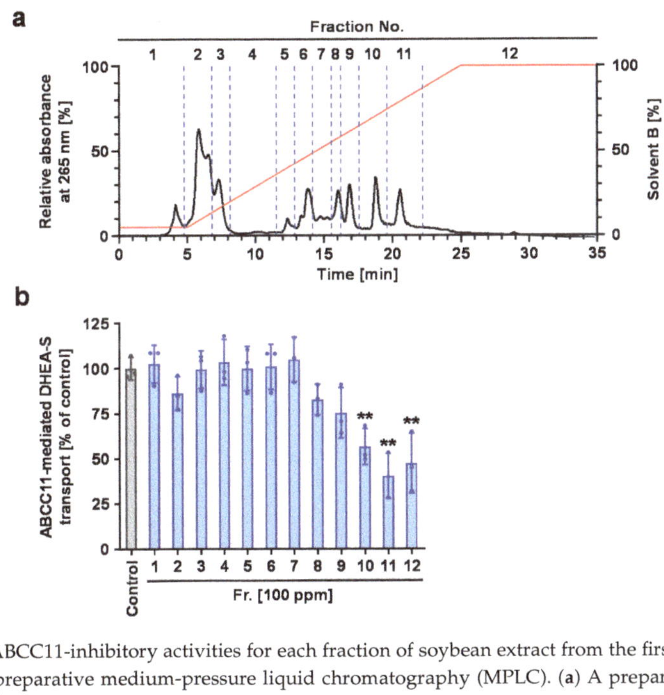

Figure 4. ABCC11-inhibitory activities for each fraction of soybean extract from the first separation step with preparative medium-pressure liquid chromatography (MPLC). (**a**) A preparative MPLC chromatogram for separating the water extract of soybeans. The chromatogram was recorded at 265 nm. Red line indicates linear gradients of solvent B (0.2% formic acid in acetonitrile). (**b**) ABCC11-inhibitory activity profile of each fraction (100 ppm) obtained from the first separation process. The effects on ABCC11-mediated [1,2,6,7-^{3}H(N)]-DHEA-S transport activity were investigated by the vesicle transport assay; 1% methanol was used for the vehicle control. Data are expressed as % of vehicle and the mean ± SD; $n = 3$. **, $p < 0.01$ vs. control (Dunnett's test).

Thirdly, to isolate the substances responsible for the ABCC11 inhibition, Fr.#11-5 was further subjected to recycling HPLC, which was repeated to afford components from peak #11-5-1 and peak #11-5-2 (denoted as Fr.#11-5-1 and Fr.#11-5-2, respectively; Figure 5a). All the wastes of this process were collected and further processed as Fr.#11-5-3. All three subfractions showed ABCC11-inhibitory activities at 20 ppm, and Fr.#11-5-2 was the most active (Figure 5b) and therefore the object of further analysis. Of note, the re-chromatography of Fr.#11-5-2 followed by LC-Q-TOF-MS and LC-DAD analyses suggested that this subfraction was mainly composed of a single substance that should be responsible for the ABCC11-inhibitory activity (Figure 5c). Indeed, a full LC-Q-TOF-MS scan of Fr.#11-5-2 revealed a constituent with a retention time of 5.83 min. Ions were detected in the positive ion mode at *m*/*z* 271.0616 and 293.0428, which corresponded to the $[M+H]^+$ and $[M+Na]^+$ of the constituent, respectively (Figure 5d).

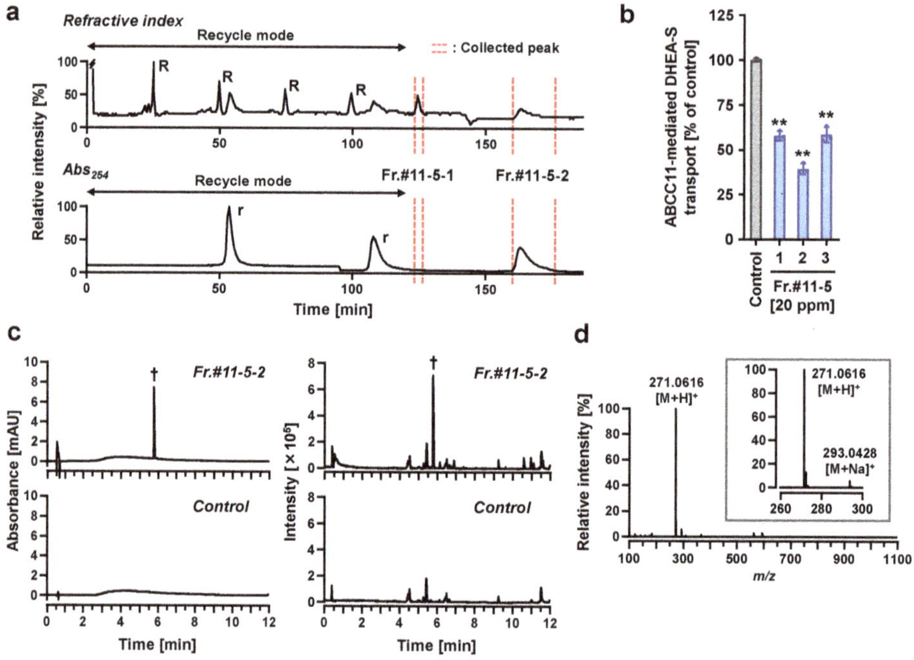

Figure 5. Isolation of an ABCC11-inhibitory ingredient by means of recycling preparative HPLC. (**a**) Recycling preparative HPLC chromatograms for the separation of fractions Fr.#11-5-1 and Fr.#11-5-2. The upper chromatogram was recorded with a refractive index detector, and the lower one was recorded with a diode array and multiple-wavelength detector at 254 nm. After separation under the recycling mode (0–120 min), the mode was changed; Fr.#11-5-1 (123–126 min) and Fr.#11-5-2 (160–176 min) were collected, and all the wastes were collected and further processed as Fr.#11-5-3. R, recycled peaks for Fr.#11-5-1; r, recycled peaks for Fr.#11-5-2. (**b**) ABCC11-inhibitory activities of each subfraction (20 ppm) in terms of ABCC11-mediated [1,2,6,7-^{3}H(N)]-DHEA-S transport activity measured by the vesicle transport assay; 1% DMSO was used for the vehicle control. Data are expressed as % of vehicle and the mean ± SD; $n = 3$. **, $p < 0.01$ vs. control (Dunnett's test). (**c**) Purity verification of the isolated ingredient in Fr.#11-5-2 by spectrometric analyses. Left: UV chromatograms recorded at 265 nm. Right: LC-quadrupole time-of-flight-MS (LC-Q-TOF-MS) base peak chromatograms, excluding peaks derived from the plasticizing materials and injected solvent. †, a specific peak in Fr.#11-5-2 with a retention time of 5.83 min. (**d**) Full scan mass spectrum obtained in the positive ion mode of this peak (indicated by † in **c**) at 5.83 min. The inset is the magnified view for ions at *m*/*z* 271.0616 and 293.0428, which corresponded to the $[M + H]^+$ and $[M + Na]^+$ of the target constituent, respectively.

3.4. Structural Characterization of the Putative ABCC11 Inhibitor Derived from Soybeans

We next conducted a series of spectrometric analyses (Figure 6) to obtain structural information about the candidate active ingredient, which was almost completely isolated from the soybean extract into Fr.#11-5-2. Based on accurate mass information from the LC-Q-TOF-MS analysis (Figure 5d), the elemental composition of the target analyte was determined as $C_{15}H_{10}O_5$ (Δ−5.51 and Δ−2.84 ppm from $[M+H]^+$ and $[M+Na]^+$, respectively). The three major soy isoflavones are genistein, daidzein, and glycitein, with the respective formulas (monoisotopic mass) of $C_{15}H_{10}O_5$ (270.0528), $C_{15}H_{10}O_4$ (254.0579), and $C_{16}H_{10}O_4$ (284.0685) [17]. Additionally, isoflavones exhibit an intense UV absorption between 240 and 280 nm associated with their benzoyl system, and the target analyte showed a similar spectrometric feature. Therefore, we hypothesized that the active ingredient would be genistein (Figure 6a). This hypothesis was tested by spectroscopic analyses, which demonstrated that the Fr.#11-5-2 and authentic genistein were identical in their retention time (Figure 6b), accurate mass of parent ion and the ratios of adduct ions (Figure 6c), photoabsorption spectrum (Figure 6d), and MS/MS spectrum (Figure 6e). Hence, the isolated substance should be genistein.

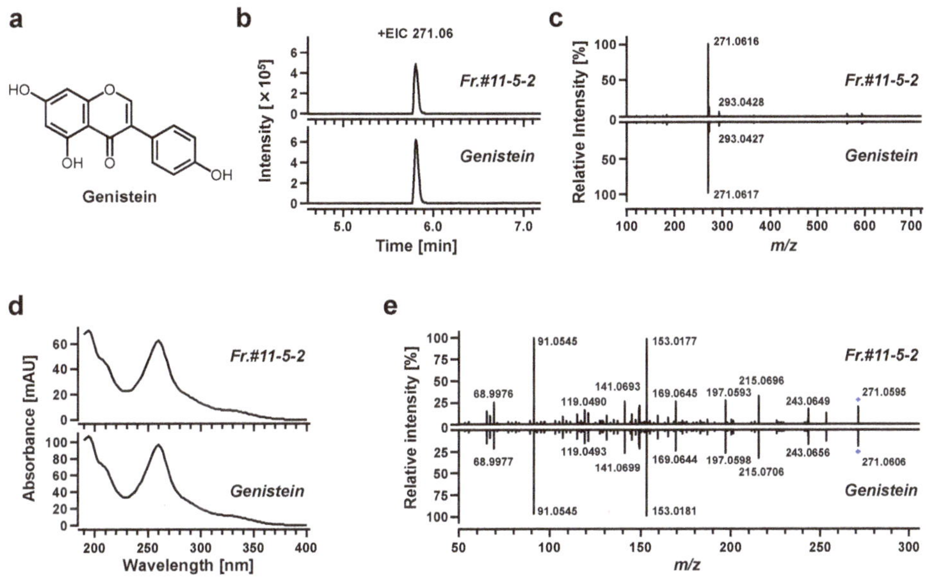

Figure 6. Chemical characterization of an ABCC11 inhibitory activity-guided fraction from soybean extract. Fraction (Fr.) #11-5-2 (upper panels) and authentic genistein (lower panels) were analyzed by a high-performance liquid chromatography instrument coupled with a diode array and multiple wavelength detector (DAD) and Q-TOF-MS system. (**a**) Chemical structure of genistein. (**b**) Extracted ion chromatograms (EICs) with a single peak at *m/z* 271.0621 in the positive ESI spectrum. (**c**) MS spectrums with a retention time of 5.83 min for the parent ion. (**d**) DAD spectrums. (**e**) Information on the fragment ions derived from MS/MS analyses.

3.5. Identification of the Active Ingredient as Genistein

To check whether genistein was indeed responsible for inhibiting the ABCC11 function, we examined the effects of genistein and the other two major soy isoflavones (daidzein and glycitein), as well as their metabolites (genistein 7-β-D-glucuronide 4′-sulfate, daidzein 7-β-D-glucuronide 4′-sulfate, and equol) on the ABCC11 function (Figure 7a,b). As expected, genistein inhibited ABCC11; its ABCC11-inhibitory activity was the highest among the tested compounds at 100 μM. The further examination of its concentration-dependent inhibitory effects revealed an IC_{50} of 61.5 μM (Figure 7c).

If the Fr.#11-5-2 of soybean extract only contained genistein, 20 ppm of this subfraction corresponded to approximately 74 μM of genistein. The detected ABCC11-inhibitory effect of Fr.#11-5-2 at 20 ppm was approximately 40% (Figure 5b), and this is consistent with the measured concentration-dependent effects of genistein (Figure 7c). After combining these results and the determined structural characters (Figure 6), we concluded that the active ingredient in the Fr.#11-5-2 was indeed genistein.

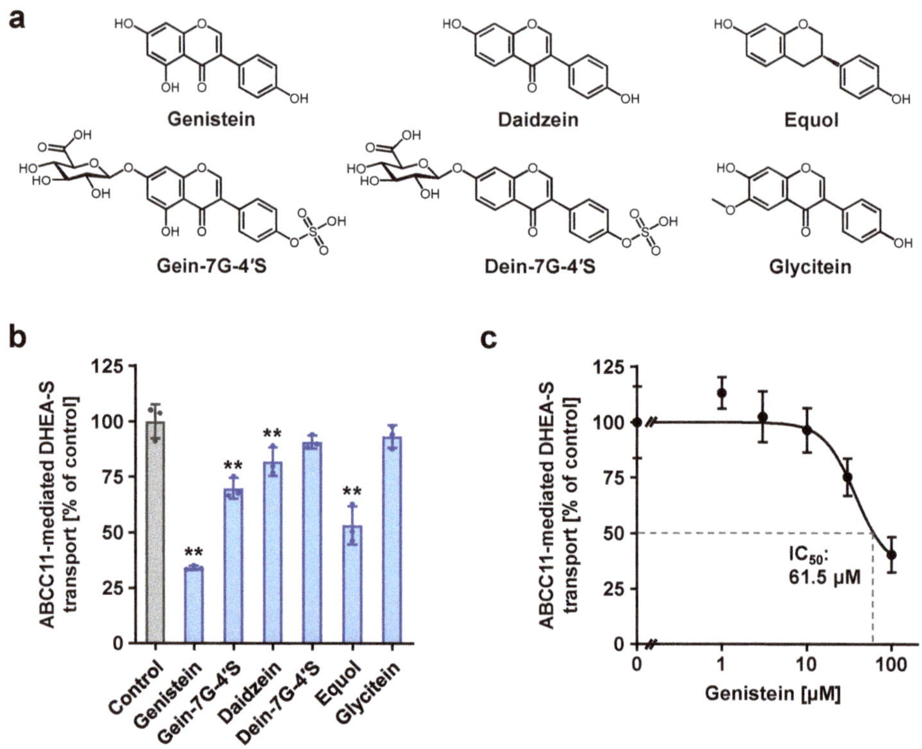

Figure 7. Effects of soybean flavonoids and their metabolites on the transport activity of ABCC11. (**a**) Chemical structures. Gein-7G-4'S, genistein 7-β-D-glucuronide 4'-sulfate; dein-7G-4'S, daidzein 7-β-D-glucuronide 4'-sulfate. (**b**) Inhibitory effects of each flavonoid (100 μM) on ABCC11-mediated [1,2,6,7-^{3}H(N)]-DHEA-S transport. (**c**) Concentration-dependent inhibition of ABCC11-mediated DHEA-S transport by genistein. Data are expressed as % of vehicle and the mean ± SD; n = 3–6. **, $p < 0.01$ vs. control (Dunnett's test).

Daidzein and glycitein only exhibited weak and minimal ABCC11-inhibitory activity, respectively. (*S*)-equol, which is a daidzein-derived metabolite produced by the intestinal bacterial flora in human intestines [18], showed a stronger effect than daidzein (Figure 7b). Moreover, compared with the non-conjugated forms, the glucuronide-sulfate diconjugates of genistein and daidzein showed lower inhibitory activities, suggesting that the polyfunctionalization-mediated structural enlargement might affect the interaction between the soy isoflavones and ABCC11 protein.

3.6. Investigation of ABCC11-Inhibitory Activities of Other Dietary Flavonoids

Finally, we investigated the effects of other dietary flavonoids of interest on the ABCC11 function. The chemical structures of the selected compounds are shown in Figure A1. As shown in Table 2, at 100 μM (the same concentration used in Figure 7b), 13 of the flavonoids lowered the ABCC11-mediated DHEA-S transport to less than 30% of that of the control. Among them, luteolin, nobiletin, myricetin, quercetagetin, isoliquiritigenin, and phloretin powerfully inhibited the transport activity of ABCC11. Additionally, hardly any ABCC11-inhibitory activity was observed for (+)-catechin, (−)-epicatechin, (−)-epigallocatechin, and (+)-gallocatechin in this study, but their galloylated forms inhibited ABCC11, thus suggesting that the gallic acid esterified with catechins would be an important chemical structure for an interaction with ABCC11. These results provide a framework for the further investigation of naturally derived ABCC11 inhibitors.

Table 2. ABCC11-inhibitory activities of dietary flavonoids.

Class	Tested Food Ingredients	ABCC11-Mediated DHEA-S Transport (% of Control)	p Value †
Flavanonol	Dihydromyricetin	56.9 ± 23.4	0.043
	Taxifolin	43.2 ± 26.0	0.032
Flavone	Apigenin	25.3 ± 12.5	0.005
	Luteolin	0 *	<0.001
	Nobiletin	0 *	<0.001
Flavanone	Hesperetin	5.2 ± 26.9	0.013
	Naringenin	38.4 ± 31.6	0.039
Flavonol	3-Hydroxyflavone	91.7 ± 1.1	0.003
	Fisetin	37.7 ± 1.0	<0.001
	Galangin	63.6 ± 5.9	0.004
	Gossypetin	46.0 ± 8.4	0.004
	Kaempferol	42.8 ± 1.0	<0.001
	Morin	41.5 ± 3.4	<0.001
	Myricetin	0 *	0.002
	Quercetin	34.6 ± 18.3	0.013
	Quercetagetin	0 *	<0.001
Chalcone	Cardamonin	73.4 ± 10.8	0.025
	Isoliquiritigenin	0 *	<0.001
	Naringenin chalcone	12.4 ± 4.2	<0.001
	Phloretin	0.6 ± 3.9	<0.001
	Xanthohumol	19.7 ± 13.6	0.005
Catechins	(+)-Catechin	91.9 ± 5.9	0.071 (NS)
	(−)-Catechin gallate	10.5 ± 6.0	0.001
	(−)-Epicatechin	85.2 ± 24.4	0.202 (NS)
	(−)-Epicatechin gallate	29.8 ± 6.8	0.002
	(−)-Epigallocatechin	83.6 ± 29.0	0.215 (NS)
	(−)-Epigallocatechin gallate	37.5 ± 7.0	0.002
	(+)-Gallocatechin	87.6 ± 27.6	0.259 (NS)
	(−)-Gallocatechin gallate	24.0 ± 2.5	<0.001

Inhibitory effects of each food ingredient (100 μM) on ABCC11-mediated [1,2,6,7-^{3}H(N)]-DHEA-S transport activity were investigated by using plasma membrane vesicles (0.5 mg/mL in the reaction mixture) prepared form ABCC11-expressing or control adenovirus-infected 293A cells. Additionally, major green tea catechins, based on a previous study [19], were tested in this study. Data are expressed as % of vehicle and the mean ± SD; $n = 3$. *, Values were calculated under 0; †, one-sample t-test (vs. vehicle control as 100%); NS, not significantly different from control ($p > 0.05$).

4. Discussion

In this study, we examined the effects of water extracts of various dietary plant materials on ABCC11-mediated DHEA-S transport activity as an indicator of their ABCC11 inhibitory function (Figure 2). Among them, the extract of soybeans exhibited the strongest inhibition. Moreover, we identified genistein as an active ingredient responsible for the activity in the extract (Figures 4–7). Hitherto, interactions between ABC proteins and phytochemicals, especially flavonoids, have attracted a lot of interest within the frameworks of multi-drug resistance (MDR) in cancer chemotherapy and the intestinal absorption of a variety of drugs, bioactive food ingredients, and/or toxins upon oral uptake because most ABC proteins are known to significantly affect the pharmacokinetic features of their substrate xenobiotics. In this way, the effects of flavonoids on MDR-related and/or intestinal ABC transporters, such as ABCB1 (also known as P-glycoprotein—P-gp), ABCG2 (breast cancer resistance protein—BCRP), and ABCC2 (multidrug resistance-associated protein 2—MRP2), have been studied, including the inhibitory effects of genistein on several ABC transporters [20–23]. However, to the best of our knowledge, no studies have examined the effects of phytochemicals on ABCC11 function. In fact, in a completely different context, the present study is the first to address and demonstrate the nutrient(s)-mediated ABCC11 inhibition by food extracts and dietary flavonoids.

Our findings may also provide a deeper understanding of the beneficial effects of flavonoids, especially soy isoflavones, that have been proposed to have a number of positive effects on human health [17,24,25]. Though the results have not been entirely consistent, there is considerable interest in using soy isoflavones to prevent cardiovascular diseases, certain types of cancer, menopausal symptoms, etc. This point of view is also supported by a recent umbrella review [26], which reported that the consumption of soy and isoflavones generally provides more benefit than harm in a series of health outcomes. While soy-based foods are traditionally consumed mainly in some Asian countries, their potential health effects have attracted growing attention from health-conscious consumers elsewhere, especially in Western countries [27]. Given this global interest, whether soy flavonoids can ameliorate the constitution causing AO or not is worth studying from the perspectives of dermatology and functional food ingredients.

Previous studies on the bioavailability and metabolism of isoflavones in humans have found that most of the circulating isoflavones are the phase II metabolites including glucuronides and sulfates [28–30], as well as that aglycons such as genistein and daidzein have good affinity for protein binding (>80%) [31,32]. Additionally, after the oral administration of isoflavones to humans (approximately 300 or 600 mg/day genistein and half this amount of daidzein), the plasma levels of aglycones were only in hundreds of nano molar range [33]. Hence, it will not be easy to achieve clinically relevant plasma concentrations of unbound isoflavones to inhibit ABCC11 expressed in the apocrine glands. On the other hand, given that the human axillary apocrine glands open onto the hair follicles that lead to the skin surface [34], the administration of natural extracts with ABCC11-inhibitory activity or their isolated active ingredients on the affected skin may inhibit ABCC11. For this to be effective, the treatment must produce appropriate levels of the active ingredients in the apocrine glands and also must be safe for humans. In this context, our findings here could contribute to the development of medical creams and cosmetic products targeting body odor.

Our results have also revealed a variety of dietary flavonoids that act as inhibitors for ABCC11 (Figure 7 and Table 2). However, how the structural components affect the inhibition needs to be elucidated. With isoflavones, a hydroxy group at C5 and a carbonyl group at C4 could possibly contribute positively and negatively to the ABCC11-inhibitory activities, respectively, as shown in Figure 7. On the other hand, it remains inconclusive whether the existence of a $C_2 = C_3$ double bond, a well-documented element for various bioactivities of flavonoids [35], might contribute to the inhibitory activity. To gain more insight into the relationship between the chemical structure of tested flavonoids and the inhibition of ABCC11-mediated DHEA-S transport activity, the quantitative structure–activity relationship underlying the ABCC11-flavonoids interactions should be investigated in the future.

Some of the limitations of our study and possible future directions are as follows. First, the present study was only an in vitro evaluation for the ABCC11-inhibitory activities of food ingredients. To further investigate the pathophysiological impact of our findings in the context of AO, in vivo evaluations in animals on the scale of academic laboratory are desirable. However, mice and rats have no putative orthologous gene corresponding to the human *ABCC11* [10,36]. On the other hand, previous studies have suggested that isoflavones are fairly safe for humans—exposure to them does not seem to negatively influence human health, at least at the investigated intake levels in reported cases [24,37]. Considering these facts, well-designed human studies are highly warranted.

Second, our data indicated that in addition to genistein, soybeans contain other ABCC11-inhibitory ingredients. One of them could be daidzein, although we could not isolate it from soybean extract in the present study. Regarding the fractions obtained in the first separation step (Fr.#1-12), qualitative evaluation with accurate mass chromatograms revealed that daidzein and genistein were separately fractionated into the Fr.#10 and Fr.#11, respectively. Besides Fr.#11, Fr.#10 and Fr.#12 also showed noticeable ABCC11-inhibitory activity (Figure 4b). Thus, the activity of Fr.#10 could be at least attributable to daidzein. On the other hand, judging from the UV absorption features, the ABCC11 inhibitor(s) in Fr.#12 must be non-flavonoid substances. Moreover, unknown active compounds were collected in the recycling HPLC fractions Fr.#11-5-1 and Fr.#11-5-3 (Figure 5b). Given that these fractions had little absorption peak at 254 nm (Figure 5a), such unknown compounds may not be isoflavones. The identification of these compounds and the verification of their ABCC11-inhibitory activities should be carried out in the future.

5. Conclusions

In conclusion, we found that the soybean extract inhibits the transport activity of ABCC11. From this extract, we successfully identified genistein, a compound known to be fairly safe for humans, as an active ingredient. Additionally, to the best of our knowledge, the present study is the first one demonstrating that some dietary flavonoids can inhibit ABCC11, at least in vitro. While human studies are needed to examine the effects of ABCC11-inhibitory phytochemicals on the AO phenotype, our findings here may provide a new clue for treating AO.

Author Contributions: Conceptualization, Y.T. (Yu Toyoda), H.H., and T.T.; data curation, H.S. (Hiroki Saito), Y.T. (Yu Toyoda), and H.H.; formal analysis, H.S. (Hiroki Saito) and Y.T. (Yu Toyoda); funding acquisition, Y.T. (Yu Toyoda); investigation, H.S. (Hiroki Saito), Y.T. (Yu Toyoda), H.H., and A.O.-K.; methodology, H.S. (Hiroki Saito), Y.T. (Yu Toyoda), and H.H.; project administration, Y.T. (Yu Toyoda) and H.H.; supervision, Y.T. (Youichi Tsuchiya) and H.S. (Hiroshi Suzuki); validation, H.S. (Hiroki Saito), Y.T. (Yu Toyoda), and T.T.; visualization, Y.T. (Yu Toyoda) and H.H.; writing—original draft, Y.T. (Yu Toyoda) in consultation with H.S. (Hiroki Saito) and H.H.; writing—review and editing, Y.T. (Yu Toyoda) and T.T. All authors have read and agreed to the published version of the manuscript.

Funding: This study was supported by JSPS KAKENHI Grant Number 19K16441 (to Y. Toyoda).

Acknowledgments: The authors thank Naoyuki Kobayashi for his continuous encouragement and support for this study, and Chie Umatani for her reviewing our manuscript and technical support for visualization.

Conflicts of Interest: H. Saito, H.H., A.O.-K., and Y. Tsuchiya were the employees of Sapporo Holdings Ltd.; H. Saito, Y. Toyoda, H.H., A.O.-K., T.T. and H. Suzuki have a patent pending related to the work reported in this article.

Appendix A

Table A1. Tested plant materials.

Descriptions in this Study	Common Names	Academic Names	Details of Material *
Arctium lappa root	Burdock root tea	*Arctium lappa*	Dried root for tea
Aspalathus linearis leaves	Rooibos tea leaves	*Aspalathus linearis*	Dried leaves for tea
Barley (*Hordeum vulgare*) miso	Barley miso	*Hordeum vulgare* #	Japanese traditional fermented product
Citrus aurantiifolia peel	Lime	*Citrus aurantiifolia*	Peel
Citrus depressa peel	Shikuwasa	*Citrus depressa*	Peel
Citrus junos peel	Yuzu	*Citrus junos*	Peel
Citrus maxima peel	Pomelo	*Citrus maxima*	Peel
Citrus maxima placenta	Pomelo	*Citrus maxima*	Inner white and soft tissue layer
Citrus natsudaidai peel	Suruga elegant	*Citrus natsudaidai*	Peel
Citrus reticulata var poonensis peel	Ponkan	*Citrus reticulata var. poonensis*	Peel
Citrus sinensis peel	Blood orange	*Citrus sinensis*	Peel
Citrus sphaerocarpa peel	Kabosu	*Citrus sphaerocarpa*	Peel
Citrus sudachi peel	Sudachi	*Citrus sudachi*	Peel
(*Citrus unshiu* × *Citrus sinensis*) × *Citrus reticulata* peel	Siranuhi (Dekopon)	*(Citrus unshiu × Citrus sinensis) × Citrus reticulata*	Peel
Citrus × *Paradisi* peel	Grapefruit	*Citrus* × *Paradisi*	Peel
Citrus × *Tangelo* peel	Mineola orange (Tangelo)	*Citrus* × *Tangelo*	Peel
Commercially available lemon juice A	Not available	Not available	Commercially available product ‡
Commercially available lemon juice B	Not available	Not available	Commercially available product ‡
Commercially available lemon juice C	Not available	Not available	Commercially available product ‡
Ditrus sinensis Osbeck var. brasiliensis Tanaka peel	Navel orange	*Ditrus sinensis Osbeck var. brasiliensis Tanaka*	Peel
Fortunella peel	Kumquat	*Fortunella*	Peel
Glycine max	Soybeans(yellow soybean)	*Glycine max*	Dried product
Hidenmame (*Glycine max*)	Soybeans(green soybean)	*Glycine max*	Dried product
Humulus lupulus L. cone	Hop	*Humulus lupulus* L.	Frozen hop cone
Ilex paraguariensis leaves (roasted)	Yerba mate tea leaves	*Ilex paraguariensis*	Dried and roasted leaves for tea
Jasminum polyanthum leaves	Jasmine tea leaves	*Jasminum polyanthum*	Dried leaves for tea
Matricaria recutita	Chamomile	*Matricaria recutita*	Dried herb product
Prunus domestica extract	Prune extract	*Prunus domestica*	Product of prune pulp extract ‡
Prunus domestica pulp	Prune	*Prunus domestica*	Product of prune pulp without seed
Psidium guajava leaves (from China)	Guava tea leaves	*Psidium guajava*	Dried leaves for tea cultivated in China
Psidium guajava leaves (from Japan)	Guava tea leaves	*Psidium guajava*	Dried leaves for tea cultivated in Japan
Rice (*Oryza sativa*) miso	Rice miso	*Oryza sativa* #	Japanese traditional fermented product
Smallanthus sonchifolius tea powder	Yacon tea powder	*Smallanthus sonchifolius*	Dried product
Soybean (*Glycine max*) miso	Soybean miso	*Glycine max* #	Japanese traditional fermented product

*, Unless otherwise indicated, fresh materials were used; #, academic name of main material of miso product; ‡, after defatting via liquid-liquid partition with an equal volume of ethyl acetate, the obtained water phase of each juice or extract was subjected to lyophilization.

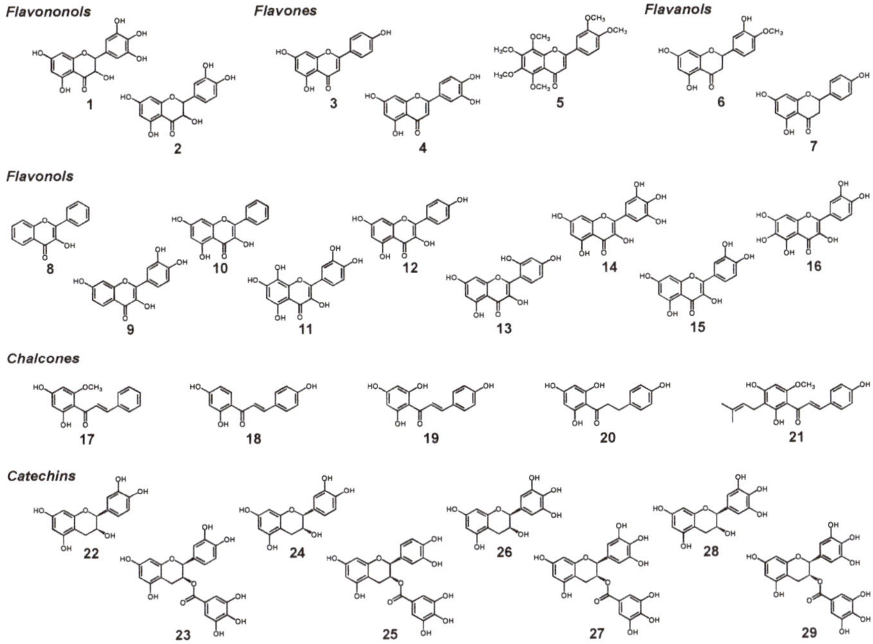

Figure A1. Chemical structures of dietary flavonoids tested in this study. **1**, dihydromyricetin; **2**, taxifolin; **3**, apigenin; **4**, luteolin; **5**, nobiletin; **6**, hesperetin; **7**, naringenin; **8**, 3-hydroxyflavone; **9**, fisetin; **10**, galangin; **11**, gossypetin; **12**, kaempferol; **13**, morin; **14**, myricetin; **15**, quercetin; **16**, quercetagetin; **17**, cardamonin; **18**, isoliquiritigenin; **19**, naringenin chalcone; **20**, phloretin; **21**, xanthohumol; **22**, (+)-catechin; **23**, (−)-catechin gallate; **24**, (−)-epicatechin; **25**, (−)-epicatechin gallate; **26**, (−)-epigallocatechin; **27**, (−)-epigallocatechin gallate; **28**, (+)-gallocatechin; and **29**, (−)-gallocatechin gallate.

References

1. Inoue, Y.; Mori, T.; Toyoda, Y.; Sakurai, A.; Ishikawa, T.; Mitani, Y.; Hayashizaki, Y.; Yoshimura, Y.; Kurahashi, H.; Sakai, Y. Correlation of axillary osmidrosis to a SNP in the ABCC11 gene determined by the Smart Amplification Process (SmartAmp) method. *J. Plast. Reconstr. Aesthet. Surg.* **2010**, *63*, 1369–1374. [CrossRef] [PubMed]
2. Toyoda, Y.; Gomi, T.; Nakagawa, H.; Nagakura, M.; Ishikawa, T. Diagnosis of Human Axillary Osmidrosis by Genotyping of the Human ABCC11 Gene: Clinical Practice and Basic Scientific Evidence. *Biomed. Res. Int.* **2016**, *2016*, 7670483. [CrossRef] [PubMed]
3. Martin, A.; Saathoff, M.; Kuhn, F.; Max, H.; Terstegen, L.; Natsch, A. A functional ABCC11 allele is essential in the biochemical formation of human axillary odor. *J. Investig. Dermatol.* **2010**, *130*, 529–540. [CrossRef] [PubMed]
4. Toyoda, Y.; Sakurai, A.; Mitani, Y.; Nakashima, M.; Yoshiura, K.; Nakagawa, H.; Sakai, Y.; Ota, I.; Lezhava, A.; Hayashizaki, Y.; et al. Earwax, osmidrosis, and breast cancer: Why does one SNP (538G>A) in the human ABC transporter ABCC11 gene determine earwax type? *FASEB J.* **2009**, *23*, 2001–2013. [CrossRef] [PubMed]
5. Nakano, M.; Miwa, N.; Hirano, A.; Yoshiura, K.; Niikawa, N. A strong association of axillary osmidrosis with the wet earwax type determined by genotyping of the ABCC11 gene. *BMC Genet.* **2009**, *10*, 42. [CrossRef] [PubMed]
6. Bortfeld, M.; Rius, M.; Konig, J.; Herold-Mende, C.; Nies, A.T.; Keppler, D. Human multidrug resistance protein 8 (MRP8/ABCC11), an apical efflux pump for steroid sulfates, is an axonal protein of the CNS and peripheral nervous system. *Neuroscience* **2006**, *137*, 1247–1257. [CrossRef]

7. Chen, Z.S.; Guo, Y.; Belinsky, M.G.; Kotova, E.; Kruh, G.D. Transport of bile acids, sulfated steroids, estradiol 17-beta-D-glucuronide, and leukotriene C4 by human multidrug resistance protein 8 (ABCC11). *Mol. Pharmacol.* **2005**, *67*, 545–557. [CrossRef]
8. Yoshiura, K.; Kinoshita, A.; Ishida, T.; Ninokata, A.; Ishikawa, T.; Kaname, T.; Bannai, M.; Tokunaga, K.; Sonoda, S.; Komaki, R.; et al. A SNP in the ABCC11 gene is the determinant of human earwax type. *Nat. Genet.* **2006**, *38*, 324–330. [CrossRef]
9. Toyoda, Y.; Takada, T.; Gomi, T.; Nakagawa, H.; Ishikawa, T.; Suzuki, H. Clinical and Molecular Evidence of ABCC11 Protein Expression in Axillary Apocrine Glands of Patients with Axillary Osmidrosis. *Int. J. Mol. Sci.* **2017**, *18*, 417. [CrossRef]
10. Toyoda, Y.; Takada, T.; Miyata, H.; Ishikawa, T.; Suzuki, H. Regulation of the Axillary Osmidrosis-Associated ABCC11 Protein Stability by N-Linked Glycosylation: Effect of Glucose Condition. *PLoS ONE* **2016**, *11*, e0157172. [CrossRef]
11. Saito, H.; Toyoda, Y.; Takada, T.; Hirata, H.; Ota-Kontani, A.; Miyata, H.; Kobayashi, N.; Tsuchiya, Y.; Suzuki, H. Omega-3 Polyunsaturated Fatty Acids Inhibit the Function of Human URAT1, a Renal Urate Re-Absorber. *Nutrients* **2020**, *12*, 1601. [CrossRef] [PubMed]
12. Toyoda, Y.; Mancikova, A.; Krylov, V.; Morimoto, K.; Pavelcova, K.; Bohata, J.; Pavelka, K.; Pavlikova, M.; Suzuki, H.; Matsuo, H.; et al. Functional Characterization of Clinically-Relevant Rare Variants in ABCG2 Identified in a Gout and Hyperuricemia Cohort. *Cells* **2019**, *8*, 363. [CrossRef] [PubMed]
13. Toyoda, Y.; Takada, T.; Suzuki, H. Inhibitors of Human ABCG2: From Technical Background to Recent Updates with Clinical Implications. *Front. Pharmacol.* **2019**, *10*, 208. [CrossRef]
14. Hirata, H.; Takazumi, K.; Segawa, S.; Okada, Y.; Kobayashi, N.; Shigyo, T.; Chiba, H. Xanthohumol, a prenylated chalcone from *Humulus lupulus* L., inhibits cholesteryl ester transfer protein. *Food Chem.* **2012**, *134*, 1432–1437. [CrossRef] [PubMed]
15. Miyata, H.; Takada, T.; Toyoda, Y.; Matsuo, H.; Ichida, K.; Suzuki, H. Identification of Febuxostat as a New Strong ABCG2 Inhibitor: Potential Applications and Risks in Clinical Situations. *Front. Pharmacol.* **2016**, *7*, 518. [CrossRef] [PubMed]
16. Chen, K.I.; Erh, M.H.; Su, N.W.; Liu, W.H.; Chou, C.C.; Cheng, K.C. Soyfoods and soybean products: From traditional use to modern applications. *Appl. Microbiol. Biotechnol.* **2012**, *96*, 9–22. [CrossRef] [PubMed]
17. Messina, M. Soy and Health Update: Evaluation of the Clinical and Epidemiologic Literature. *Nutrients* **2016**, *8*, 754. [CrossRef]
18. Mayo, B.; Vazquez, L.; Florez, A.B. Equol: A Bacterial Metabolite from The Daidzein Isoflavone and Its Presumed Beneficial Health Effects. *Nutrients* **2019**, *11*, 2231. [CrossRef]
19. Taylor, P.W.; Hamilton-Miller, J.M.; Stapleton, P.D. Antimicrobial properties of green tea catechins. *Food Sci. Technol. Bull.* **2005**, *2*, 71–81. [CrossRef]
20. Sjostedt, N.; Holvikari, K.; Tammela, P.; Kidron, H. Inhibition of Breast Cancer Resistance Protein and Multidrug Resistance Associated Protein 2 by Natural Compounds and Their Derivatives. *Mol. Pharm.* **2017**, *14*, 135–146. [CrossRef]
21. Tan, K.W.; Li, Y.; Paxton, J.W.; Birch, N.P.; Scheepens, A. Identification of novel dietary phytochemicals inhibiting the efflux transporter breast cancer resistance protein (BCRP/ABCG2). *Food Chem.* **2013**, *138*, 2267–2274. [CrossRef] [PubMed]
22. Li, Y.; Revalde, J.L.; Reid, G.; Paxton, J.W. Interactions of dietary phytochemicals with ABC transporters: Possible implications for drug disposition and multidrug resistance in cancer. *Drug Metab. Rev.* **2010**, *42*, 590–611. [CrossRef] [PubMed]
23. Brand, W.; Schutte, M.E.; Williamson, G.; van Zanden, J.J.; Cnubben, N.H.; Groten, J.P.; van Bladeren, P.J.; Rietjens, I.M. Flavonoid-mediated inhibition of intestinal ABC transporters may affect the oral bioavailability of drugs, food-borne toxic compounds and bioactive ingredients. *Biomed. Pharm.* **2006**, *60*, 508–519. [CrossRef] [PubMed]
24. Xiao, Y.; Zhang, S.; Tong, H.; Shi, S. Comprehensive evaluation of the role of soy and isoflavone supplementation in humans and animals over the past two decades. *Phytother. Res.* **2018**, *32*, 384–394. [CrossRef] [PubMed]
25. Zaheer, K.; Humayoun Akhtar, M. An updated review of dietary isoflavones: Nutrition, processing, bioavailability and impacts on human health. *Crit. Rev. Food Sci. Nutr.* **2017**, *57*, 1280–1293. [CrossRef]

26. Li, N.; Wu, X.; Zhuang, W.; Xia, L.; Chen, Y.; Zhao, R.; Yi, M.; Wan, Q.; Du, L.; Zhou, Y. Soy and Isoflavone Consumption and Multiple Health Outcomes: Umbrella Review of Systematic Reviews and Meta-Analyses of Observational Studies and Randomized Trials in Humans. *Mol. Nutr. Food. Res.* **2020**, *64*, e1900751. [CrossRef]
27. Cao, Z.H.; Green-Johnson, J.M.; Buckley, N.D.; Lin, Q.Y. Bioactivity of soy-based fermented foods: A review. *Biotechnol. Adv.* **2019**, *37*, 223–238. [CrossRef]
28. Rodriguez-Morato, J.; Farre, M.; Perez-Mana, C.; Papaseit, E.; Martinez-Riera, R.; de la Torre, R.; Pizarro, N. Pharmacokinetic Comparison of Soy Isoflavone Extracts in Human Plasma. *J. Agric. Food Chem.* **2015**, *63*, 6946–6953. [CrossRef]
29. Rowland, I.; Faughnan, M.; Hoey, L.; Wahala, K.; Williamson, G.; Cassidy, A. Bioavailability of phyto-oestrogens. *Br. J. Nutr.* **2003**, *89* (Suppl. S1), S45–S58. [CrossRef]
30. Setchell, K.D.; Brown, N.M.; Desai, P.; Zimmer-Nechemias, L.; Wolfe, B.E.; Brashear, W.T.; Kirschner, A.S.; Cassidy, A.; Heubi, J.E. Bioavailability of pure isoflavones in healthy humans and analysis of commercial soy isoflavone supplements. *J. Nutr.* **2001**, *131*, 1362S–1375S. [CrossRef]
31. Yang, Z.; Kulkarni, K.; Zhu, W.; Hu, M. Bioavailability and pharmacokinetics of genistein: Mechanistic studies on its ADME. *Anticancer Agents Med. Chem.* **2012**, *12*, 1264–1280. [CrossRef] [PubMed]
32. Csanady, G.A.; Oberste-Frielinghaus, H.R.; Semder, B.; Baur, C.; Schneider, K.T.; Filser, J.G. Distribution and unspecific protein binding of the xenoestrogens bisphenol A and daidzein. *Arch. Toxicol.* **2002**, *76*, 299–305. [CrossRef] [PubMed]
33. Fischer, L.; Mahoney, C.; Jeffcoat, A.R.; Koch, M.A.; Thomas, B.E.; Valentine, J.L.; Stinchcombe, T.; Boan, J.; Crowell, J.A.; Zeisel, S.H. Clinical characteristics and pharmacokinetics of purified soy isoflavones: Multiple-dose administration to men with prostate neoplasia. *Nutr. Cancer* **2004**, *48*, 160–170. [CrossRef] [PubMed]
34. Wilke, K.; Martin, A.; Terstegen, L.; Biel, S.S. A short history of sweat gland biology. *Int. J. Cosmet. Sci.* **2007**, *29*, 169–179. [CrossRef] [PubMed]
35. Wang, T.Y.; Li, Q.; Bi, K.S. Bioactive flavonoids in medicinal plants: Structure, activity and biological fate. *Asian J. Pharm. Sci.* **2018**, *13*, 12–23. [CrossRef] [PubMed]
36. Shimizu, H.; Taniguchi, H.; Hippo, Y.; Hayashizaki, Y.; Aburatani, H.; Ishikawa, T. Characterization of the mouse Abcc12 gene and its transcript encoding an ATP-binding cassette transporter, an orthologue of human ABCC12. *Gene* **2003**, *310*, 17–28. [CrossRef]
37. Huser, S.; Guth, S.; Joost, H.G.; Soukup, S.T.; Kohrle, J.; Kreienbrock, L.; Diel, P.; Lachenmeier, D.W.; Eisenbrand, G.; Vollmer, G.; et al. Effects of isoflavones on breast tissue and the thyroid hormone system in humans: A comprehensive safety evaluation. *Arch. Toxicol.* **2018**, *92*, 2703–2748. [CrossRef]

nutrients

MDPI

Review

Potential of *Chlorella* as a Dietary Supplement to Promote Human Health

Tomohiro Bito [1], Eri Okumura [2], Masaki Fujishima [2] and Fumio Watanabe [1,*]

[1] Department of Agricultural, Life and Environmental Sciences, Faculty of Agriculture, Tottori University, Tottori 680-8553, Japan; bito@tottori-u.ac.jp
[2] Sun Chlorella Corporation, Kyoto 600-8177, Japan; okumura@sunchlorella.co.jp (E.O.); mfujishima@sunchlorella.co.jp (M.F.)
* Correspondence: watanabe@tottori-u.ac.jp; Tel.: +81-857-31-5412

Received: 26 May 2020; Accepted: 17 August 2020; Published: 20 August 2020

Abstract: *Chlorella* is a green unicellular alga that is commercially produced and distributed worldwide as a dietary supplement. *Chlorella* products contain numerous nutrients and vitamins, including D and B_{12}, that are absent in plant-derived food sources. *Chlorella* contains larger amounts of folate and iron than other plant-derived foods. *Chlorella* supplementation to mammals, including humans, has been reported to exhibit various pharmacological activities, including immunomodulatory, antioxidant, antidiabetic, antihypertensive, and antihyperlipidemic activities. Meta-analysis on the effects of *Chlorella* supplementation on cardiovascular risk factors have suggested that it improves total cholesterol levels, low-density lipoprotein cholesterol levels, systolic blood pressure, diastolic blood pressure, and fasting blood glucose levels but not triglycerides and high-density lipoprotein cholesterol levels. These beneficial effects of *Chlorella* might be due to synergism between multiple nutrient and antioxidant compounds. However, information regarding the bioactive compounds in *Chlorella* is limited.

Keywords: antioxidants; *Chlorella*; dietary fibers; dietary supplements; folate; lutein; vitamin B_{12}; vitamin D_2

1. Introduction

Microalgae are primarily found in aquatic ecosystems, living in both seawater and freshwater, and are photosynthetic eukaryotic organisms that contain chloroplasts and nuclei, similar to plants. Microalgae more efficiently yield biomass than land-based plants owing to their higher performance in utilizing sunlight and CO_2, leading to their extremely high growth rates [1]. Therefore, microalgae have been used in the food, pharmaceutical, and cosmetic industries, and their pigments, nutrients, bioactive compounds and whole biomass are already in use worldwide. Recently, various bioactive compounds and nutrients have been detected in both seawater and freshwater microalgae, including cyanobacteria. These compounds and nutrients have been reported to promote human health [1,2]. However, there is limited information regarding the bioactive compounds of freshwater-living *Chlorella* species, which are classified as green algae.

Chlorella species can be mass-cultured, and their dietary supplement products are commercially available worldwide. However, the commercial cultivation of their biomass has started only several years ago. *Chlorella vulgaris* was discovered and reported in 1890 by Dr. Martinus Willem Beijerinck, a famous microbiologist and botanist [3]. Another *Chlorella* species, distinguished by the presence pyrenoids in chloroplasts, was identified and accordingly named *C. pyrenoidosa* in 1903 [4]. Since then, more than 20 *Chlorella* species have been characterized, with over 100 strains described [5]. At present, *Chlorella* species are divided into three varieties: *C. vulgaris, C. lobophora, and C. sorokiniana* [6]. *C. sorokiniana* is a sub-species first isolated in 1953 by Sorokin and originally thought to be a thermotolerant mutant of *C. pyrenoidosa* [7,8]. *C. pyrenoidosa*, the subject of many scientific studies, is now called *C. sorokiniana*.

Investigations of the dietary value of *Chlorella* in human health began in the early 1950s, when the use of *Chlorella* as a food source was initiated in the midst of a global food crisis [9]. *Chlorella* was first produced and consumed in Asia, mainly in Japan, and then used as a dietary supplement worldwide [10]. *Chlorella* is produced commercially for use in foods and as a source of its intrinsic compounds. Using large-scale cultivation technology, *C. vulgaris* and *C. pyrenoidosa* are prepared as commercial sources for dietary supplements [11]. Studies have shown that *Chlorella* cells contain a variety of nutrients and bioactive compounds that promote human health and prevent certain diseases [10,12], suggesting that *Chlorella*-derived natural compounds might provide substitutes for synthetic compounds or drugs. The content of natural compounds in *Chlorella* differs greatly between culture conditions and *Chlorella* species [13,14].

Here, we present updated information on the *Chlorella* content of nutrients and bioactive compounds that promote human health. However, at present, there is limited information available regarding the bioactive compounds responsible for its pharmacological activities, which might be due to the synergistic effects of various nutrients and antioxidant compounds in *Chlorella*.

2. Nutrients in Commercial *Chlorella* Products

2.1. Macronutrients

The macronutrient content of 13 commercially available *Chlorella* products, based on information provided on the packaging label, are summarized in Table 1. Humans cannot digest *Chlorella* cells in their natural state because their cell walls are made of cellulose. Therefore, *Chlorella* cell walls are mechanically broken down in most dietary supplements. An animal study has shown that more than 80% of *Chlorella* proteins are digestible [15].

Table 1. Nutrient content of 13 commercially available *Chlorella* products.

Macronutrients (Per 100 g Dry Weight)	A	B	C	D	E	F	G	H	I	J	K	L	M
Proteins (g)	50–65	61	63	65	57	50	56–72	50–67	50–70	62	60	58	57
Fats (g)	7–14	10	13	12	11		7–20	5–15	8–15	11	10	10	12
Carbohydrates (g)			15	11			5–23	8–42	8–20		20	18	
Sugars (g)	5–21	7	5	0–1			0–5	2–23		1–10			11
Dietary fibers (g)	7–14	11	10	11	11		5–18	7–18		8–16			10
Remarks	*1 *2 78%	*1 *4	*2	*2			*1 *2 77–82% *3	*2 75–85% *3	*1 *2 83% *3	*2 82% *3		*4	*1 *3 *4

*1 Cell walls disrupted; *2 digestibility of proteins; *3 contains *Chlorella* extract; *4 *C. pyrenoidosa*.

These *Chlorella* products contain a large amount of proteins (approximately 59% based on dry weight), coinciding with the analytical data of the protein contents of *C. pyrenoidosa* (57%) [16] and *C. vulgaris* (51–58%) [17]. This protein content is higher than that of soybeans (approximately 33%, dry weight). The amino acid composition of *Chlorella* products C and M are shown in Table 2. These amino acid profiles indicate that all essential amino acids for humans (isoleucine, leucine, lysine, methionine, phenylalanine, threonine, tryptophan, valine, and histidine) are present in substantial concentrations in these products. According to the essential amino acid index (EAAI) used to evaluate protein quality for human nutrition, the quality of *C. pyrenoidosa* (EAAI, 1.35) [18] and a commercially available *Chlorella* product (EAAI, 0.92) [19] are higher than that of soybean protein (EAAI, 0.66) [18]. These results indicate that proteins in *Chlorella* products are of high or good quality. Notably, *Chlorella* products contain a considerable amount of arginine (approximately 3200 mg/100 g dry weight), which serves as a substrate for the production of NO, a potent intracellular signaling molecule that influences every mammalian system [20]. Arginine also serves as a potent modulator of immune functions [21].

Table 2. Amino acid content of commercially available *Chlorella* products C and M.

Amino Acids (mg/100 g Dry Weight)	C	M
Essential		
Isoleucine	1820	2030
Leucine	4180	4480
Lysine	4659	3140
Methionine	1009	1240
Phenylalanine	2230	2580
Threonine	2209	2490
Tryptophan	1030	1090
Valine	2780	3090
Histidine	1141	1040
Non-essential		
Tyrosine	1720	1940
Cystine	659	650
Aspartic acid	4469	4710
Serine	1930	2120
Glutamic acid	6209	6030
Proline	2320	2560
Glycine	2859	2990
Alanine	4009	4170
Arginine	3109	3260

Approximately 17% (dry weight) of carbohydrates are found in the commercially available *Chlorella* products. Similar results have been reported for *C. vulgaris* [17]. As shown in Table 1, more than 65% of the carbohydrate is dietary fiber, which appears to be derived from the *Chlorella* cell wall. Various polysaccharides have been extracted and characterized [22–25]. *Chlorella* polysaccharides exhibited a variety of biologically active compounds, including antioxidants [24] and stimulators of plant growth [25]. Tabarsa et al. [26] characterized an immune-enhancing water-soluble α-glucan prepared from *C. vulgaris*.

Commercially available *Chlorella* products contain a small amount of fats (approximately 11%, dry weight) (Table 1), which coincides with the analytical data of the fat content of *C. vulgaris* (14–22%) [17]. *Chlorella* products contain α-linolenic acid (approximately 10–16% of total fatty acids) and linoleic acid (approximately 18% of total fatty acids) but not eicosapentaenoic acid, docosahexenoic acid, or arachidonic acid [19,27]. Approximately 65–70% of the total fatty acids found in commercially available *Chlorella* products are derived from polyunsaturated fatty acids [19,27].

Different growth conditions, such as temperature, nutrient composition, and light availability, can readily alter the levels of biomass, macro- and micronutrients, and other valuable bioactive compounds, including antioxidants, in *Chlorella* cells [28–30].

2.2. Micronutrients

2.2.1. Vitamins

As shown in Table 3, commercially available *Chlorella* products contain all the vitamins required by humans, i.e., B_1, B_2, B_6, B_{12}, niacin, folate, biotin, pantothenic acid, C, D_2, E, and K, and α- and β-carotenes. *Chlorella* products contain substantial amounts of vitamins D_2 and B_{12}, both of which are well known to be absent in plants. Commercially available *Chlorella* (*C. vulgaris*) products contain higher amounts of folate (approximately 2.5 mg/100 g dry weight) than spinach [31]. Vitamin B_{12} and folate deficiencies induce the accumulation of serum homocysteine, which is involved in cardiovascular diseases. In this section, we discuss vitamin D_2, vitamin B_{12}, and folate.

Table 3. Content of vitamins and related compounds in 13 commercially available *Chlorella* products.

Vitamins (Per 100 g Dry Weight)	A	B	C	D	E	F	G	H	I	J	K	L	M
Vitamin B_1 (mg)		1.9	2.5		6.5		1.0–3.0		1.0–3.0	1.8			1.6
Vitamin B_2 (mg)	3–8	5.6	5.0	5.7	5.5		2.0–9.0	4.0–9.0	4.0–8.0	5.0		5.0	4.8
Vitamin B_6 (mg)		0.9	2.5		1.7		1.0–3.0		1.0–3.0	1.0–3.0			1.8
Vitamin B_{12} (μg)		20.0							6.0–30.0	200.0–500.0			230.0
Niacin (mg)		20.4	50.0				40.0–80.0		20.0–50.0	10.0–40.0			45.9
Folate (mg)		0.3	2.0				1.2–3.6						1.4
Biotin (μg)													227.0
Pantothenic acid (mg)										1.0–6.0			1.8
Vitamin C (mg)		7.0	50.0		30.0		10.0–200.0						14.0
Vitamin D_2 (mg)													1.4
Vitamin E (mg)		3.0	25.0				10.0–45.0						6.2
Vitamin K (mg)		1.4	1.1	0.3						0.5–3.5 *1			1.2 *1
Carotenoids (mg)				25.0 *2			36.0–150.0 *2			100.0–500.0			31.5 *3

*1 Vitamin K_1 (mg), *2 β-Carotene (mg), *3 α-Carotene + β-Carotene (mg).

Vitamin D, a major regulator of calcium absorption, reduces the risk of osteomalacia in adults and rickets in children [32]. The two main dietary forms of vitamin D are vitamin D_2 and D_3, which are found in fungi such as mushrooms [33,34] and animal-derived foods such as fish and fish products [35], respectively. Mushrooms have the ability to synthesize ergosterol (known as provitamin D_2), which is converted into ergocalciferol as vitamin D_2 upon ultraviolet irradiation [34,36]. Thus, ultraviolet-irradiated mushrooms are suitable for use as vitamin D_2 sources in strict vegetarians [36]. Cell walls of mushrooms contain high concentrations of ergosterol, which plays a physiological role in modulating cell membrane strength and fluidity similar to cholesterol in animals [37]. Sun-dried, commercially available mushrooms reportedly contain approximately 17 μg of vitamin D_2 per g dry weight [38]. The bioavailability of vitamin D_2 from mushrooms has been studied in humans [39,40].

Ergosterol was first reported as the main sterol compound in *C. pyrenoidosa* in the early 1950s [41]. *C. vulgaris* also contains a substantial amount of ergosterol [42,43]. Our unpublished data show that one commercially available *Chlorella* product contains both ergosterol (1.68 mg/g dry weight) and vitamin D_2 (15.2 μg/g dry weight), similar in amounts to those in sun-dried mushrooms. The vitamin D_2 in this *Chlorella* product is synthesized from ergosterol upon exposure to sunlight during cultivation (Figure 1). Although it has been reported that vitamin D_3 is more effective than vitamin D_2 at increasing the concentration of circulating 25-hydroxyvitamin D [44], *Chlorella* products and sun-dried mushrooms could become sources of vitamin D for vegetarians.

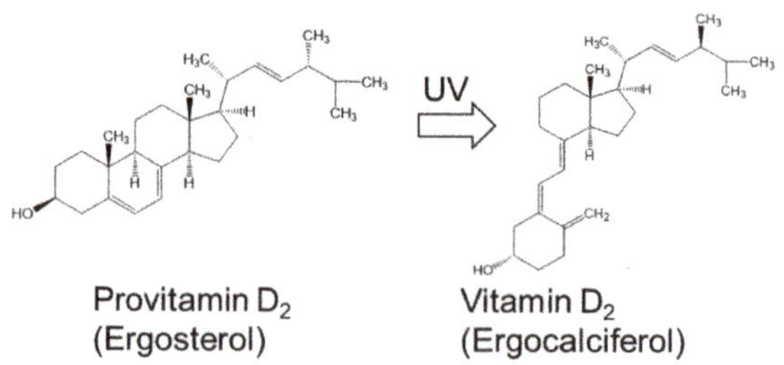

Figure 1. Structures of provitamin D_2 and vitamin D_2 found in commercially available *Chlorella* products.

Serum homocysteine (Hcy) is an established biomarker of cardiovascular disease in humans [45,46]. Hcy is a non-protein forming amino acid (Figure 2) produced as an intermediate compound of methionine metabolism and is further metabolized to cystathionine via cystathionine β-synthetase, a vitamin B_6-dependent enzyme [46]. Alternatively, Hcy can be remethylated back to methionine by methionine synthase, a vitamin B_{12}-dependent enzyme. Folate is also required for the remethylation of Hcy by providing 5-methyltetrahydrofolate. Deficiencies in vitamin B_{12} [47], vitamin B_6 [48], and folate [49] cause hyper-homocysteinemia. Several clinical studies report a correlation between atherosclerosis and deficiencies in vitamin B_{12} and folate [50,51]. Folate deficiency in women before and during pregnancy is associated with neural tube defects in newborns [52]. Plants can synthesize folate compounds de novo, but animals cannot [53]. Thus, plant-derived foods are sources of dietary folates for humans. High concentrations of folate (approximately 1.69–2.45 mg/100 g dry weight) are reported in commercially available *Chlorella* (*C. vulgaris*) products [31], with concentrations similar to those of the products shown in Table 3 (0.3–3.6 mg/100 g dry weight). The main folate compounds identified in *Chlorella* products are 5-CHO-H_4 folate (60–62%) and 5-CH_3-H_4 folate (24–26%) and the minor folate compounds are 10-CHO-folate (5–7%), H_4 folate (4%), and fully oxidized folate (3–6%) [31]. The chemical structures of the *Chlorella* folate compounds are shown in Figure 3. The main dietary sources of folates are vegetables (25%), bread and cereal products (22%), dairy products (10%), fruit (10%), and oils and fats (5%) [31]. Spinach has high folate content (165 μg/100 g fresh weight; 1.7 mg/100 g dry weight) [31,54], which is similar to that of *Chlorella* products. Thus, *Chlorella* products are an excellent source of folate for humans.

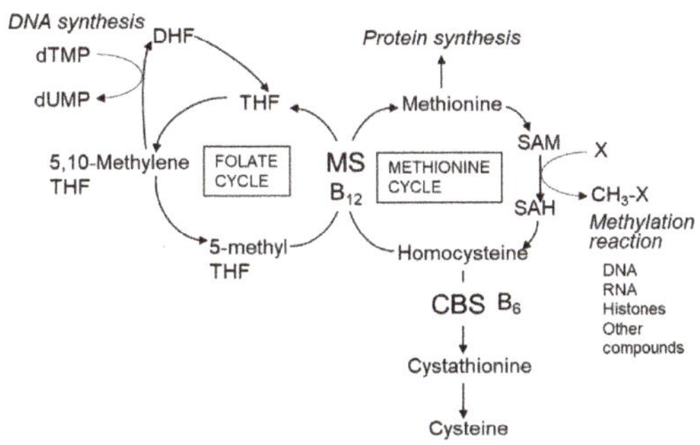

Figure 2. Homocysteine metabolic pathway in mammals. Abbreviations: B_6, vitamin B_6; B_{12}, vitamin B_{12}; CBS, cystathionine β-synthetase; DHF, dihydrofolate; MS, cobalamin-dependent methionine synthase; SAM, S-adenosyl methionine; SAH, S-adenosyl homocysteine; THF, tetrahydrofolate.

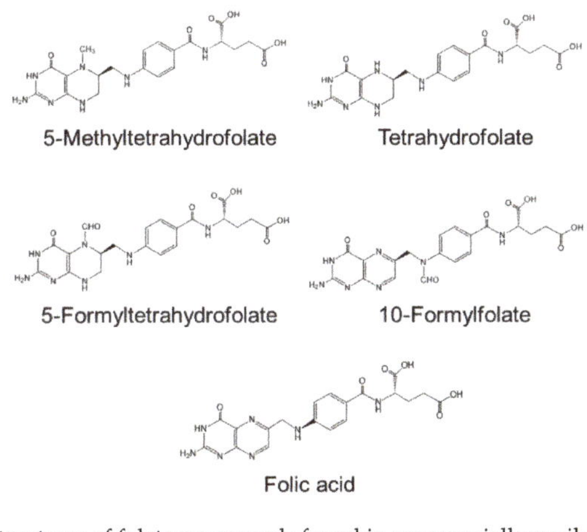

Figure 3. Chemical structures of folate compounds found in commercially available *Chlorella* products.

Vitamin B_{12} (B_{12}) is synthesized by certain bacteria and archaea but not by plants [55]. Animal-derived foods, such as meats, milk, fish, and shellfish, are the major dietary sources of B_{12} for humans [56]. B_{12} content is high in beef, pork, and chicken livers (approximately

25–53 μg/100 g fresh weight) [56] and in edible bivalves such as clams (approximately 60 μg/100 g fresh weight) [57]. The reported B_{12} content of *Chlorella* products varies from <0.1 to 400 μg per 100 g of dry weight [58,59], consistent with that of the products shown in Table 3 (6–500 μg/100 g dry weight). Among *Chlorella* species, the B_{12} content is much higher in *C. pyrenoidosa* than in *C. vulgaris* when grown under open culture conditions [59]. B_{12} is not essential for the growth of these *Chlorella* species [59,60], suggesting that *Chlorella* cells absorb and accumulate large amounts of exogenous B_{12}. Some of the high B_{12}-containing *Chlorella* products contain inactive corrinoid compounds such as 5-methoxybenzimidazolylcobamide and cobalt-free corrinoid (Figure 4). Thus, if *Chlorella* products with high B_{12} are consumed as a sole B_{12} source, accurate content estimation requires the identification of B_{12} compounds using liquid chromatography–tandem mass spectrometry [59].

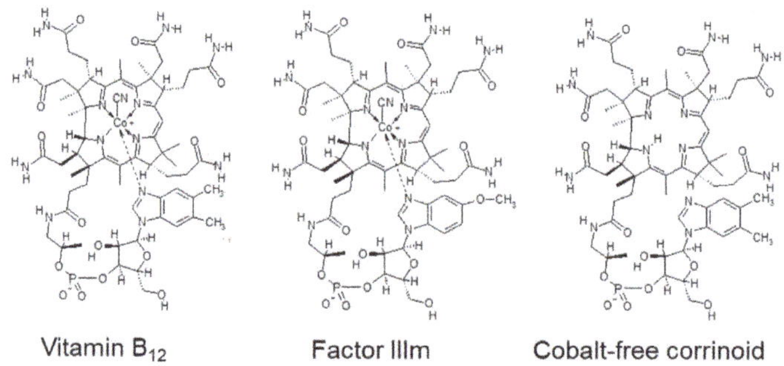

Figure 4. Chemical structures of vitamin B_{12} and related compounds found in commercially available *Chlorella* products. Abbreviations: Factor IIIm, 5-methoxybenzimidazolylcobamide.

Rauma et al. [61] demonstrated that substantial consumption of *Chlorella* products can supply adequate amounts of B_{12}. Another study of strict vegetarians (vegans) with an elevated baseline of serum methylmalonic acid (as an index of B_{12} deficiency) showed that ingestion of 9 g of *C. pyrenoidosa* daily for 60 days resulted in significant decreases in serum methylmalonic acid in 88% of the subjects [62]; serum Hcy decreased and serum B_{12} tended to increase, although the mean corpuscular volume, hemoglobin, and hematocrit levels were unchanged. These results suggest that *Chlorella* products with high B_{12} and without inactive corrinoid compounds are suitable for use as B_{12} sources in humans, particularly vegans.

2.2.2. Minerals

As shown in Table 4, commercially available *Chlorella* products contain a variety of minerals that are required in humans. In particular, *Chlorella* products contain substantial amounts of iron (104 mg/100 g dry weight) and potassium (986 mg/100 g dry weight), of which adequate intake prevents anemia [63] and hypertension [64], respectively. Iron plays physiological roles in respiration, energy production, DNA synthesis, and cell proliferation [65]. The phytates in grains potently inhibit the intestinal absorption of iron because they chelate iron to form an insoluble complex [66]. Thus, people on vegan and vegetarian diets may be at risk for iron-deficiency anemia [63]. Studies in rats and humans have investigated whether *Chlorella* supplementation can prevent iron-deficiency anemia [67,68]. In a cohort of 32 women in the second and third trimester of pregnancy, oral *Chlorella* supplementation (6 g/day) for 12–18 weeks decreased markers of anemia as compared to the control group [68], suggesting that *Chlorella* supplementation significantly reduces the risk of pregnancy-associated anemia.

Table 4. Mineral content of 13 commercially available *Chlorella* products.

Minerals (Per 100 g Dry Weight)	A	B	C	D	E	F	G	H	I	J	K	L	M
Sodium (mg)	5–75	65			40		80–220	10–45		5–30	80	65	47
Iron (mg)	10–130	160	121	62			350–1600		100–200	50–100		110	113
Calcium (mg)		650	513		850		500–1500		100–300				433
Potassium (mg)		970	1075		350		200–500			500–1500			1020
Magnesium (mg)		350	250							23–420			298
Zinc (mg)		2											1
Copper (mg)		1											1
Phosphorus (mg)		1600											1320
Manganese (mg)													5

Selenium (Se) is an essential trace mineral that serves as a fundamental nutrient to human health. It is a component of selenoproteins, such as thioredoxin reductase and glutathione peroxidases, and protects against intercellular oxidative damage [69–71]. Therefore, low levels of Se show various pharmaceutical activities, including antitumor and antiaging effects; however, high levels of Se induce the generation of reactive oxygen species. Generally, the organic forms of Se are more bioavailable and less toxic than the inorganic forms of Se. Selenite (SeO_3^{2-}) and selenite (SeO_4^{2-}) are the predominant forms of Se in freshwater. Microalgae act as a major transporter of Se from water to filter-feeders and other organisms. Although most plant species accumulate less than 25 μg Se/g dry weight [72], some microalgae species can accumulate Se at high concentrations (100 μg Se/g dry weight) [73]. Se is essential for many algae and functions to protects them from oxidative damage. Sun et al. [74] indicated that *C. vulgaris* can accumulate Se at high concentrations (857 μg/g dry weight) when grown under Se concentrations of 0–200 mg/L in a growth medium and that relatively low Se concentrations (75 mg selenite/L medium) positively promotes *C. vulgaris* growth and acts as an antioxidant by inhibiting lipid peroxidation and intracellular reactive oxygen species. The maximum accumulation of organic Se was found at 316 μg/g dry weight under relatively low Se (75 mg selenite/L medium) conditions [75], indicating that *C. vulgaris* is an efficient Se accumulator and that Se-enriched *Chlorella* cells might be useful for human supplementation.

2.3. Pigments

Carotenoids are secondary metabolites in the most abundant naturally occurring pigments that participate in various biological processes in plants, including photosynthesis, photomorphogenesis, photoprotection, and development [76]. They also serve as colorants and critical components of the human diet, such as antioxidants and provitamin A [76]. More than 400 carotenoids have been identified in living organisms [77], and β-carotene, astaxanthin, lutein, zeaxanthin, and lycopene are widely known as the major carotenoids among them. The green microalgae *Dunaliella salina* produces high concentrations of β-carotene of up to 14% of algal dry weight [78]. *Haematococcus pluvialis* increases astaxanthin concentration up to 4–5% of algal dry weight [79] under stressful conditions. *Chlorella* products contain lower contents of total carotenoids (approximately 1.3%) [80], compared with the above-mentioned green algae. *C. vulgaris* reportedly produces lutein as the primary carotenoid [81,82]. However, *C. zofingiensis* reportedly accumulates significant amounts of astaxanthin, and it might be a suitable organism for the mass production of astaxanthin [83].

3. Pharmacological Activities of *Chlorella* Products

Because *Chlorella* cells contain various nutrients and biologically active compounds, the effects of *Chlorella* supplementation on preventing the development of various diseases has been studied in rats and mice, including disease-specific model animals. These animal studies have been useful for elucidating the specific health effects of *Chlorella* supplementation. Moreover, the effects of *Chlorella* supplementation on mitigating a variety of diseases in humans have been investigated. These studies have used either *C. vulgaris* or *C. pyrenoidosa* because these species are commercially available as *Chlorella* products.

3.1. Antihypertensive Effects

Hypertension increases the risk of cardiovascular disease [84]. Antihypertensive compounds in foods have been identified using a stroke-prone spontaneously hypertensive (SHRSP) rat model, which is genetically predisposed to hypertension and cerebral stroke [85]. Sansawa et al. [86] investigated the effects of dried *Chlorella* powder (*C. regularis*) on blood pressure, cerebral stroke lesions, and the life span of SHRSP rats. In 12-week-old SHRSP rats fed *Chlorella* (5%, 10%, and 20%) for 13 weeks, elevated blood pressure significantly decreased in the 10% and 20% *Chlorella* groups compared with the untreated controls. Serum total cholesterol levels were significantly lower in all *Chlorella* groups, and their average life span was more than that of the controls. To characterize the antihypertensive compounds in *Chlorella*, *Chlorella* powder was fractionated into hot-water-soluble, lipid-soluble, and residual fractions. Blood pressure was significantly lower in rats fed the lipid or residual fraction but not in those fed the hot-water-soluble fraction. The lipid fraction contained substantial amounts of carotenoids, which are potent antioxidants, and phospholipids, which mediate aorta collagen and elastin metabolism. The residual fraction contained a high level of arginine, which increases the production of endothelium-derived relaxing factor. These beneficial effects of *Chlorella* powder on SHRSP rats might result from synergism between its numerous bioactive compounds.

To evaluate whether daily *Chlorella* supplementation can reduce blood pressure in subjects with mild to moderate hypertension, a pilot study was conducted in 24 participants administered *C. pyrenoidosa* (10 g of *Chlorella* tablets and 100 mL *Chlorella* extract) [87]. After two months of *Chlorella* supplementation, the average heart rate and sitting systolic and diastolic blood pressures only slightly changed. On the other hand, for some subjects with mild to moderate hypertension, *Chlorella* supplementation reduced or maintained their sitting diastolic blood pressure.

Arterial stiffness is a well-established risk factor of cardiovascular disease [88]. Previous studies have reported that antioxidants [89], potassium [90], and *n*-3 unsaturated fatty acids [91] decrease arterial stiffness. Nitric oxide (NO), derived from arginine in the vascular endothelium, is an important modulator of arterial stiffness [92]. *Chlorella* products contain antioxidants, vitamins, potassium, arginine, and *n*-3 unsaturated fatty acids. To evaluate the effects of *Chlorella* supplementation on arterial stiffness, a single-blinded, placebo-controlled crossover study was conducted in 14 young participants who were administered *C. pyrenoidosa* (6 g/day) or placebo for four weeks, with a 12-week washout period between trials, in a randomized order [93]. No differences were observed in blood pressure or heart rate before and after supplementation in both the placebo and *Chlorella* groups. Brachial-ankle pulse wave velocity, a measure of arterial stiffness, decreased in the *Chlorella* group but not in the placebo group [93]. A similar trial in 32 middle-aged and older subjects reports that the brachial-ankle pulse wave velocity decreased after *Chlorella* supplementation but not after placebo supplementation [94]. These changes in brachial-ankle pulse wave velocity with *Chlorella* supplementation correlated with the plasma NOx level. These results suggest that *Chlorella* supplementation decreases arterial stiffness in both younger and older subjects.

The efficacy of *Chlorella* supplementation in reducing cardiovascular risk factors was assessed in a meta-analysis of 19 randomized controlled trials including 797 subjects [95]. This study concluded that *Chlorella* supplementation improves total cholesterol levels, low-density lipoprotein cholesterol levels, systolic blood pressure, diastolic blood pressure, and fasting blood glucose levels but not triglyceride levels, high-density lipoprotein cholesterol levels, and body mass index.

3.2. Antihypercholesterolemic and Antihyperlipemic Effects

Elevated total cholesterol and triglycerides and abnormal metabolism of lipoproteins and apolipoproteins are responsible for an increased risk of cardiovascular disease [96–98]. The indigestible components of foods, such as dietary fiber, decrease serum cholesterol by inhibiting the intestinal absorption of neutral steroids [99]. *Chlorella* supplementation reportedly decreases serum cholesterol levels in model animals [100]. To identify the bioactive compounds responsible for this effect, the indigestible fraction of *C. regularis* powder was isolated and characterized, revealing a content of

43% crude protein, 37.3% dietary fiber, 6.9% carbohydrate, 5.4% moisture, 4.3% crude fat, and 2.7% ash [101]. Rats fed a diet with 5.3% of this indigestible fraction demonstrated lower serum and liver cholesterol levels and higher fecal neutral steroid levels as compared with those fed a diet of 12.7% *Chlorella* powder. Both *Chlorella* powder and the indigestible fraction exhibited a high bile-acid binding capacity in vitro. Furthermore, the indigestible fraction increased the hepatic mRNA levels of cholesterol 7α-hydroxylase, which is the rate limiting enzyme for cholesterol catabolism and bile-acid synthesis [102]. These results indicate that the indigestible fraction of *Chlorella* possesses hypercholesteromic activity, which improves cholesterol catabolism via the upregulation of hepatic cholesterol 7α-hydroxylase expression.

Chlorella supplementation is also reported to decrease serum cholesterol levels in hyperlipemia and mild hypercholesterolemic patients in a small, open-label trial [103]. To evaluate the preventive role of *Chlorella* in maintaining serum cholesterol levels against excess dietary cholesterol intake, a double-blind, randomized, placebo-control study was conducted in 63 mildly hypercholesterolemic subjects treated with either *C. vulgaris* (5 g/day) or placebo for four weeks [104]. A similar trial investigated cholesterol levels in 34 participants administered 510 mg of dietary cholesterol from three eggs concomitantly with either *Chlorella* (*C. vulgaris*) (5 g/day) or a matched placebo for 4 weeks [105]. Participants on the three-egg diet alone exhibited significant elevation in serum total cholesterol, low-density lipoprotein cholesterol, and high-density lipoprotein cholesterol levels. The administration of *Chlorella* in addition to the three-egg diet significantly suppressed these increases in total cholesterol and low-density lipoprotein cholesterol levels and significantly increased serum lutein and α-carotene levels [105]. In mildly hypercholesterolemic subjects, *Chlorella* administration resulted in marked changes in total cholesterol, triglycerides, lutein/zeaxanthin, and α-carotene levels as well as a significant decrease in very low-density lipoprotein cholesterol, apolipoprotein B, non-high-density lipoprotein, and high-density lipoprotein/triglyceride levels [104]. These results suggest that *Chlorella* might inhibit the intestinal absorption of dietary and endogenous lipids. In addition, the observed changes in serum lipids may be associated with changes in serum carotenoids. These results suggest that daily consumption of *Chlorella* provides potential health benefits by reducing the levels of serum lipid risk factors, such as triglycerides and total cholesterol, in mild hypercholesterolemic subjects.

3.3. Antidiabetic Effect

Type 2 diabetes, accounting for 90–95% of all diabetes cases, is a severe health problem affecting over 380 million people worldwide [106]. Elevated blood glucose levels, insulin resistance, and low insulin sensitivity are the main characteristics of patients with type 2 diabetes [107], resulting in serious conditions, including arteriosclerosis, renal damage, and retinopathy [108]. In a streptozotocin-induced animal model of diabetes, several studies have been conducted to elucidate the mechanisms underlying the antidiabetic activity of *Chlorella* supplementation [109–111]. Shibata et al. [109] evaluated the effects of *Chlorella* supplementation on antioxidant status and cataracts by feeding a diet containing 7.3% (*w*/*w*) *Chlorella* powder (*C. regularis*) to 11-week old rats with streptozotocin-induced diabetes. After 11 weeks of supplementation, serum lipid peroxide levels (an index of oxidative status) and blood glycated hemoglobin were lower in *Chlorella*-supplemented rats than in control rats; however, the serum glucose level did not differ between groups. *Chlorella* supplementation delayed the development of lens opacities. These results indicate that *Chlorella* supplementation might be beneficial for preventing diabetes complications such as cataracts, possibly due to the activity of its antioxidant compounds.

Cherng and Shih reported potential hypoglycemic effects of *Chlorella* supplementation in streptozotocin-induced diabetic mice [110]. Oral administration of *Chlorella* 60 min before glucose administration (0.5 g/kg body weight) resulted in a transient hypoglycemic effect at 90 min after glucose administration without an increase in insulin secretion. *Chlorella* supplementation increased 2-deoxyglucose uptake in the liver and soleus muscles of streptozotocin-treated mice and was likely the cause of the observed hypoglycemic effects [111].

The prophylactic effect of *Chlorella* (*C. vulgaris*) supplementation on diabetes was studied by Vecina et al. [112], who investigated body weight, lipid profile, blood glucose, and insulin signaling in liver, skeletal muscle, and adipose tissue in high-fat diet-induced obese mice. *Chlorella* supplementation improves glycemic control in obesity and diabetes because it decreases insulin resistance caused by increased expression of glucose transporter 4 via the activation of protein kinase B phosphorylation in skeletal muscle. *Chlorella* supplementation combined with aerobic exercise training showed more pronounced effects on the improvement of glycemic control via increased activation of muscle phosphorylation signaling in type 2-diabetic rats [113].

A randomized, double-blind, placebo-controlled human study was conducted in 28 borderline-diabetic participants treated with either *Chlorella* (8 g/day) or placebo for 12 weeks [114]. The expression levels of 252 genes, including six associated with type 2 diabetes, differed between the two groups. Notably, the mRNA expression level of resistin, an insulin resistance inducer, was significantly lower in the *Chlorella* group than in the placebo group and correlated with the expression levels of hemoglobin A_{1c}, tumor necrosis factor-a, and interleukin-6 [114], all of which are involved in glucose metabolism and/or inflammation.

3.4. Hepatoprotective Effect

Li et al. [115] demonstrated that *C. vulgaris* extract has a potent hepatoprotective effect on carbon tetrachloride-induced acute hepatic injury in mice. *Chlorella* extract of 50, 100, or 200 mg/kg of diet, was administered to mice every other day for four weeks, and carbon tetrachloride was administrated intraperitoneally 3 h after the final *Chlorella* supplement. Carbon tetrachloride treatment increased serum alanine and aspartate aminotransferases levels, lipid peroxidation, and cytochrome P450 expression and decrease in reduced glutathione and cellular antioxidant defense enzyme levels; all of these changes were significantly lower in the *Chlorella* (100 and 200 mg/kg diet) groups. Although hepatocyte necrosis was mildly diminished in the 50 mg/kg *Chlorella*-treated group, it was absent in the 100 and 200 mg/kg *Chlorella*-treated groups. These results indicate that *Chlorella* extract has a protective effect on carbon tetrachloride-induced acute hepatic injury in mice, presumably due to the inhibition of carbon tetrachloride-induced cytochrome P450 activation and the activation of antioxidant enzymes and free radical scavengers.

Non-alcoholic fatty liver disease (NAFLD) is a group of metabolic disorders that involving abnormal fat accumulation of more than 5–10% in hepatocytes [116]. It affects 10–35% of the world population [117]. NAFLD includes steatosis, non-alcoholic steatohepatitis, fibrosis, cirrhosis, and hepatocellular carcinoma [118]. Most NAFLD patients have at least one characteristic metabolic syndrome, including insulin resistance, hypertension, dyslipidemia, diabetes, and obesity [119]. Seventy NAFLD patients were randomly administered *C. vulgaris* (1.2 g/day) or placebo for eight weeks [120]. The mean body weight and serum concentrations of liver enzymes were significantly lower in the *Chlorella* group than in the placebo group, and the serum insulin concentration was significantly higher in the *Chlorella* group than in the placebo group. Therefore, *Chlorella* supplementation may have beneficial effects on reducing weight and serum glucose levels and improving inflammatory biomarkers as well as liver function in NAFLD patients [120,121].

To evaluate the safety and efficacy of *Chlorella* (*C. pyrenoidosa*) in patients chronically infected with hepatitis C virus genotype 1, patients received daily oral supplement of *Chlorella* (both of *Chlorella* extract and tablets) for 12 weeks [122]. The majority (approximately 85%) of the patients exhibited a significant decrease in alanine aminotransferase levels from Week 0 to Week 12. Patients with decreased alanine aminotransferase level showed a tendency toward decreased hepatitis C virus load.

3.5. Detoxification Effect

Dioxins are a group of polychlorinated dibenzo-*p*-dioxin and dibenzofuran-related compounds that are industrial contaminants and ubiquitous environmental pollutants [123]. These compounds are easily absorbed in the mammalian gastrointestinal tract [124] and then stored in the liver, adipose tissue, and breast milk due to their lipophilic properties [125]. An incident involving the consumption of cooking oil contaminated with dioxins had tragic effects [126]. To investigate the effects of *Chlorella* supplementation on fecal excretion of dioxins, rats were administered dioxin-contaminated rice oil [127]. The rats were fed 4 g of a 10% (*w*/*w*) *Chlorella* (*C. vulgaris*) diet or a control diet (without *Chlorella*) once during the five-day experimental period, and the amounts of fecal dioxins were measured. The fecal dioxin levels were significantly greater in the *Chlorella* group than in the control group. In addition, *Chlorella* supplementation significantly inhibited the gastrointestinal absorption of dioxins (approximately 2–53% decrease). These results indicate that *Chlorella* supplementation might be useful in promoting dioxin excretion.

Heterocyclic amines have been established as carcinogenic chemicals that form when amino acids, sugars, and creatine in muscle meats (beef, pork, fish, and poultry) react with one another during cooking at high temperatures [128]. To evaluate the effect of *Chlorella* supplementation on the detoxification of carcinogenic heterocyclic amines, a randomized, double-blind, placebo-controlled crossover study with *Chlorella* supplementation (100 mg/day) for two weeks was conducted [129]. *Chlorella* supplementation decreased urinary excretion of the predominant metabolite of carcinogenic heterocyclic amines [129], suggesting that *Chlorella* either inhibits the intestinal absorption of heterocyclic amines or inactivates carcinogenic compounds.

Methylmercury is a neurotoxic metal compound that is converted from inorganic mercury by microorganisms in aquatic environments and is then accumulated in fish and shellfish through marine food chains [130]. Therefore, the major route of human exposure to methylmercury is the consumption of seafood [130]. In many countries, pregnant women are cautioned against consuming large fish, such as tuna, to prevent fetal exposure [131]. As *Chlorella* consumption is reported to increase the excretion of methylmercury and lower tissue mercury levels in methylmercury-treated mice [132], an open-label clinical trial was performed to estimate the effects of *Parachlorella beijerinckii* supplementation (9 g/day) for three months on mercury concentrations in the hair and blood of healthy subjects [133]. *Chlorella* supplementation reduced mercury levels in both the hair and blood [133]. Fecal excretion is the major route of methylmercury elimination (90%) in humans [134]. Most of the methylmercury in the liver is secreted as a glutathione complex via the bile duct, with a small portion excreted in the feces [135]. The dietary fiber in *Chlorella* cells increases the amount of feces excreted by humans [136]. Dietary fiber has been shown to absorb some methylmercury in vitro [132]. These observations suggest that the observed lowering of hair and blood mercury levels in *Chlorella*-treated participants may result from the promotion of fecal methylmercury excretion via accelerated bile secretion, the binding of methylmercury to dietary fiber in the intestinal tract, and increased feces production.

3.6. Immunomodulatory Effects

Allergic disease is a prevalent aberrant immune responsive against innocuous environmental proteins (antigens) [137]. Allergen-specific CD4^{+} T cells involved in the initiation of allergic reactivity can develop into either type 1 or type 2 helper T cells [138]. CD4^{+} T cells stimulated in the presence of interleukin-12 and γ-interferon can develop into type 1 helper T cells [138], while interleukin-4 promotes the development of type 2 helper T cells and inhibits the generation of type 1 helper T cells [139]. Since type 1 and 2 helper T cells regulate each other, interleukin-12 functions not only to induce the type 1 helper T-cell response but also to regulate the type 2 helper T-cell response [140]. Interleukin-12 strongly suppresses the production of IgE by preventing type 2 helper T-cell development [141]. Allergen-specific IgE induces the pathogenesis of allergic disorder [142].

Hasegawa et al. [143] described the effects of a *Chlorella* (*C. vulgaris*) hot-water extract on antigen specific response in mice. A 2% (*w*/*w*) *Chlorella* hot-water extract diet or control diet (without *Chlorella* extract) was given to mice for two weeks before intraperitoneal administration of casein/complete Freund's adjuvant (an immunostimulant). Mice that received the hot-water extract exhibited suppressed IgE production and mRNA expression of interleukin-6 involved in the type 2 helper T-cell response. They also exhibited increased levels of interleukin-12 and g-interferon mRNA, increasing the type 1 helper T-cell response and suppressing the type 2 helper T-cell response. These results suggest that *Chlorella* hot-water extract supplementation might be useful for suppressing allergic responses with a predominant type 2 helper T-cell response. To clarify the mechanisms underlying the immunomodulatory activity of *Chlorella* hot-water extract, soluble polysaccharides were isolated from *C. pyrenoidosa* hot-water extract and characterized [144]. GC-MS analysis indicated that the major monosaccharide components of the soluble polysaccharides are rhamnose (31.8%), glucose (20.4%), galactose (10.3%), mannose (5.2%), and xylose (1.3%). These soluble polysaccharides were intraperitoneally administrated (100 mg/kg of body weight) to 6–8-week-old mice. After 24 h, lipopolysaccharide as an antigen was administrated to mice, and their serum was collected 1.5 h later [144]. The soluble polysaccharides induced interleukin-1b secretion in macrophages via the toll-like receptor protein kinase signaling pathway. Interleukin-1β is one of the most important mediators of inflammation and host responses to infection [145]. These results suggest that *Chlorella* hot-water-soluble polysaccharides could be used as an agent source to stimulate anti-microorganism activity.

Halperin et al. [146] evaluated the effect of *C. pyrenoidosa* supplementation (200 or 400 mg) on the immune response to influenza vaccination. After 28 days of *Chlorella* supplementation, the antibody response to the influenza vaccine was not elevated in the overall study population but was increased in participants aged 50–55 years.

Salivary secretory immunoglobulin A (SIgA) plays a crucial role in mucosal immune function and is the first line of defense against pathogenic microbial invasion in human [147]. To evaluate whether *Chlorella* supplementation increases salivary SIgA secretion in humans, a blind, randomized, crossover study was conducted in participants administered *Chlorella* (*C. pyrenoidosa*) (6 g/day) or placebo for four weeks [148]. Although no difference was observed in salivary SIgA levels before and after placebo ingestion, salivary SIgA levels were significantly elevated after *Chlorella* ingestion than at baseline. The SIgA secretion rate increased significantly after *Chlorella* supplementation. These results suggest that four-week *Chlorella* supplementation increases salivary SIgA secretion and improves mucosal immune function in humans.

Natural killer cells are the predominant innate lymphocyte subsets that mediate antitumor and antiviral responses [149]. To evaluate the effect of *Chlorella* supplementation on natural killer cell activity and early inflammatory response in humans, a randomized, double-blinded, placebo-controlled trial was conducted in healthy adults ingested with *Chlorella* (*C. vulgaris*) (5 g/day) or placebo [150]. After eight weeks of supplementation, serum interferon-γ and interleukin-1β levels were significantly elevated and that of interleukin-12 tended to increase in the *Chlorella* group. Natural killer cell activities were significantly elevated in the *Chlorella* group. These results suggest a beneficial immunostimulatory effect of short-term *Chlorella* supplementation that increases natural killer cell activity and produces interferon-γ, interleukin-12, and interleukin-1β.

3.7. Antioxidant Effects

C. vulgaris hot-water extract [151] and acetone extract [152] are reported to have antitumor activity. A *Chlorella* aqueous extract containing substantial amounts of antioxidants also exhibit antiproliferative activity in human hepatoma cells [153]. Lipophilic pigments, including carotenoids antheraxanthin, zeaxanthin, and lutein, extracted from *Chlorella* cells were observed to significantly inhibit the growth of human colon cancer cells [154]. These results suggest that the antitumor activity of *Chlorella* might be the synergistic effect of multiple bioactive compounds. Romos et al. [155] reported that *Chlorella* supplementation can modulate immunomyelopoietic activity and disengage tumor-induced suppression of various cytokines and related cell activities in tumor-bearing mice. Interestingly, a 63.1-kD antitumor glycoprotein was isolated from the culture supernatant of *C. vulgaris* strain CK22 [156,157], and its chemical and antitumor properties were characterized [158], suggesting possible contribution of this glycoprotein toward the observed antitumor activity.

Alzheimer's disease is a severe neurodegenerative condition affecting humans [159]. The erythrocytes of Alzheimer's disease patients are known to be in an excessively oxidized state [160]. α-Tocopherol and carotenoids such as lutein are important lipophilic antioxidants in human erythrocytes [161]. Erythrocyte lutein levels were found to be significantly lower in Alzheimer's disease patients than in normal subjects [162]. Oral intake of lutein capsules increases lutein levels and prevents phospholipid hydroperoxide accumulation in human erythrocytes [163], suggesting that dietary lutein has the potential to act as an important antioxidant in erythrocytes and thus may have beneficial effects in Alzheimer's disease patients. According to the labels on *Chlorella* products D and M, the products contain substantial amounts of lutein (approximately 200 mg/100 g dry weigh). A randomized, double-blind, placebo-controlled human study was conducted to evaluate the effects of *Chlorella* supplementation (8 g *Chlorella*/day/person; equivalent to 22.9 mg lutein/day/person) on phospholipid hydroperoxide and lutein levels in erythrocytes [164]. After two months of *Chlorella* supplementation, erythrocyte lutein levels increased 4.6-fold, but tocopherol levels did not change [164], suggesting that daily *Chlorella* intake may be effective for improving and maintaining erythrocyte antioxidant status and lutein levels in humans. These results suggest that *Chlorella* supplementation contributes to maintaining the normal function of erythrocytes and has beneficial effects on Alzheimer's disease-related dementia in humans.

Major depressive disorder is a widespread mental disorder that greatly impairs the quality of life of humans. Approximately 12% of people experience at least one episode of depression during their lifetime [165]. Although various antidepressant drugs are available for treating depression, a considerable proportion of patients are not responsive to these drugs and some experience side effects [166,167]. Therefore, alternative antidepressant drugs with adequate efficacy and safety are needed. The therapeutic effect of dried *C. vulgaris* extract administration (1.8 g/day) for six weeks was evaluated in patients with major depressive disorder [168]. After treatment, the participants exhibited improvements in physical and cognitive symptoms of depression [168]. As oxidative stress is an important pathophysiological mechanism underlying major depressive disorder, major depressive disorder has been effectively reversed via antioxidant therapy [169,170]. These observations suggest that the therapeutic effectiveness of *Chlorella* supplementation may result from the action of its antioxidant nutrients and compounds [171].

3.8. Other Effects

Stress is well known to disturb homeostasis, impairing immunological functions. *Chlorella* supplementation reportedly stimulates the pool of hematopoietic stem cells and activates leukocytes [172]. To further understand the influence of *Chlorella* (*C. vulagaris*) supplementation on hematopoiesis, hematopoietic cell populations in the bone marrow of mice subjected to a single or repeated stressor were measured [173]. Reduced numbers of hematopoietic progenitors in the bone marrow were observed after treatment with either stressor. Both stressors induced a decrease in mature myeloid and lymphoid populations but did not affect pluripotent hematopoietic progenitors. Both stressors reduced the levels of interleukin-1α and interleukin-6. *Chlorella* supplementation prevented the changes produced by both stressors in all of the parameters tested, suggesting that *Chlorella* supplementation is an effective tool for the prophylaxis of myelosuppression caused by single or repeated stressors.

Stressors are processed in the brain through the activation of several types of neurons. Immediate early genes such as *c-fos* are extensively used to map brain areas involved in stress responses [174]. Using *c-fos* expression, Oueiroz et al. [175] evaluated the effect of acute pretreatment with *Chlorella* (*C. vulgaris*) on the peripheral and central responses to forced swimming stress in rats. *Chlorella* supplementation produced a significant reduction in stress-related hypothalamic–pituitary–adrenal axis activation due to decreased corticotrophin releasing factor gene expression in the hypothalamic paraventricular nucleus and a lower adrenocorticotropic hormone response. Hyperglycemia induced by the stressor was similarly reduced. These results suggest that *Chlorella* supplementation might reduce the impact of stressors.

A hot-water extract of *Chlorella* (*C. pyrenoidosa*) increased the lifespan of superoxide dismutase-1 mutant adults of *Drosophila melanogaster* in a dose-dependent manner (200–800 μg/mL) [176]. An active compound was purified and identified as phenethylamine, an aromatic amine, which exhibited no superoxide dismutase-like activity. Treatment with this compound extended the lifespan of the mutant flies at very low concentration (60 ng/g diet) [176]. Similarly, supplementation of *C. sorokiniana* (4 mg/mL) reportedly increased the lifespan of *D. melanogaster* by 10% increase as compared to a control diet, likely due to the increased mRNA expression of antioxidative enzymes (Cu/Zn-superoxide dismutase and catalase) [177].

However, for the beneficial effects described above, no human study has been conducted.

4. Conclusions

Commercially available *Chlorella* products contain a variety of nutrients essential for humans, as well as a large amount of good quality protein, dietary fibers, and polyunsaturated fatty acids, including α-linolenic and linoleic acids. In particular, *Chlorella* products contain vitamins D_2 and B_{12}, which are absent from plant-derived food sources, and larger amounts of folate and iron than other plant-derived foods. Mounting scientific evidence of the health benefits of daily *Chlorella* consumption has been presented in animal and human studies. The pharmacological activities reported in *Chlorella* studies include immunomodulation, antioxidative activity, and effects against diabetes, hypertension, and hyperlipidemia. The beneficial effects of *Chlorella* might involve synergism between multiple nutrient and antioxidant compounds. Overall, the information regarding bioactive compounds in *Chlorella* is limited. Thus, new bioactive compounds responsible for its pharmacological activities may be identified in future studies.

Author Contributions: T.B., E.O., M.F., and F.W. conceived of and designed the study. T.B. and F.W. wrote the original draft. E.O. and M.F. reviewed and edited the manuscript. All authors have read and agreed to the published version of the manuscript.

Funding: This research received no external funding.

Conflicts of Interest: The authors declare no conflict of interest.

References

1. Yan, N.; Fan, C.; Chen, Y.; Hu, Z. The potential for microalgae as bioreactors to produce pharmaceuticals. *Int. J. Mol. Sci.* **2016**, *17*, 962. [CrossRef] [PubMed]
2. Barkia, I.; Saari, N.; Manning, S.R. Microalgae for high-value products towards human health and nutrition. *Mar. Drugs* **2019**, *17*, 304. [CrossRef] [PubMed]
3. Beijerinck, M.W. Culturversuche mit Zoochlorellen, Lichenengonidien und anderen niederen Algen. *Botanische Zeitung* **1890**, *47*, 725–739.
4. Chick, H. A study of a unicellular green alga, occurring in polluted water, with especial reference to its nitrogenous metabolism. *Proc. Royal Soc. Biol. Sci. Ser. B* **1903**, *71*, 458–476.
5. Wu, H.L.; Hseu, R.S.; Lin, L.P. Identification of *Chlorella* spp. isolates using ribosomal DNA sequences. *Bot. Bull. Acad. Sin.* **2001**, *42*, 115–121.
6. Krienitz, L.; Hegewald, E.H.; Hepperle, D.; Huss, V.A.R.; Rohr, T.; Wolf, M. Phylogenetic relationship of *Chlorella* and *Parachlorella* gen. nov. (Chloropyta, Trebouxiophyceae). *Phycologia* **2004**, *43*, 529–542. [CrossRef]
7. Sorokin, C.; Myers, J. A high-temperature strain of *Chlorella*. *Science* **1953**, *117*, 330–331. [CrossRef]
8. Lizzul, A.M.; Lekuona-Amundarain, A.; Purton, S.; Campos, L.C. Characterization of *Chlorella sorokiniana*, UTEX 1230. *Biology* **2018**, *7*, 25. [CrossRef]
9. Montoya, E.Y.O.; Casazza, A.A.; Aliakbarian, B.; Perego, P.; Converti, A.; Carvalho, M. Production of *Chlorella vulgaris* as a source of essential fatty acids in a tubular photobioreactor continuously fed with air enriched with CO_2 at different concentrations. *Biotechnol. Prog.* **2014**, *30*, 916–922. [CrossRef]
10. Rani, K.; Sandal, N.; Sahoo, P.K. A comprehensive review on chlorella-its composition, health benefits, market and regulation scenario. *Pharma Innov. J.* **2018**, *7*, 583–589.

11. De Ortega, A.R.; Roux, J.C. Production of *Chlorella* biomass in different types of Flat bioreactors in temperate zones. *Biomass* **1986**, *10*, 141–156. [CrossRef]
12. Ru, I.T.K.; Sung, Y.Y.; Jusoh, M.; Wahid, M.E.A. *Chlorella vulgaris*: A perspective on its potential for combining high biomass with high value bioproducts. *App. Phycol.* **2020**, *1*, 2–11. [CrossRef]
13. Shukla, S.P.; Kvíderová, J.; Tríska, J.; Elster, J. *Chlorella mirabilis* as a potential species for biomass production in low-temperature environment. *Front. Microbiol.* **2013**, *4*, 97. [CrossRef] [PubMed]
14. Ward, V.C.A.; Rehmann, L. Rast media optimization for mixotrophic cultivation of *Chlorella vulgaris*. *Sci. Rep.* **2019**, *9*, 19262. [CrossRef]
15. Komaki, H.; Yamashita, M.; Niwa, Y.; Tanabe, Y.; Kamiya, N.; Ando, Y.; Furuse, M. The effect of processing of *Chlorella vulgaris*: K-5 on in vitro and in vivo digestibility in rats. *Anim. Feed Sci. Technol.* **1998**, *70*, 363–366. [CrossRef]
16. Chisti, Y. Biodiesel from microalgae. *Biotechnol. Adv.* **2007**, *25*, 294–306. [CrossRef]
17. Becker, E.W. Micro-algae as a sourse of protein. *Biotechnol. Adv.* **2007**, *25*, 207–210. [CrossRef]
18. Waghmare, A.G.; Salve, M.K.; LeBlanc, J.G.; Arya, S.S. Concentration and characterization of microalgae proteins from *Chlorella pynenoidosa*. *Bioresour. Bioprocess.* **2016**, *3*, 16. [CrossRef]
19. Kent, M.; Welladsen, H.M.; Mangott, A.; Li, Y. Nutritional evaluation of Australian microalgae as potential human health supplements. *PLoS ONE* **2015**, *10*, e0118985. [CrossRef]
20. Morris, S.M., Jr. Recent advances in arginine metabolism: Roles and regulation of the arginases. *Br. J. Pharm.* **2009**, *157*, 922–930. [CrossRef]
21. Bansal, V.; Ochoa, J.B. Arginine availability, arginase, and the immune response. *Curr. Opin. Clin. Nutr. Metab. Care* **2003**, *6*, 223–228. [CrossRef] [PubMed]
22. Olaitan, S.A.; Northcote, D.H. Polysaccharides of *Chlorella pyrenoidosa*. *Biochem. J.* **1962**, *82*, 509–519. [CrossRef] [PubMed]
23. Sui, Z.; Gizaw, Y.; BeMiller, J.N. Extraction of polysaccharides from a species of *Chlorella*. *Carbohydr. Polym.* **2012**, *90*, 1–7. [CrossRef] [PubMed]
24. Yu, M.; Chen, M.; Gui, J.; Huang, S.; Liu, Y.; Shentu, H.; He, J.; Fang, Z.; Wang, W.; Zhag, Y. Preparation of *Chlorella vulgaris* polysaccharides and their antioxidant activity in vitro and in vivo. *Int. J. Biol. Macromol.* **2019**, *137*, 139–150. [CrossRef] [PubMed]
25. El-Naggar, N.E.A.; Hussein, M.H.; Shaaban-Dessuuki, S.A.; Dalal, S.R. Production, extraction and characterization of *Chlorella vulgaris* soluble polysacharaides and their applications in AgNPs biosynthesis and biostimulation of plant growth. *Sci. Rep.* **2020**, *10*, 3011. [CrossRef]
26. Tabarsa, M.; Shin, I.S.; Lee, J.H.; Surayot, U.; Park, W.J.; You, S.G. An immune-enhancing water-soluble a-glucan from *Chlorella vulgaris* and structural characteristics. *Food Sci. Biotechnol.* **2015**, *24*, 1933–1941. [CrossRef]
27. Ötles, S.; Pire, R. Fatty acid composition of *Chlorella* and Spirulina microalgae species. *J. AOAC Int.* **2001**, *84*, 1708–1714. [CrossRef]
28. Panahi, Y.; Khosroshahi, A.Y.; Sahebkar, A.; Heidari, H.R. Impact of cultivation condition and media content on *Chlorella vulgaris* composition. *Adv. Pharm. Bull.* **2019**, *9*, 182–194. [CrossRef]
29. Chiu, S.Y.; Kao, C.Y.; Chen, T.Y.; Chang, Y.B.; Kuo, C.M.; Lin, C.S. Cultivation of microalgal *Chlorella* for biomass and lipid production using wastewater as nutrient resource. *Bioresour. Technol.* **2015**, *184*, 179–189. [CrossRef]
30. Mao, X.; Wu, T.; Sun, D.; Zhang, Z.; Chen, F. Differential responses of the green microalga *Chlorella zofingiensis* to the starvation of various nutrients for oil and astaxanthin production. *Bioresour. Technol.* **2018**, *249*, 791–798. [CrossRef]
31. Woortman, D.V.; Fuchs, T.; Striegel, L.; Fuchs, M.; Weber, N.; Brück, T.B.; Rychlik, M. Microalae a superior source of folates: Quantification of folates in halophile microalgae by stable isotope dilution assay. *Front. Bioeng. Biotechnol.* **2020**, *7*, 481. [CrossRef] [PubMed]
32. Lips, P. Vitamin D physiology. *Prog. Biophys. Mol. Biol.* **2006**, *92*, 4–8. [CrossRef] [PubMed]
33. Taofiq, O.; Fernandes, A.; Barreiro, M.F.; Ferreira, I.C.F.R. UV-irradiated mushrooms as a source of vitamin D_2: A review. *Trends Food Sci. Technol.* **2017**, *70*, 82–94. [CrossRef]
34. Holman, E.H.; Martin, B.R.; Lackcik, P.L.; Godeon, D.T.; Fleet, J.C.; Weaver, C.M. Bioavailability and efficacy of vitamin D_2 from UV-irradiated yeast in growing, vitamin D-deficient rats. *J. Agric. Food Chem.* **2011**, *59*, 2341–2346.

35. Ložnjak, P.; Jakobsen, J. Stability of vitamin D_3 and vitamin D_2 in oil, fish and mushrooms after household cooking. *Food Chem.* **2018**, *254*, 144–149. [CrossRef]
36. Cardwell, G.C.; Bornman, J.F.; James, A.P.; Black, L.J. A review of mushrooms as a potential source of dietary vitamin D. *Nutrients* **2018**, *10*, 1498. [CrossRef]
37. Weete, J.D.; Abril, M.; Blackwell, M. Phylogenetic distribution of fungal sterols. *PLoS ONE* **2010**, *5*, e10899. [CrossRef]
38. Huang, G.; Cai, W.; Xu, B. Vitamin D_2, ergosterol, and vitamin B_2 content in commercially dried mushrooms. *Int. J. Vitam. Nutr. Res.* **2016**, *1*, 1–10.
39. Urbain, P.; Singler, F.; Ihorst, G.; Biesalski, H.K.; Bertz, H. Bioavailability of vitamin D_2 from UV-B-irradiated button mushrooms in healthy adults deficient in serum 25-hydroxyvitamin D: A randomized controlled trial. *Eur. J. Clin. Nutr.* **2011**, *65*, 965–971. [CrossRef]
40. Cashman, K.D.; Kiely, M.; Seamans, K.M.; Urbain, P. Effect of Ultraviolet light-exposed mushrooms on vitamin D status: Liquid chromatography-tandem mass spectrometry reanalysis of biobanked sera from a randomized controlled trial and a systematic review plus meta-analysis. *J. Nutr.* **2016**, *146*, 565–575. [CrossRef]
41. Klosty, M.; Bergmann, W. Sterols of algae. III. The occurrence of ergosterol in *Chlorella pyranoidosa*. *J. Am. Chem. Soc.* **1952**, *74*, 1601. [CrossRef]
42. Patterson, G.W. Sterols of *Chlorella*. II. The occurrence of an unusual sterol mixture in *Chlorella vulgaris*. *Plant. Physiol.* **1967**, *42*, 1457–1459. [CrossRef] [PubMed]
43. Wu, J.; Liu, C.; Lu, Y. Preparative separation of phytosterol analogues from green alga *Chlorella vulgaris* using recycling counter-current chromatography. *J. Separ. Sci.* **2017**, *40*, 2326–2334. [CrossRef] [PubMed]
44. Wilson, L.R.; Tripkovic, L.; Hart, K.H.; Lanham-New, S.A. Vitamin D deficiency as a public health issue: Using vitamin D_2 or vitamin D_3 in future fortification strategies. *Proc. Nutr. Soc.* **2017**, *106*, 481–490. [CrossRef]
45. Selhub, J.; Jacques, P.F.; Bostom, A.G.; D'Agostino, R.B.; Wilson, P.W.F.; Belanger, A.J.; O'Leary, D.H.; Wolf, P.A.; Schaffer, E.J.; Rosenberg, I.H. Association between plasma homocysteine concentrations and extracranial carotid-artery stenosis. *N. Engl. J. Med.* **1995**, *332*, 286–291. [CrossRef]
46. Ganguly, P.; Alam, S.F. Role of homocysteine in the develop of cardiovascular disease. *Nutr. J.* **2015**, *14*, 6. [CrossRef]
47. Green, R.; Allen, L.H.; Bjorke-Mercado, A.L.; Brto, A.; Gueant, J.L.; Miller, J.W.; Molloy, A.M.; Nexo, E.; Stabler, S.; Toh, B.H.; et al. Vitamin B_{12} deficiency. *Nat. Rev. Dis. Prim.* **2017**, *3*, 17040. [CrossRef]
48. Miller, J.W.; Ribaya-Meercado, J.D.; Russell, R.M.; Shepard, D.C.; Morrow, F.D.; Cochary, E.F.; Sadowski, J.A.; Gershoff, S.N.; Selhub, J. Effect of vitamin B_6-deficiency on fasting plasma homocysteine concentrations. *Am. J. Clin. Nutr.* **1992**, *55*, 1154–1160. [CrossRef]
49. Forges, T.; Monnier-Barbarino, P.; Alberto, J.M.; Gueant-Rodriguez, R.M.; Daval, J.L.; Gueant, J.L. Impact of folate and homocysteine metabolism on human reproductive health. *Hum. Reprod. Update* **2007**, *13*, 225–238. [CrossRef]
50. Celik, S.F.; Celik, E. Subclinical atherosclerosis and impaired cardiac autonomic control in pediatric patients with vitamin B_{12} deficiency. *Niger. J. Clin. Pract.* **2018**, *21*, 1012–1016.
51. Bunout, D.; Petermann, M.; Hirsch, S. Low serum folate but normal homocysteine levels in patients with atherosclerotic vascular disease and matched healthy controls. *Nutrition* **2000**, *16*, 434–438. [CrossRef]
52. Czeizel, A.E.; Dudás, I.; Vereczkey, A.; Bránhidy, F. The prevention of neural-tube defects and congenital heart defects. *Nutrients* **2013**, *5*, 4760–4775. [CrossRef] [PubMed]
53. Gorelova, V.; Ambach, L.; Rébeillé, F.; Stove, C.; Van der Straeten, D. Fotes in plants: Research advances and progress in crop biofortification. *Front. Chem.* **2017**, *5*, 21. [CrossRef] [PubMed]
54. Delchier, N.; Herbig, A.L.; Rychlik, M.; Renard, M.G.C. Folates in fruits and vegetables: Contents, processing and stability. *Comp. Rev. Food Sci. Food Saf.* **2016**, *15*, 506–528. [CrossRef]
55. Watanabe, F.; Bito, T. Corrinoids in food and biological samples. In *Frontiers in Natural Product Chemistry*; Atta-ur-Rahman, Ed.; Bentham Science Publishers: Dubai, UAE, 2016; Volume 2, pp. 229–244.
56. Watanabe, F.; Bito, T. Vitamin B_{12} sources and microbial interaction. *Exp. Biol. Med.* **2018**, *243*, 148–158. [CrossRef]
57. Bito, T.; Tanioka, Y.; Watanabe, F. Characterization of vitamin B_{12} compounds from marine foods. *Fish. Sci.* **2018**, *84*, 747–755. [CrossRef]

58. Kittaka-Katsura, H.; Fujita, T.; Watanabe, F.; Nakano, Y. Purification and characterization of a corrinoid compound from *Chlorella* Tablets as an algal health food. *J. Agric. Food Chem.* **2002**, *50*, 4994–4997. [CrossRef]
59. Bito, T.; Bito, M.; Asai, Y.; Takenaka, S.; Yabuta, Y.; Tago, K.; Ohnishi, M.; Mizoguchi, T.; Watanabe, F. Characterization and quantitation of vitamin B_{12} compounds in various *Chlorella* supplements. *J. Agric. Food Chem.* **2016**, *64*, 8516–8524. [CrossRef]
60. Watanabe, F.; Abe, K.; Takenaka, S.; Tamura, Y.; Maruyama, I.; Nakano, Y. Occurrence of cobalamin coenzymes in the photosynthetic green alga, *Chlorella vulgaris*. *Biosci. Biotechnol. Biochem.* **1997**, *61*, 896–897. [CrossRef]
61. Rauma, A.L.; Törrönen, R.; Hänninen, O.; Mykkänen, H. Vitamin B_{12} status of long-term adherents of strict uncooked vegan diet ("living food diet") is comprised. *J. Nutr.* **1995**, *125*, 2511–2515.
62. Merchant, R.E.; Phillips, T.W.; Udani, J. Nutritional supplementation with *Chlorella pyrenoidosa* lowers serum methylmalonic acid in vegans and vegetarians with a suspected vitamin B_{12} deficiency. *J. Med. Food* **2015**, *18*, 1357–1362. [CrossRef] [PubMed]
63. Camaschella, C. Iron-deficiency anemia. *N. Engl. J. Med.* **2015**, *372*, 1832–1843. [CrossRef] [PubMed]
64. Adrogué, H.J.; Madias, N.E. Sodium and potassium in pathogenesis of hypertension. *N. Engl. J. Med.* **2007**, *356*, 1966–1978. [CrossRef] [PubMed]
65. Hentze, M.W.; Muckenthaler, M.U.; Galy, B.; Camaschella, C. Two to tango: Regulation of mammalian iron metabolism. *Cell* **2010**, *142*, 24–38. [CrossRef] [PubMed]
66. Gibson, R.S.; Raboy, V.; King, J.C. Implication of phytate in plant-based foods for iron and zinc bioavailability, setting dietary requirements, and formulating programs and policies. *Nutr. Rev.* **2018**, *76*, 793–804. [CrossRef] [PubMed]
67. Gao, F.; Guo, W.; Zeng, M.; Feng, Y.; Feng, G. Effect of microalgae as iron supplements on ion-deficiency anemia in rats. *Food Funct.* **2019**, *10*, 723–732. [CrossRef]
68. Nakano, S.; Takekoshi, H.; Nakano, M. *Chlorella pyrenoidosa* supplementation reduces the risk of anemia, proteinuria and edema in pregnant women. *Plant Food Hum. Nutr.* **2010**, *65*, 25–30. [CrossRef]
69. Rayman, M.P. The importance of selenium to human health. *Lancet* **2000**, *356*, 233–241. [CrossRef]
70. Allen, C.B.; Lacourciere, G.M.; Stadtman, T.C. Responsiveness of selenoproteins to dietary selenium. *Ann. Rev. Nutr.* **1999**, *19*, 1–16. [CrossRef]
71. Sunde, R.A. Handbook of nutritionally essential mineral elements. In *Handbook of Nutritionally Essential Mineral Elements*; O'Dell, B.L., Sunde, R.A., Eds.; Marcel Dekker Inc: New York, NY, USA, 1997; pp. 493–556.
72. White, P.J.; Brown, H.C.; Parmaguru, P.; Fritz, M.; Spracklen, W.P.; Spiby, R.E.; Meacham, M.C.; Trueman, L.J.; Smith, B.M.; Thomas, B.; et al. Interactions between selenium and Sulphur nutrition in Arabidopsis thaliana. *J. Exp. Bot.* **2004**, *55*, 1927–1937. [CrossRef]
73. Doucha, J.; Lívanský, K.; Kotrbáček, V.; Zachleder, V. Production of *Chlorella* biomass enriched by selenium and its use in animal nutrition: A review. *Appl. Microbiol. Biotechnol.* **2009**, *83*, 1001–1008. [CrossRef] [PubMed]
74. Sun, X.; Zhong, Y.; Huang, Z.; Yang, Y. Selenium accumulation in unicellular green alga *Chlorella vulgaris* and its effects on antioxidant enzymes and content of photosynthetic pigments. *PLoS ONE* **2014**, *9*, e112270. [CrossRef] [PubMed]
75. Chen, T.F.; Zheng, W.J.; Wong, Y.S.; Yang, F. Selenium-induced changes in activities of antioxidant enzymes and content of photosynthetic pigments in *Spirulina platensis*. *J. Integr. Plant. Biol.* **2008**, *50*, 40–48. [CrossRef] [PubMed]
76. Nisar, N.; Li, L.; Lu, S.; Khin, N.C.; Pogson, B.J. Carotenoid metabolism in plants. *Mol. Plant.* **2015**, *8*, 68–82. [CrossRef]
77. Sathasivam, R.; Radhakrishnan, R.; Hashem, A.; Abd_Allah, E.F. Microalgae metabolites: A rich source for food and medicine. *Saudi J. Biol. Sci.* **2019**, *26*, 709–722. [CrossRef]
78. Raja, R.; Hemaiswarya, S.; Rengasamy, R. Exploitation of *Dunaliella* for β-carotene production. *Appl. Microbiol. Biotechnol.* **2007**, *74*, 517–523. [CrossRef]
79. Boussiba, S.; Bing, W.; Yuan, J.P.; Zarka, A.; Chen, F. Changes in pigments profile in the green alga *Haeamtococcus pluvialis* exposed to environmental stresses. *Biotechnol. Lett.* **1999**, *20*, 601–604. [CrossRef]
80. O'Sullivan, A.M.; O'Callaghan, Y.C.; O'Connor, T.P.; O'Brien, N.M. The content and bioaccessibility of carotenoids from selected commercially available health supplements. *Proc. Nutr. Soc.* **2011**, *70*, E62. [CrossRef]

81. Guedes, A.C.; Amaro, H.M.; Malcata, F.X. Microalgae as sources of carotenoids. *Mar. Drugs* **2011**, *9*, 625–644. [CrossRef]
82. Plaza, M.; Herrero, M.; Cifuentes, A.; Ibáñez, E. Innovative natural functional ingredients from microalgae. *J. Agric. Food Chem.* **2009**, *57*, 7159–7170. [CrossRef]
83. Lui, J.; Sun, Z.; Gerken, H.; Liu, S.; Jiang, Y.; Chen, F. *Chlorella zofingiensis* as an alternative microalgal producer of astaxanthin: Biology and industrial potential. *Mar. Drugs* **2014**, *12*, 3487–3515.
84. McFarlane, S.I.; Jean-Lours, G.; Zizi, F.; Whaley-Connel, A.T.; Ogedegbe, O.; Nakaryus, A.N.; Maraj, I. Hypertension in the high-cardiovascular-risk populations. *Int. J. Hypertens.* **2012**, *2011*, 746369. [CrossRef] [PubMed]
85. Yamori, Y. Experimental evidence for dietary prevention of cardiovascular disease. *Clin. Exp. Pharm. Physiol.* **1989**, *16*, 303–307. [CrossRef] [PubMed]
86. Sansawa, H.; Takahashi, M.; Tsuchikura, S.; Endo, H. Effect of *Chlorella* and its fractions on blood pressure, cerebral stroke lesions, and life-span in stroke-prone spontaneously hypertensive rats. *J. Nutr. Sci. Vitam.* **2006**, *52*, 457–466. [CrossRef] [PubMed]
87. Merchant, R.E.; Andre, C.A.; Sica, D.A. Nutritional supplementation with *Chlorella pyrenoidosa* for mild to moderate hypertension. *J. Med. Food* **2002**, *5*, 141–152. [CrossRef] [PubMed]
88. Vlachopoulos, C.; Aznaouridis, K.; Terentes-Printzios, D.; Ioakeimidis, N.; Stefanadis, C. Prediction of cardiovascular events and all-cause mortality with brachial-ankle elasticity index: A systematic review and meta-analysis. *Hypertension* **2012**, *60*, 556–562. [CrossRef]
89. Plantinga, Y.; Ghiadoni, L.; Magagna, A.; Giannarelli, C.; Franzoni, F.; Salvetti, A. Supplementation with vitamins C and E improves arterial stiffness and endothelial function in essential hypertensive patients. *Am. J. Hypertens.* **2007**, *20*, 392–397. [CrossRef]
90. Nestel, P.J.; Pomeroy, S.E.; Sasahara, T.; Yamashita, T.; Ling, Y.L.; Dart, A.M.; Jennings, G.L.; Abbey, M.; Cameron, J.D. Arterial compliance in obese subjects is improved with dietary plant n-3 fatty acid from flaxseed oil despite increased LDL oxidizability. *Arter. Thromb. Vasc. Biol.* **1997**, *17*, 1163–1170. [CrossRef]
91. He, F.J.; Marciniak, M.; Carney, C.; Markandu, N.D.; Anand, V.; Fraser, W.D.; Dalton, R.N.; Kaski, J.C.; MacGregor, G.A. Effects of potassium chloride and potassium bicarbonate on endothelial function, cardiovascular risk factors, and bone turnover in mild hypertensives. *Hypertension* **2010**, *55*, 681–688. [CrossRef]
92. Kinlay, S.; Creager, M.A.; Fukumoto, M.; Hikita, H.; Fang, J.C.; Selwyn, A.P.; Ganz, P. Endothelium-derived nitric oxide regulates arterial elasticity in human arteries in vivo. *Hypertension* **2001**, *38*, 1049–1053. [CrossRef]
93. Osuki, T.; Shimizu, K.; Iemitsu, M.; Kono, I. Multicomponent supplement containing *Chlorella* decreases arterial stiffness in healthy young men. *J. Clin. Biochem. Nutr.* **2013**, *53*, 166–169. [CrossRef] [PubMed]
94. Otuski, T.; Shimizu, K.; Maeda, S. Changes in arterial stiffness and nitric oxide production with *Chlorella*-derived multicomponent supplementation in middle-aged and older individuals. *J. Clin. Biochem. Nutr.* **2015**, *57*, 228–232. [CrossRef] [PubMed]
95. Fallah, A.A.; Sarmast, E.D.; Dehkordi, S.H.; Engardeh, J.; Mahmoodnia, L.; Khaledifar, A.; Jafari, T. Effect of *Chlorella* supplementation on cardiovascular risk factors: A meta-analysis of randomized controlled trials. *Clin. Nutr.* **2018**, *37*, 1892–1901. [CrossRef] [PubMed]
96. Mora, S.; Otvos, J.D.; Rifai, N.; Rosenson, R.S.; Buring, J.E.; Ridker, P.M. Lipoprotein particle profiles by nuclear magnetic resonance compared with standard lipids and apolipoproteins in predicting incident cardiovascular disease in women. *Circulation* **2009**, *119*, 931–939. [CrossRef]
97. Nordestgaard, B.G.; Benn, M.; Schnohr, P.; Tybjaerg-Hansen, A. Nonfasting triglycerides and risk of myocardial infarction, ischemic heart disease, and death in men and women. *J. Am. Med Assoc.* **2007**, *298*, 299–308. [CrossRef]
98. Sacks, F.M.; Alaupovic, P.; Moye, L.A.; Cole, T.G.; Sussex, B.; Stampfer, M.J.; Pfeffer, M.A.; Braunwald, E. VLDL, apolipoproteins B, CIII and E, and risk of recurrent coronary events in the cholesterol and recurrent events (CARE) trial. *Circulation* **2000**, *102*, 1886–1892. [CrossRef]
99. Adnerson, J.W.; Deakins, D.A.; Bridges, S.R. Soluble fiber: Hypocholesteromic effects and proposed mechanisms. In *Dietary Fiber: Chemistry, Physiology and Health Effects*; Kritchevsky, D., Bonfield, C., Anderson, J.W., Eds.; Plenum Press: New York, NY, USA, 1990; pp. 339–363.
100. Cherng, J.Y.; Shih, M.F. Preventing dyslipidemia by *Chlorella pyrenoidosa* in rats and hamsters after chronic high fat diet treatment. *Life Sci.* **2005**, *76*, 3001–3013. [CrossRef]

101. Shibata, S.; Oda, K.; Onodera-Masuoka, N.; Matsubara, S.; Kikuchi-Hayakawa, H.; Ishikawa, F.; Iwabuchi, A.; Sansawa, H. Hypocholesteromic effect of indigestible fraction of *Chlorella regularis* in cholesterol-fed rats. *J. Nutr. Sci. Vitam.* **2001**, *47*, 373–377. [CrossRef]
102. Shibata, S.; Hayakawa, K.; Egashira, Y.; Sanada, H. Hypocholesteromic mechanism of *Chlorella*: *Chlorella* and its indigestible fraction enhance hepatic cholesterol catabolism through up-regulation of cholesterol 7α-hydroxylase in rats. *Biosci. Biotechnol. Biochem.* **2007**, *71*, 916–925. [CrossRef]
103. Sansawa, H.; Inoue, K.; Shirai, T. Effect of *Chlorella* tablet ingestion on mild hypercholesterolemic patients. *J. Jpn. Soc. Food Sci. Technol.* **2002**, *49*, 167–173. [CrossRef]
104. Ryu, N.H.; Lim, Y.; Park, J.E.; Kim, J.; Kim, J.Y.; Kwon, S.W.; Kwon, O. Impact of daily *Chlorella* consumption on serum lipid and carotenoid profiles in mild hypercholesterolemic adults: A double-blinded, randomized, placebo-controlled study. *Nutr. J.* **2014**, *13*, 57. [CrossRef] [PubMed]
105. Kim, S.; Kim, J.; Lim, Y.; Kim, Y.J.; Kim, J.Y.; Kwon, O. A dietary cholesterol challenge study to assess *Chlorella* supplementation in maintaining healthy lipid levels in adults: A double-blinded, randomized, placebo-controlled study. *Nutr. J.* **2016**, *15*, 54. [CrossRef] [PubMed]
106. Caruso, R.; Magon, A.; Baroni, I.; Dellafiore, F.; Arrigoni, C.; Pittella, F.; Ausili, D. Health literacy in type 2 diabetes patients: A systematic review of systematic reviews. *Acta Diabetol.* **2018**, *55*, 1–2. [CrossRef] [PubMed]
107. Schwingshackl, L.; Chaimani, A.; Hoffmann, G.; Schwedhelm, C.; Boeing, H. Impact of different dietary approaches on glycemic control and cardiovascular risk factors in patients with type 2 diabetes: A protocol for a systematic review and network meta-analysis. *Sys. Rev.* **2017**, *6*, 57. [CrossRef]
108. De Groot, M.; Anderson, R.; Freedland, K.E.; Clouse, R.E.; Lustman, P.J. Association for depression and diabetes complications: A meta-analysis. *Psychosom. Med.* **2001**, *63*, 619–630. [CrossRef]
109. Shibata, S.; Natori, Y.; Nishihara, T.; Tomisaka, K.; Matsumoto, K.; Sansawa, H.; Nguyen, V.C. Antioxidant and anti-cataract effects of *Chlorella* on rats with streptozotocin-induced diabetes. *J. Nutr. Sci. Vitam.* **2003**, *49*, 334–339. [CrossRef]
110. Cherng, J.Y.; Shih, M.F. Potential hypoglycemic effects of *Chlorella* in streptozotocin-induced diabetic mice. *Life Sci.* **2005**, *77*, 980–990.
111. Cherng, J.Y.; Shih, M.F. Improving glycogenesis in streptozotocin (STZ) diabetic mice after administration of green algae *Chlorella*. *Life Sci.* **2006**, *78*, 1181–1186. [CrossRef]
112. Vecina, J.F.; Oliveira, A.G.; Araujo, T.G.; Baggio, S.R.; Torello, C.O.; Saad, M.J.A.; Queiroz, M.L.S. *Chlorella* modulates insulin signaling pathway and prevents high-fat diet-induced insulin resistance in mice. *Life Sci.* **2014**, *95*, 45–52. [CrossRef]
113. Horii, N.; Hasegawa, N.; Fujii, S.; Uchida, M.; Iemitsu, K.; Inoue, K.; Iemitsu, M. Effect of combination of chlorella intake and aerobic exercise training on glycemic control in type 2 diabetic rats. *Nutrition* **2019**, *63–64*, 45–50. [CrossRef]
114. Itakura, H.; Kobayashi, M.; Nakamura, S. *Chlorella* ingestion suppresses resistin gene expression in peripheral blood cells of borderline diabetics. *Clin. Nutr. ESPEN* **2015**, *10*, e95–e101. [CrossRef] [PubMed]
115. Li, L.; Kim, Y.; Lee, Y.W. *Chlorella vulgaris* extract ameliorates carbon tetrachloride-induced acute hepatic injury in mice. *Exp. Toxicol. Pathol.* **2013**, *65*, 73–80. [CrossRef] [PubMed]
116. Paradis, V.; Bedossa, P. Definition and natural history of metabolic steatosis: Histology and cellular aspects. *Diabetes Metab.* **2008**, *34*, 638–642. [CrossRef]
117. Schwimmer, J.B.; Deutsch, R.; Kahen, T.; Lavine, J.E.; Stanley, C.; Behling, C. Prevalence of fatty liver in children and adolescents. *Pediatrics* **2006**, *118*, 1388–1393. [CrossRef] [PubMed]
118. Ben, M.D.; Polimeni, L.; Baratta, F.; Pastori, D.; Loffredo, L.; Angelico, F. Modern approach to the clinical management of non-alcoholic fatty liver disease. *World J. Gastroentrol.* **2014**, *20*, 8341–8350. [CrossRef]
119. Yki-Järvinen, H. Non-alcoholic fatty liver disease as a cause and a consequence of metabolic syndrome. *Lancet Diabetes Endocrinol.* **2014**, *2*, 901–910. [CrossRef]
120. Ebrahimi-Mameghani, M.; Sadeghi, Z.; Farhangi, M.A.; Vaghef-Mehrabany, E.; Aliashrafi, S. Glucose homeostasis, insulin resistance and inflammatory biomarkers in patients with non-alcoholic fatty liver disease: Beneficial effects of supplementation with microalgae Chlorella vulgaris: A double-blind placebo-controlled randomized clinical trial. *Clin. Nutr.* **2017**, *36*, 1001–1006. [CrossRef]

121. Ebrahimi-Mameghani, M.; Aliashrafi, S.; Javadzadeh, Y.; AsghariJafarabadi, M. The effect of *Chlorella vulgaris* supplementation on liver enzymes, serum glucose and lipid profile in patients with non-alcoholic fatty liver disease. *Health Promot. Perspect.* **2014**, *4*, 107–115.
122. Azocar, J.; Diaz, A. Efficacy and safety of *Chlorella* supplementation in adults with chronic hepatitis C virus infection. *World J. Gastroenterol.* **2013**, *19*, 1085–1090. [CrossRef]
123. Schecter, A.; Startin, J.; Wright, C.; Kelly, M.; Papke, O.; Lis, A.; Ball, M.; Olson, J.R. Dioxins in, U.S. food and estimated daily intake. *Chemosphere* **1994**, *29*, 2261–2265. [CrossRef]
124. Ven den Berg, M.; Jongh, J.; Poiger, H.; Olson, J.R. The toxicokinetics and metabolism of polychlorinated dibenzo-p-dioxins (PCDDs) and dibenzofurans (PCDFs) and their relevance for toxicity. *Crit. Rev. Toxicol.* **1994**, *24*, 1–74. [CrossRef] [PubMed]
125. Schecter, A.; Papke, O.; Lis, A.; Ball, M.; Ryan, J.J.; Olson, J.R.; Li, L.; Kessler, H. Decrease in milk and blood dioxin levels over two years in a mother nursing twins: Estimates of decreased maternal and increased infant dioxin body burden from nursing. *Chemosphere* **1996**, *32*, 543–549. [CrossRef]
126. Masuda, Y.; Kuroki, H.; Haraguchi, H.; Nagayama, J. PCB and PCDF congeners in the blood and tissues of Yusho and Yu-cheng patients. *Envion. Health Perspect.* **1985**, *59*, 53–58.
127. Morita, K.; Matsueda, T.; Iida, T.; Hasegawa, T. *Chlorella* accelerates dioxin excretion in rats. *J. Nutr.* **1999**, *129*, 1731–1736. [CrossRef] [PubMed]
128. Kinze, M.G.; Dolbere, F.A.; Carroll, K.L.; Moore, D.H., II; Felton, J.S. Effect of cooking time and temperature on the heterocyclic amine content of fried beef patties. *Food Chem. Toxicol.* **1994**, *32*, 595–603.
129. Lee, I.; Tran, M.; Evans-Nguyen, T.; Stickle, D.; Kim, S.; Han, J.; Park, J.Y.; Yang, M. Detoxification of chlorella supplement on heterocyclic amines in Korean young adults. *Environ. Toxicol. Pharm.* **2015**, *39*, 441–446. [CrossRef]
130. Harding, G.; Dalziel, J.; Vass, P. Bioaccumulation of methylmercury within the marine food web of the outer bay of Fundy, gulf of maine. *PLoS ONE* **2018**, *13*, e0197220. [CrossRef]
131. His, H.C.; Hsu, Y.W.; Chang, T.C.; Chien, L.C. Methylmercury concentration in fish and risk-benefit assessment of fish intake among pregnant versus infertile women in Taiwan. *PLoS ONE* **2016**, *11*, e0155704.
132. Uchikawa, T.; Yasutake, A.; Kumamoto, Y.; Maruyama, I.; Kumamoto, S.; Ando, Y. The influence of Parachlorella beijerinckii CK-5 on the absorption and excretion of methylmercury (MeHg) in mice. *J. Toxicol. Sci.* **2010**, *35*, 101–105. [CrossRef]
133. Maruyama, I.; Uchikawa, T.; Kanno, T.; Ando, Y.; Kitsuki, H.; Yasutake, A. *Chlorella* supplementation decreases methylmercury concentrations of hair and blood in healthy volunteers. *Fundam. Toxicol. Sci.* **2018**, *5*, 117–122. [CrossRef]
134. Miettinen, J.K.; Rahola, T.; Hattula, T.; Rissanen, K.; Tillander, M. Elimination of ^{203}Hg-methylmercury in man. *Ann. Clin. Res.* **1971**, *3*, 116–122. [PubMed]
135. Ballatori, N.; Clarkson, T.W. Biliary transport of glutathione and methylmercury. *Am. J. Physiol* **1983**, *244*, G435–G441. [CrossRef] [PubMed]
136. Fujiwara, Y.; Sinpo, K.; Imae, Y.; Nonomura, M.; Hirakawa, K. Effect of *Chlorella vulgaris* strain CK-5 on the frequency of bowel movement in humans. *Ipn. J. Nutr.* **1998**, *56*, 253–263.
137. Nauta, A.; Engels, F.; Knippples, L.M.; Garssen, J.; Nijkamp, F.; Redegeld, F.A. Mechanisms of allergy and asthma. *Eur. J. Pharmcol.* **2008**, *585*, 354–360. [CrossRef] [PubMed]
138. Zhu, J.; Paul, W.E. CD4 T cells: Fates, functions, and faults. *Blood* **2008**, *112*, 1557–1569. [CrossRef] [PubMed]
139. Luckheeram, R.V.; Zhou, R.; Verma, A.D.; Xia, B. CD4+ cells: Differentiation and functions. *Clin. Dev. Immunol.* **2012**, *2012*, 925135. [CrossRef]
140. Bancroft, G.J.; Schreiber, R.D.; Unanue, E.R. Natural immunity: A T-cell-independent pathway of macrophage activation, defined in the scid mouse. *Immunol. Rev.* **1991**, *124*, 5–24. [CrossRef]
141. Wynn, T.A.; Jankovic, D.; Hieny, S.; Cheever, A.W.; Sher, A. IL-12 enhances vaccine induced immunity to Schistosomiasia mansoni in mice and decreases T helper 2 cytokine expression, IgE production, and tissue eosinophilia. *J. Immumol.* **1995**, *154*, 4701–4709.
142. Galli, S.J.; Tsai, M. IgE and mast cells in allergic disease. *Nat. Med.* **2012**, *18*, 693–704. [CrossRef]
143. Hasegawa, T.; Ito, K.; Ueno, S.; Kumamoto, S.; Ando, Y.; Yamada, A.; Nomoto, K.; Yasunobu, Y. Oral administration of hot water extracts of *Chlorella vulgaris* reduces IgE production against milk casein in mice. *Int. J. Immunopharmacol.* **1999**, *21*, 311–323. [CrossRef]

144. Hsu, H.Y.; Jeyashoke, N.; Yeh, C.H.; Song, Y.J.; Hua, K.F.; Chao, L.K. Immunostimulatory bioactivity of algal polysaccharides from *Chlorella pyrenoidosa* activates macrophages via Toll-like receptor 4. *J. Agric. Food Chem.* **2010**, *58*, 927–936. [CrossRef] [PubMed]
145. Eder, C. Mechanisms of interkeukin-1β release. *Immnobiology* **2009**, *214*, 543–553. [CrossRef] [PubMed]
146. Halperin, S.A.; Smith, B.; Nolan, C.; Shay, J.; Kralovec, J. Safety and immunoenhancing effect of a *Chlorella*-derived dietary supplement in healthy adults undergoing influenza vaccination: Randomized, double-bind, placebo-controlled trial. *Can. Med Assoc. J.* **2003**, *169*, 111–117.
147. Lamm, M.E.; Nedrud, J.G.; Kaetzel, C.S.; Mazanec, M.B. IgA and mucosal defense. *Acta Pathol. Microbiol. Immunol. Scand.* **1995**, *103*, 241–246. [CrossRef] [PubMed]
148. Otsuki, T.; Shimizu, K.; Iemitsu, M.; Kono, I. Salivary secretory immunoglobulin a secretion increases after 4-weeks ingestion of chlorella-derived multicomponent supplement in humans: A randomized cross over study. *Nutr. J.* **2011**, *10*, 91. [CrossRef] [PubMed]
149. Abel, A.M.; Yang, C.; Thakar, M.S.; Malarkannan, S. Natural killer cells: Development, maturation, and clinical utilization. *Front. Immunol.* **2018**, *9*, 1869. [CrossRef]
150. Kwak, J.H.; Baek, S.H.; Woo, Y.; Han, J.K.; Kim, B.G.; Kim, O.Y. Beneficial immunostimulatory effect of short-term *Chlorella* supplementation: Enhancement of Natural Killer cell activity and early inflammatory response (randomized, double-blinded, placebo-controlled trial). *Nutr. J.* **2012**, *11*, 53. [CrossRef]
151. Konishi, F.; Tanaka, K.; Himeno, K.; Taniguchi, K.; Nomoto, K. Antitumor effect induced by a hot water extract of *Chlorella vultaris* (CE): Resistance to Meth-A tumor growth mediated by CE-induced polymorphonuclear leukocytes. *Cancer Immunol. Immunother.* **1985**, *19*, 73–78. [CrossRef]
152. Tanaka, K.; Tomita, Y.; Tsuruta, M.; Konishi, F.; Okuda, M.; Himeno, K. Oral administration of *Chlorella vulgaris* augments concomitant antitumor immunity. *Immunopharmacol. Immunotoxicol.* **1990**, *12*, 277–291. [CrossRef]
153. Wu, L.C.; Ho, J.A.; Shieh, M.C.; Lu, I.W. Antioxidant and antiproliferatives of Spirulina and *Chlorella* water extracts. *J. Agric. Food Chem.* **2005**, *53*, 4207–4212. [CrossRef]
154. Cha, K.H.; Koo, S.Y.; Lee, D.U. Antiproliferative effects of carotenoids extracted from *Chlorella ellipsoidea* and *Chlorella vulgaris* on human colon cancer cells. *J. Agric. Food Chem.* **2008**, *56*, 10521–10526. [CrossRef] [PubMed]
155. Romos, A.L.; Torello, C.O.; Queiroz, M.L.S. *Chlorella vulgaris* modulates immunomyelopoietic activity and enhances the resistance of tumor-bearing mice. *Nutr. Cancer* **2010**, *62*, 1170–1180. [CrossRef] [PubMed]
156. Noda, K.; Tanaka, K. A water-soluble glycoprotein from *Chlorella vulgaris*. *Planta Med.* **1996**, *62*, 423–426. [CrossRef] [PubMed]
157. Noda, K.; Ohno, N.; Tanaka, K. A new type of biological response modifier from *Chlorella vulgaris* which needs protein moiety to show antitumor activity. *Phytother. Res.* **1998**, *12*, 309–319. [CrossRef]
158. Hasegawa, T.; Matsuguchi, T.; Noda, K.; Tanaka, K.; Kumamoto, S.; Shoyama, Y.; Yoshikai, Y. Toll-like receptor 2 is at least partly involved in the antitumor activity of glycoprotein from *Chlorella vulgaris*. *Int. Immunopharmacol.* **2002**, *2*, 579–589. [CrossRef]
159. Selkoe, D.J. Alzheimer's disease: Genes, proteins, and therapy. *Physiol. Rev.* **2001**, *81*, 741–766. [CrossRef]
160. Bosman, G.J.; Barholomeus, I.G.; de Man, A.J.; van Kalmthout, P.J.; de Grip, W.J. Erythrocyte membrane characteristics indicate abnormal cellular aging in patients with Alzheimer's disease. *Neurobiol. Aging.* **1991**, *12*, 13–18. [CrossRef]
161. Nakagawa, K.; Kiko, T.; Hatade, K.; Asai, A.; Kumra, F.; Sookwong, P.; Tsuduki, T.; Arai, H.; Miyazawa, T. Development of a high-performance liquid chromatography-based assay for carotenoids in human red blood cells: Application to clinical studies. *Anal. Biochem.* **2008**, *381*, 129–134. [CrossRef]
162. Kiko, T.; Nakagawa, K.; Tsuduki, T.; Suzuki, T.; Arai, H.; Miyazawa, T. Significance of lutein in red blood cells of Alzheimer's disease patients. *J. Alzheimers Dis.* **2012**, *28*, 593–600. [CrossRef]
163. Nakagawa, K.; Kiko, T.; Hatade, K.; Sookwong, P.; Arai, H.; Miyazawa, T. Antioxidant effect of lutein towards phospholipid hydroperoxidation in human erythrocytes. *Br. J. Nutr.* **2009**, *102*, 1280–1284. [CrossRef]
164. Miyazawa, T.; Nakagawa, K.; Takekoshi, H.; Higuchi, O.; Koto, S.; Kondo, M.; Kimura, F.; Miyazawa, T. Ingestion of *Chlorella* reduced the oxidation of erythrocyte membrance lipids in senior Japanese subjects. *J. Oleo Sci.* **2013**, *62*, 873–881. [CrossRef] [PubMed]

165. Andrade, L.; Caraveo-Anduaga, J.J.; Berglung, P.; Bijl, R.V.; de Graaf, R.; Vollebergh, W.; Dragomirecka, E.; Kohn, R.; Keller, M.; Kessler, R.; et al. The epidemiology of major depressive episodes: Results from the international consortium of psychiatric epidemiology (ICPE) surveys. *Int. J. Methods Psychiatry Res.* **2003**, *12*, 3–21. [CrossRef] [PubMed]
166. Nemeroff, C.B. Prevalence and management of treatment-resistant depression. *J. Clin. Psychiatry* **2007**, *68*, s17–s25.
167. Khawam, E.; Laurence, G.; Malone, J.R.D. Side effects of antidepressants: An overview. *Cleve Clin. J. Med.* **2006**, *73*, 351–353. [CrossRef]
168. Panahi, Y.; Badeli, R.; Karami, G.R.; Badeli, Z.; Sahebkar, A. A randomized controlled trial of 6-week *Chlorella vulgaris* supplementation in patients with major depressive disorder. *Comp. Ther. Med.* **2015**, *23*, 598–602. [CrossRef]
169. Halliwell, B. Oxidative stress and neurodegeneration: Where are we now? *J. Neurochem.* **2006**, *97*, 1634–1658. [CrossRef]
170. Valko, M.; Leibfritz, D.; Moncol, J.; Cronin, M.T.; Mazur, M.; Telser, J. Free radicals and antioxidants in normal physiological functions and human disease. *Int. J. Biochem. Cell Biol.* **2007**, *39*, 44–84. [CrossRef]
171. Vijayavel, K.; Anbuselvam, C.; Balasubramanian, M.P. Antioxidant effect of the marine algae *Chlorella vulgaris* against naphthalene-induced oxidative stress in the albino rats. *Mol. Cell Biochem.* **2007**, *303*, 39–44. [CrossRef]
172. Konishi, F.; Mitsuyama, M.; Okuda, M.; Tanaka, K.; Hasegawa, T.; Nomoto, K. Protective effect of an acidic glycoprotein obtained from culture of *Chlorella vulgaris* against myelosuppression by 5-fluorouraci. *Cancer Immunol. Immunother.* **1996**, *42*, 268–274. [CrossRef]
173. De Souza Queiroz, J.; Barbosa, C.M.V.; da Rocha, M.; Bincoletto, C.; Paredes-Gamero, E.J.; de Souza Queiroz, M.L.; Neto, J.P. *Chlorella vulgaris* treatment ameliorates the suppressive effects of single and repeated stressors on hematopoiesis. *Brain Behav. Immun.* **2013**, *29*, 39–50. [CrossRef]
174. Armario, A. The contribution of immediate early genes to the understanding of brain processing of stressors. In *Immediate Early Genes in Sensory Processing Cognitive Performance and Neurological Disorders*; Pinaud, R., Termere, L., Eds.; Springer: Berlin, Germany, 2006; pp. 199–221.
175. Queiroz, J.S.; Blasco, I.M.; Gagliano, H.; Daviu, N.; Román, A.G.; Belda, X.; Carrasco, J.; Rocha, M.C.; Neto, J.P.; Armario, A. *Chlorella vulgaris* reduces the impact of stress on hypothalamic-pituitary-adrenal axis and brain c-fos expression. *Psychoneuroendocrinology* **2016**, *65*, 1–8. [CrossRef] [PubMed]
176. Zheng, Y.; Inoue, Y.H.; Kohno, N.; Fujishima, M.; Okumura, E.; Sato, K. Phenethylamine in hot water extract of *Chlorella pyrenoidosa* expands lifespan of *SOD1* mutant adults of *Drosophila melanogaster* at very low dose. *J. Food Bioact.* **2020**, *9*, 52–57. [CrossRef]
177. Qiu, S.; Shen, Y.; Zhang, L.; Ma, B.; Amadu, A.A.; Ge, S. Antioxidant assessment of wastewater-cultivated *Chlorella sorokiniana* in *Drosophila melanogaster*. *Algal. Res.* **2020**, *46*, 101795. [CrossRef]

nutrients

MDPI

Article

Anti-Obesity Effect of *DKB-117* through the Inhibition of Pancreatic Lipase and α-Amylase Activity

Do Hoon Kim [1,†], Yu Hwa Park [1,†], Jung Suk Lee [2], Hyun Il Jeong [2], Kye Wan Lee [2] and Tong Ho Kang [1,*]

1 Department of Oriental Medicine Biotechnology, College of Life Sciences and Graduate School of Biotechnology, Kyung Hee University, Global Campus, Gyeonggi 17104, Korea; hoony3914@gmail.com (D.H.K.); best1rd@gmail.com (Y.H.P.)

2 R&D Center, Dongkook Pharm. Co., Ltd., Gyeonggi 16229, Korea; ljs@dkpharm.co.kr (J.S.L.); jhi@dkpharm.co.kr (H.I.J.); lkw1@dkpharm.co.kr (K.W.L.)

* Correspondence: panjae@khu.ac.kr

† These authors contributed equally to this work.

Received: 25 August 2020; Accepted: 2 October 2020; Published: 6 October 2020

Abstract: This study sought to evaluate the effects of *Phaseolus multiflorus* var. albus Bailey extract (PM extract) and *Pleurotus eryngii* var. ferulae extract (PF extract) on the inhibition of digestive enzymes and to confirm the anti-obesity effect of DKB-117 (a mixture of PM extract and PF extract) in digestive enzyme inhibition in a mouse model of obesity induced by a high-fat diet. In in vitro studies, PM extract and PF extract have increased dose-dependent inhibitory activity on α-amylase (Inhibitory concentration (IC_{50} value: 6.13 mg/mL)) and pancreatic lipase (IC_{50} value; 1.68 mg/mL), respectively. High-fat diet–induced obese mice were orally administered DKB-117 extracts at concentrations of 100, 200, and 300 mg/kg/day, while a positive control group was given orlistat (pancreatic lipase inhibitor) and *Garcinia cambogia* (inhibiting the enzymes needed to synthesize carbohydrates into fat) at concentrations of 40 and 200 mg/kg/day, respectively, for eight weeks. As a result, body weight, fat mass (total fat mass, abdominal fat, and subcutaneous fat) detected with microcomputed tomography, fat mass (abdominal fat and inguinal fat) after an autopsy, and liver triglyceride levels were decreased significantly in the DKB-117 (300 mg/kg/day) group compared to those in the HFD control group. Additionally, we obtained results indicating that the presence of carbohydrates was found more in the DKB-117-300 (300 mg/kg/day) group than in the HFD control group. These data clearly show that DKB-117 extracts are expected to have an anti-obesity effect through a complex mechanism that promotes carbohydrate release through the inhibition of carbohydrate-degrading enzymes while blocking lipid absorption through lipase inhibition.

Keywords: *Phaseolus multiflorus* var. albus Bailey (PM); *Pleurotus eryngii* var. ferulae (PF); anti-obesity; pancreatic lipase; α-amylase

1. Introduction

Obesity is a metabolic disorder characterized by an excess accumulation of fat in the body due to one's energy intake exceeding one's energy expenditure [1]. Obesity is a very common global health problem. It was reported by the World Health Organization in 2016 that more than 1.9 billion adults were overweight and, of these, more than 650 million were obese [2]. Obesity is a risk factor for metabolic syndrome and can lead to hypertension, type 2 diabetes (T2DM), dyslipidemia, cardiovascular disease (CVD), and stroke [3,4].

Currently, four weight-loss medicines (orlistat marketed as Xenical®, Roche Holding AG, Basel Switzerland or Alli®, GlaxoSmithKline, Brentford, UK; Contrave® from Nalpropion Pharmaceuticals,

San Diego, CA, USA; Belviq® from Eisai, Tokyo, Japan; and Qsymia® from Vivus, Campbell, CA, USA) have been approved by the United States Food and Drug Administration, with users' body weight regulated by pancreatic lipase inhibition, increased energy consumption, and appetite suppression [5,6]. However, existing synthetic drugs have been reported to cause heart attack and stroke as well as liver damage [7].

Accordingly, there have been a number of studies conducted on the development of anti-obesity materials that are effective in decreasing appetite and reducing weight by using natural substances with long histories of use [8,9]. According to a recent report, many herbal extracts (*Garcinia cambogia, Plantago psyllium, Morus alba*) have been suggested to act on fat and carbohydrate metabolism to regulate body weight [10].

However, a lot of safety issues related to hepatic insufficiency, hepatitis, and heart disease have been reported in the case of *Garcinia cambogia* extract, which is the most widely sold and health functional food material in the world. Accordingly, it is required that new natural materials with good safety and efficacy be developed.

White kidney beans (*Phaseolus multiflorus* var. albus Bailey; PM) are native to Italy and belong to the Leguminosae family. Abdulwahid et al. reported that the white kidney bean–treated group in a diabetic-induced mouse model showed glucose and weight loss effects compared to the control group [10]. In addition, many studies on anti-obesity-related clinical trials have reported that white kidney bean ingestion is effective in weight loss, waist circumference reduction, and weight loss through α-amylase [11].

Pleurotus eryngii var. ferulae (PF) is a mushroom of the family Pleurotaceae that is rich in protein and dietary fiber [12]. Wang et al. and Alam et al. reported on the antioxidative, anti-inflammatory, and hypotensive effects of PF. In addition, PF water extracts reduced body weight, white adipose tissue weight, and liver weight in a mouse model of obesity induced by a high-fat diet (HFD) while improving glucose tolerance [1,13,14].

One of the most important strategies in the treatment of obesity includes the development of nutrient digestion and absorption inhibitors in an attempt to reduce the degree of energy intake through gastrointestinal mechanisms without altering any central mechanisms. The inhibition of digestive enzymes is one of the most widely studied mechanisms used to determine the potential efficacy of natural products as anti-obesity agents [15].

This study was carried out to develop a dietary supplement ingredient that improves convenience of use and has good safety and obesity effects by respectively mixing natural plants having α-amylase inhibitory and pancreatic lipase inhibitory effects. In the literature, α-amylase inhibitors are well documented to be effective in reducing postprandial hyperglycemia by slowing the digestion of carbohydrates and absorbing postprandial glucose [10]. Reducing postprandial hyperglycemia prevents glucose uptake into adipose tissue to inhibit the synthesis and accumulation of triacylglycerol. Lipase is a hydrolytic enzyme from the pancreas that changes triglycerides (TGs) to glycerol and fatty acids. Thus, the inhibition of lipase has an important role in the treatment of obesity by inhibiting fat absorption. Previously reported studies have shown the α-amylase–inhibitory effect of PM and the pancreatic lipase–inhibitory effect of PF [12,13].

Research is actively underway to find effective anti-obesity drugs or anti-obesity health functions. The authors of this study want to confirm the anti-obesity effect by mixing *Phaseolus multiflorus* var. albus Bailey with α-amylase inhibitory effect and *Pleurotus eryngii* var. ferulae reported to have a pancreatic lipase inhibitory effect. In this study, the optimum mixing ratio of PF extract and PF extract was selected by an in vitro test, and we sought to determine the anti-obesity effect of DKB-117 in digestive enzyme inhibition in a mouse model of obesity induced by HFD.

2. Materials and Methods

2.1. Plant Material Collection and Extract Preparation

DKB-117, which is a mixture of PM extract and PF extract, was provided by Dongkook Pharm. Co., Ltd. (Suwon, Korea). The lot number was DKB-117. PM (*Phaseolus multiflorus* var. albus Bailey) used in this study was cultivated from Egypt and purchased through Solim trading Co., Ltd. in Korea. PF (*Pleurotus eryngii* var. ferulae) was obtained from the DDLE A CHE Co., Ltd. (Cheonan, Korea). A voucher specimen number (DK0117) has been deposited at the R&D Center, Dongkook Pharm. Co., Ltd.

PM and PF were cut into small pieces and extracted with a 5-fold volume of 30% ethanol (v/v) at 80 °C for 10 h, in order to obtain two ethanolic extracts, namely PM and PF, respectively. After extraction, each solution was concentrated on a rotary evaporator (BUCHI Labortechnik AG, Flawil, Switzerland) until constant weight, and the PM extract was dried by freeze-drying and the PF extract was dried using a spray dryer. DKB-117 was prepared by mixing the previously prepared PM extract and PF extract at a weight ratio of 3:1.

2.2. α-Amylase Inhibition Assay

The measurement of α-amylase inhibitory activity was carried out via the iodine reaction method as described by Wilson et al. with a slight modification [16]. Briefly, α-amylase (Sigma-Aldrich, St. Louis, MO, USA) derived from human saliva was dissolved in phosphate-buffered saline (PBS) at a concentration of 20 unit/mL. As a substrate for α-amylase, soluble starch was dissolved in PBS at a concentration of 1%. To measure the inhibitory activity against α-amylase, 290 μL of PBS, 10 μL of α-amylase solution (20 unit/mL), and 50 μL of the test substance were mixed and preincubation was performed at 37 °C for 10 min. After preincubation, 350 μL of 1% soluble starch as substrate was added and reacted at 37 °C for 30 min. To determine the amount of soluble starch remaining after the reaction, 300 μL of iodine solution (0.1% KI + 0.01% I_2/0.05 N HCl) was added to the reaction solution and the absorbance was measured at 620 nm using an enzyme-linked immunosorbent assay (ELISA) reader (Infinite 200 Pro; Tecan Austria GmBH, Grödig, Austria).

2.3. Pancreatic Lipase Inhibition Assay

Pancreatic lipase inhibitory activity was measured using the substrate p-nitrophenyl butyrate (PNPB) as described by Eom et al. with slight modification [17]. Briefly, an enzyme buffer was prepared by adding 30 μL of porcine pancreatic lipase (Sigma-Aldrich, St. Louis, MO, USA) in 10 mM of morpholinepropane sulfonic acid and 1 mM of ethylene diamine tetra acetic acid (pH: 6.8) to 850 mL of Tris buffer (100 mM of Tris-HCl and 5 mM of $CaCl_2$; pH: 7.0). Then, 100 μL of DKB-117 or orlistat was mixed with 880 mL of the enzyme buffer and incubated for 15 min at 37 °C. After incubation, we added 20 μL of the substrate solution (10 mM of PNPB in dimethyl formamide) and the enzymatic reactions were allowed to proceed for 30 min at 37 °C. Pancreatic lipase inhibitory activity was determined by measuring the hydrolysis of PNPB to p-nitrophenol at 405 nm with the use of an ELISA reader (Infinite 200 Pro; Tecan Austria GmBH, Grödig, Austria). The activities of the negative control were reviewed with and without the inhibitor. The inhibitory activity (%) was calculated according to the formula below:

$$\text{Lipase inhibition (\%)} = [1 - (B - b)/(A - a)] \times 100$$

where A is the activity of the enzyme without the inhibitor, a is the negative control without the inhibitor, B is the activity of the enzyme with the inhibitor, and b is the negative control with the inhibitor.

2.4. Animal Experiments

The protocol for animal study was approved by the Department of Biofood Research, KNOTUS Life Science Co., Ltd. Animal Ethics Committee (17-KE-265). Male C57BL/ JJmsSlc mice, five weeks old (18–20 g), were purchased from Central Lab Animal Inc. (Seoul, Republic of Korea). The animals were housed in polycarbonate cages (less than five mice/cage) under controlled temperatures (23 ± 3 °C), relative humidity (55% ± 5%), and lighting conditions (lights on from 08:00 to 20:00 h), with food and water made available ad libitum. Mice were fed either a regular diet (ND group; lab rodent chow; Cargill-Agri Purina, Seongnam-Si, Republic of Korea) or a 60% HFD (HFD group; Saeronbio, Uiwang-si, Republic of Korea). On the 19th day of HFD feeding, the body weight of the HFD group was 10% higher than that of the ND group. At this time, the HFD group was divided into six additional groups: an HFD control group, a DKB-117-100-treated group (100 mg/kg/day), a DKB-117-200-treated group (200 mg/kg/day), a DKB-117-300-treated group (300 mg/kg/day), an orlistat-treated group (40 mg/kg/day), and a *Garcinia cambogia*–treated group (200 mg/kg/day), respectively. For eight weeks, the mice were daily treated with test article and HFD, with the exception of the ND group.

After eight weeks of treatment, the mice were not fed for 15 h and their body weight was measured. Then, the mice were anesthetized with ether. Blood was drawn from the postcaval vein for serum biochemical analysis. Extracted liver, abdominal fat, and inguinal fat tissue were weighed.

2.5. *Blood Analysis*

Separated serum was examined for total cholesterol (TCHO), triglycerides (TG), high-density lipoprotein (HDL), and low-density lipoprotein (LDL) levels using a blood biochemical analyzer (model 7180; Hitachi Corp., Tokyo, Japan).

2.6. *Micro-CT Analysis*

At the end of the treatment, micro-CT (viva CT80; SCANCO Medical, Switzerland) was conducted, and body fat mass was analyzed from the basal region of second lumbar to the distal end of the fifth lumbar.

2.7. *ELISA Assay*

After eight weeks of treatment, carbohydrate levels in collected fecal samples and TG levels in right-lobule liver tissue were measured. The analysis was carried out using a commercial ELISA kit (Biochemical, MI, USA, USA, Triglyceride Assay kit, Asanpharm, Korea, Triglyceride Colorimetric Assay Kit, Cayman, Chemical, USA).

2.8. *Histopathological Analysis*

Left-lobule liver tissue dissected from mice was fixed in 10% neutral-buffered formalin solution. The dehydrated liver tissue was then embedded in paraffin wax, and sections were cut from the paraffin-embedded tissues. These sections stained with Oil-Red-O. Histopathological changes were assessed using an optical microscope (BX53, Olympus, Tokyo, Japan) and image analyzer (Zen 2.3 blue edition; Carl Zeiss, Jena, Germany).

2.9. *Statistical Analysis*

The assumption of homogeneity was tested using Levene's test. If the overall analysis of variance was significant and the assumption of homogeneity of variance was met, Duncan's multiple range test was performed. If the assumption of homogeneity of variance was not met, Dunnett's T3 test was used as the post hoc test. In nonparametric multiple analysis, the Kruskal–Wallis H-test was adopted. If a statistically significant difference was observed between groups, the Mann–Whitney U-test was used to identify the groups. SPSS Statistics 18.0K (IBM Corp., Armonk, NY, USA) was used for all statistical analysis and the level of significance was set at $p < 0.05$.

3. Result

In the α-amylase inhibitory activity test, the amount of soluble starch decreased by the enzyme reaction was measured by the iodine reaction method. The inhibitory activity of PM extract against pancreatic α-amylase was determined using different concentrations (1.875, 3.75, 7.5, 15, and 30 mg/mL). As shown in Table 1, the PM extract inhibited the enzyme activities in a dose-dependent way.

Table 1. α-Amylase inhibitory activity of PM (*Phaseolus multiflorus* var. albus Bailey) extract.

Sample	Concentration (mg/mL)	Inhibition (%)	Inhibitory Concentration IC_{50} Value (mg/mL)
PM extract	30	77.6 ± 1.36	6.13
	15	82.6 ± 3.20	
	7.5	55.0 ± 3.70	
	3.75	17.6 ± 2.52	
	1.875	5.4 ± 1.10	
Acarbose	1	85.1 ± 1.66	-

Data are presented as medians ± standard deviations (n = 3).

Pancreatic lipase is the most important enzyme for the digestion of dietary triacylglycerols. Pancreatic lipase is a key enzyme that hydrolyzes 50% to 70% of total dietary fat in the digestive system, converting TGs to monoglycerides and free fatty acids [18]. Inhibiting pancreatic lipase is an important strategy for treating obesity and other metabolic disorders [19]. In this study, PF extract inhibited pancreatic lipase activity in a concentration-dependent manner, with an Inhibitory concentration (IC_{50}) value of 1.68 mg/mL (Table 2). The lipase inhibitory activity of PF extract may be able to suppress dietary fat absorption in vivo as well.

Table 2. Pancreatic lipase (PL) inhibitory activity of PF (*Pleurotus eryngii* var. ferulae) extract.

Sample	Concentration (mg/mL)	Inhibition (%)	IC50 Value (mg/mL)
PF extract	10	74.8 ± 2.29	1.68
	5	76.4 ± 0.85	
	2.5	57.3 ± 1.46	
	1.25	23.9 ± 2.50	
	0.625	4.4 ± 0.76	
Orlistat	1	89.3 ± 0.95	-

Data are presented as medians ± standard deviations (n = 3).

As a result of in vitro tests, the effect of inhibiting alpha amylase of PM extra and the effect of inhibiting pancreatic lipase of PF extra were the highest at 15 and 5 mg/mL, respectively. Based on the in vitro test results, a DKB-117 extract mixed with a PM:PF = 3:1 (w/w) ratio was prepared to perform in vivo tests by concentration (100, 200, and 300 mg/kg).

The results of the eight weeks of body weight measurement showed that the HFD-treated group had significantly higher values than the ND-treated group, while the DKB-117-300-, orlistat-, and *Garcinia cambogia*–treated groups had significantly decreased weight as compared with the HFD-treated group ($p < 0.001$). The administration of DKB-117-300 was considered to have affected the reduction of body weight in obese models induced by HFD (Figure 1).

The concentrations of TCHO, TGs, and low-density lipoprotein (LDL) in the serum were also remarkably higher in the HFD group compared to those in the normal diet (ND) group, which is also a typical symptom of obesity [20].

As shown in Table 3, the HFD induced hyperlipidemia, with increases of plasma triacylglycerol and cholesterol levels. The serum levels of total cholesterol (TCHO), TGs, high-density lipoprotein (HDL), and LDL sharply increased in the HFD group compared to those in the ND group. LDL level was significantly lower ($p < 0.05$) than in the HDF group and reduced by 19.7% after eight weeks of DKB-117-300 administration. No significant alteration was observed in the serum levels of TCHO and TGs. However, these conditions were improved after eight weeks of DKB-117-300 administration (Table 3). These data clearly indicate that DKB-117 intake can effectively mitigate hyperlipidemia induced by HFD by reducing the lipid content in the blood. Furthermore, it was shown that the effects of weight gain reduction by DKB-117-300 treatment had an overall effect on the blood lipid metabolism index.

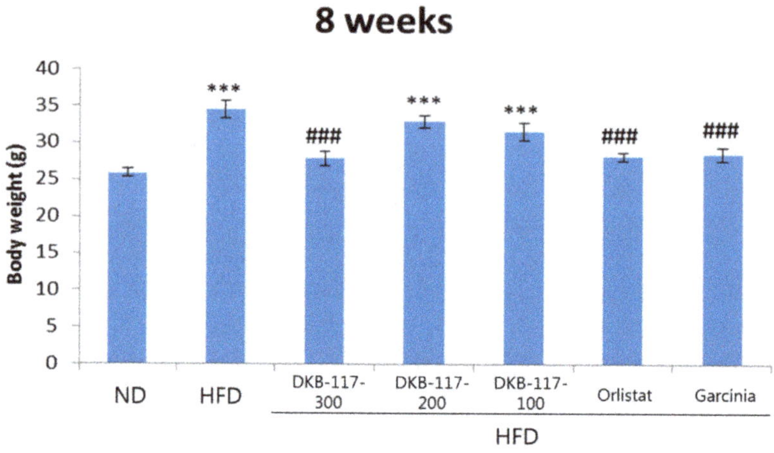

Figure 1. Effects of DKB-117 extracts on body weight in high-fat diet (HFD)–induced obese mice. The values are presented as means ± standard error of the means (SEMs). *** A significant difference at the $p < 0.001$ level was observed versus the normal diet group (ND). ### A significant difference at the $p < 0.001$ level versus the HFD. ND, normal diet control group; HFD, high-fat diet control group; DKB-117-300, DKB-117 300 mg/kg; DKB-117-200, DKB-117 200 mg/kg; DKB-117-100, DKB-117 100 mg/kg; Orlistat, orlistat 40 mg/kg; Garcinia, *Garcinia cambogia* 200 mg/kg.

Table 3. Effect of DKB-117 extracts on serum lipid profiles in HFD-induced obese mice after fasting for 15 h at the end of the study.

Experimental Group	Serum Lipid Profiles (mg/dL)			
	TCHO	TG	HDL	LDL
ND	108.6 ± 3.6	60.5 ± 8.5	72.1 ± 1.9	5.2 ± 0.6
HFD	197.6 ± 9.9 ***	148.7 ± 7.5 ***	132.2 ± 4.4 ***	16.1 ± 0.8 ***
HFD+DKB-117-300	173.5 ± 6.9 ***	133.2 ± 7.7 ***	120.2 ± 5.1 ***	12.3 ± 0.6 ***, #
HFD+DKB-117-200	189.7 ± 7.3 ***	128.5 ± 9.5 ***	124.1 ± 5.3 ***	13.9 ± 0.7 ***
HFD+DKB-117-100	195.4 ± 4.7 ***	131.6 ± 9.2 ***	120.6 ± 5.4 ***	14.2 ± 1.1 ***
Orlistat	172.5 ± 5.4 ***	133.9 ± 10.8 ***	117.6 ± 2.6 ***	12.5 ± 0.9 ***, #

The values are presented as means ± SEMs. *** A significant difference at the $p < 0.001$ level was observed versus the ND. # A significant difference at the $p < 0.05$ level was observed versus the HFD. ND, normal diet control group; HFD, high-fat diet control group; DKB-117-300, DKB-117 300 mg/kg; DKB-117-200, DKB-117 200 mg/kg; DKB-117-100, DKB-117 100 mg/kg; Orlistat, orlistat 40 mg/kg; Garcinia, *Garcinia cambogia* 200 mg/kg, TG: triglyceride; HDL, high-density lipoprotein, LDL: low-density lipoprotein, TCHO: total cholesterol.

Sung et al. reported that extracts exhibiting lipase inhibitory activity effectively lower cholesterol and TG levels that were elevated due to HFD intake. In this experiment, the lipase inhibitor (orlistat) effectively lowered TCHO, TGs, and LDL, which are increased by HFD intake [19]. Orlistat is known to inhibit the hydrolysis of dietary fat to free fatty acids so that the fat is not absorbed in the intestine but rather is excreted directly in the feces, thereby reducing a person's weight and improving their blood lipid levels and glucose metabolism [21]. The DKB-117-300-treated group appeared to present improvements in the lipid metabolism index effectively through lipase inhibitory activity.

Micro-CT scans showed that the total fat, abdominal fat, and subcutaneous fat were all increased in the HFD-induced obese mice compared with those in the ND mice. However, these fats were significantly decreased when DKB-117-100 administration was concomitant. Especially, DKB-117-100 showed an equal or superior effect to that of orlistat and *Garcinia cambogia* used as positive controls. In the case of orlistat, only the visceral fat amount was significantly higher than that in the HFD-treated group (Figure 2).

The experiment was carried out during eight weeks. We measured the weight of inguinal and abdominal fat. After finishing the eight-week test, the weights of inguinal fat and abdominal fat, respectively, were measured after the autopsy. The orlistat group showed a statistically significant decrease in the inguinal fat mass compared to the HFD group, while the DKB-117-100 group showed a statistically significant decrease in the absolute amount of abdominal fat and inguinal fat compared to the HFD group (Figure 3).

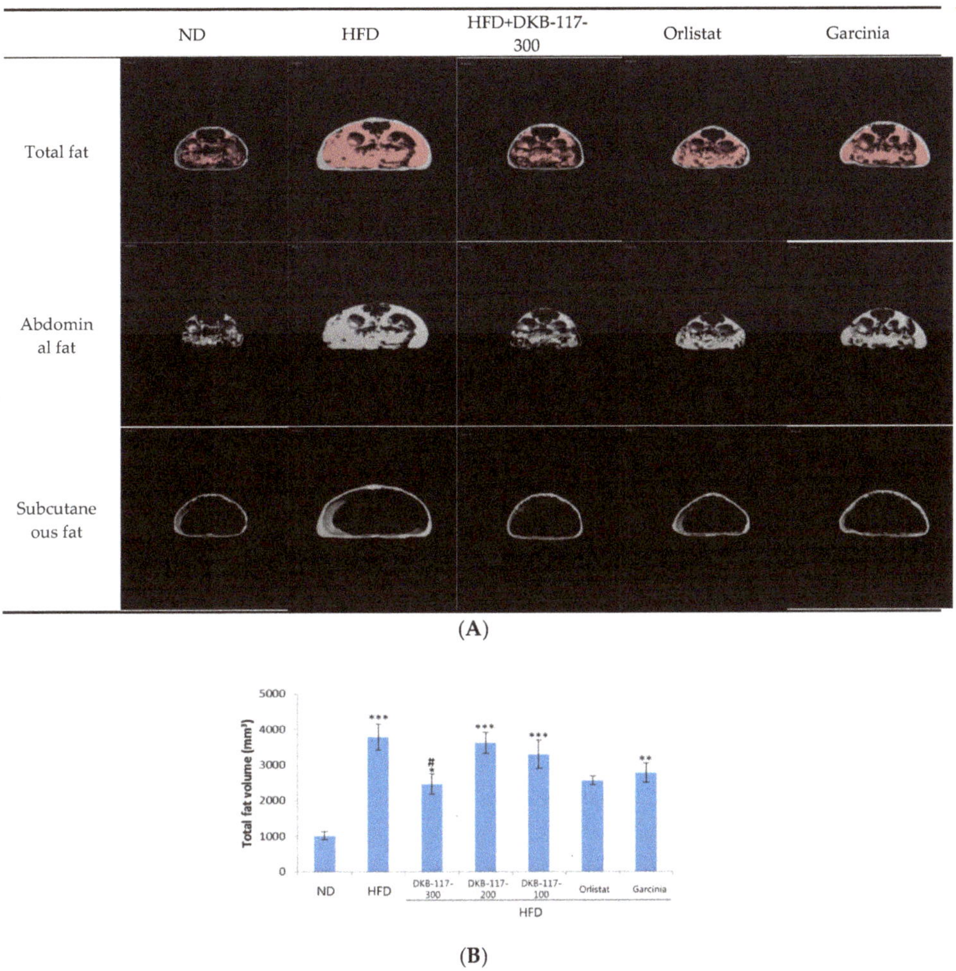

Figure 2. *Cont.*

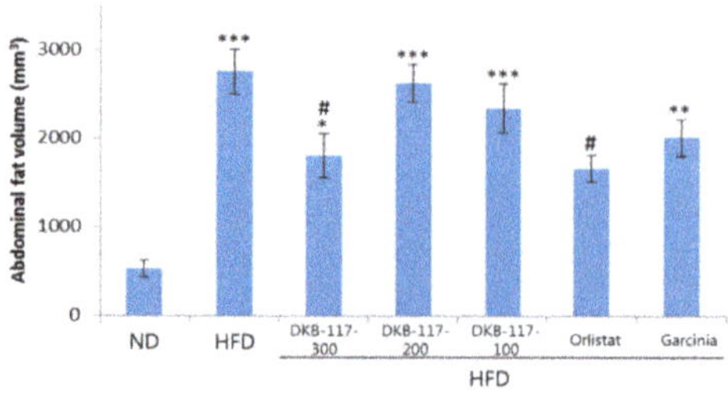

(C)

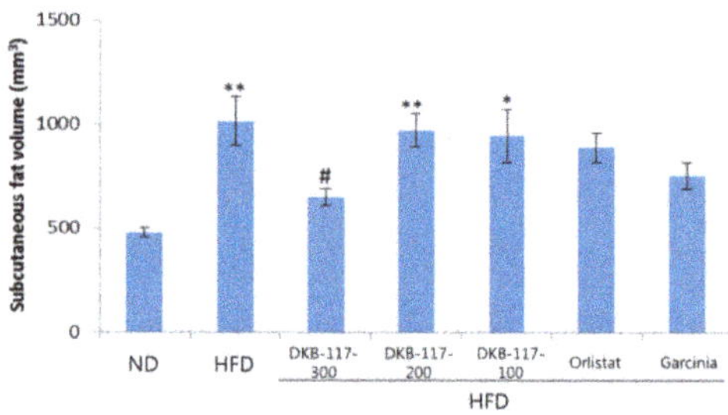

(D)

Figure 2. Effects of DKB-117 extracts on fat volume in HFD-induced obese mice. C57BL/6J mice consumed a HFD or ND for eight weeks. (**A**) Transverse microcomputed tomography (micro-CT) images. (**B**) Total fat volume, (**C**) abdominal fat volume, and (**D**) subcutaneous fat volume were measured using micro-CT. The values are presented as means ± SEMs. ***/**/* A significant difference at the $p < 0.001/p < 0.01/p < 0.05$ level was observed versus the ND. # A significant difference at the $p < 0.05$ level was observed versus the HFD. ND, normal diet control group; HFD, high-fat diet control group; DKB-117-300, DKB-117 300 mg/kg; DKB-117-200, DKB-117 200 mg/kg; DKB-117-100, DKB-117 100 mg/kg; Orlistat, orlistat 40 mg/kg; Garcinia, *Garcinia cambogia* 200 mg/kg.

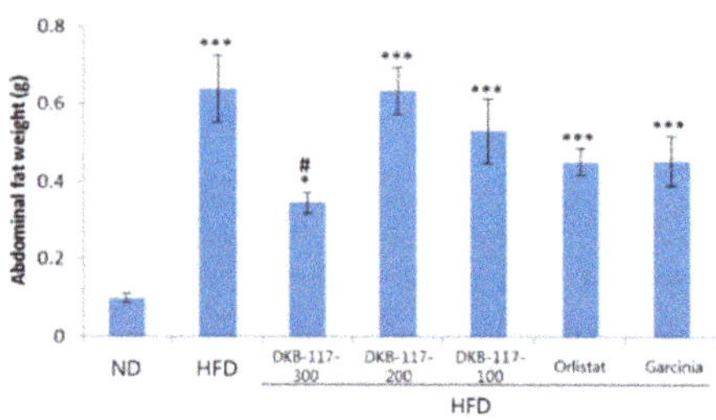

(A)

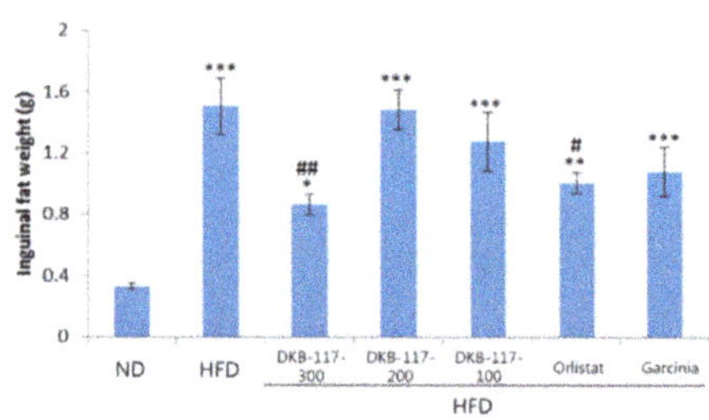

(B)

Figure 3. Effect of DKB-117 extracts on (**A**) abdominal and (**B**) inguinal fat weight in HFD-induced obese mice. The values are presented as means ± SEMs. ***/**/* A significant difference at the $p < 0.001/p < 0.01/p < 0.05$ level versus the ND. ##/# A significant difference at the $p < 0.01/p < 0.05$ level versus the HFD. ND, normal diet control group; HFD, high-fat diet control group; DKB-117-300, DKB-117 300 mg/kg; DKB-117-200, DKB-117 200 mg/kg; DKB-117-100, DKB-117 100 mg/kg; Orlistat, orlistat 40 mg/kg; Garcinia, *Garcinia cambogia* 200 mg/kg.

As a result of the measurement of fat content in the liver using Oil-Red-O staining, the HFD-treated group revealed a significant increase in fat area in the liver compared to the ND-treated group ($p < 0.001$).

In the DKB-117-treated group (100, 200, or 300 mg/kg), the dose-dependent area of fat staining decreased, and the DKB-117-100-treated group showed a significant decrease compared to the HFD-treated group ($p < 0.05$). In addition, the DKB-117-100-treated group showed statistically equivalent efficacy results compared to the *Garcinia cambogia*–treated group (Figure 4).

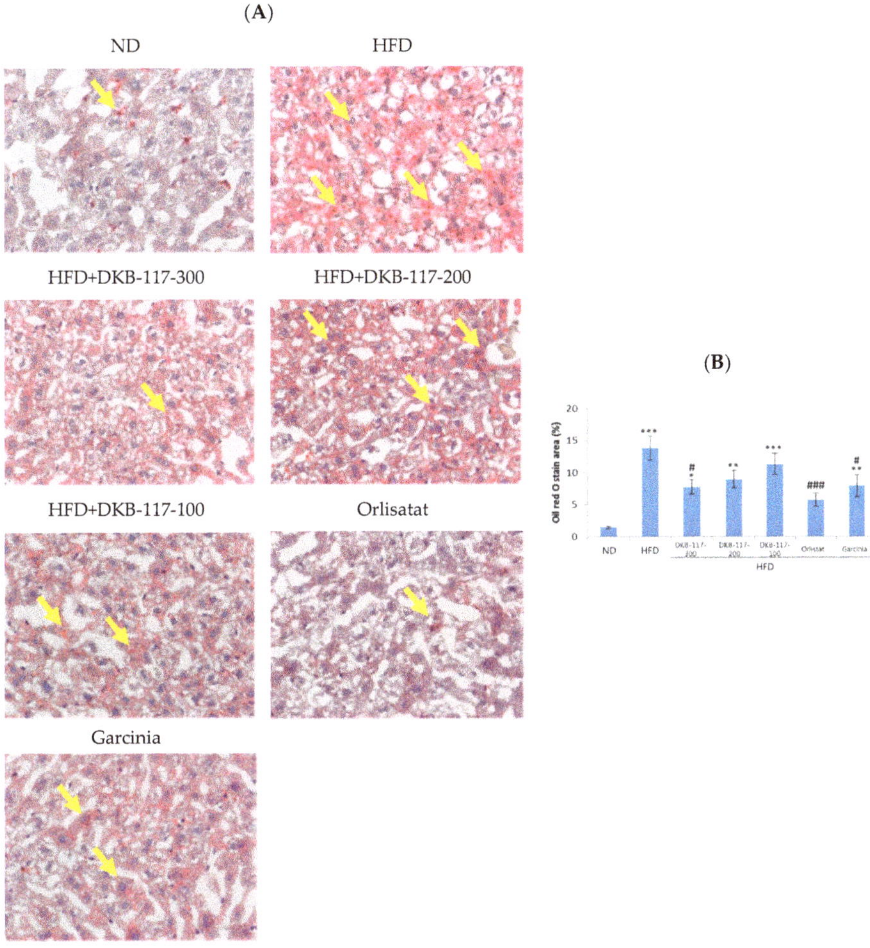

Figure 4. Effect of DKB-117 extracts on lipid accumulation in HFD-induced obese mice. (**A**) The Oil-Red-O staining in liver sections (×400, magnification) and (**B**) quantification of the positively stained area. The values are presented as means ± SEMs. ***/**/* A significant difference at the $p < 0.001/p<0.01/p<0.05$ level compared to the ND. ###/# A significant difference at the $p < 0.001/p < 0.05$ level versus the HFD. ND, normal diet control group; HFD, high-fat diet control group; DKB-117-300, DKB-117 300 mg/kg; DKB-117-200, DKB-117 200 mg/kg; DKB-117-100, DKB-117 100 mg/kg; Orlistat, orlistat 40 mg/kg; Garcinia, *Garcinia cambogia* 200 mg/kg.

The liver is an important organ responsible for lipid metabolism, along with fat tissue [22]. Free fatty acids are the most basic elements of the energy metabolism in the body and are transferred to other organs in the form of TGs [23]. An imbalance in lipid metabolism causes an intracellular accumulation of TGs. We confirmed that liver fat was significantly accumulated by eating HFDs.

As shown in Table 3 and Figure 5, the HFD induced hyperlipidemia, with increases in plasma triacylglycerol and cholesterol levels and induced fatty liver along with an accumulation of triacylglycerol in the liver.

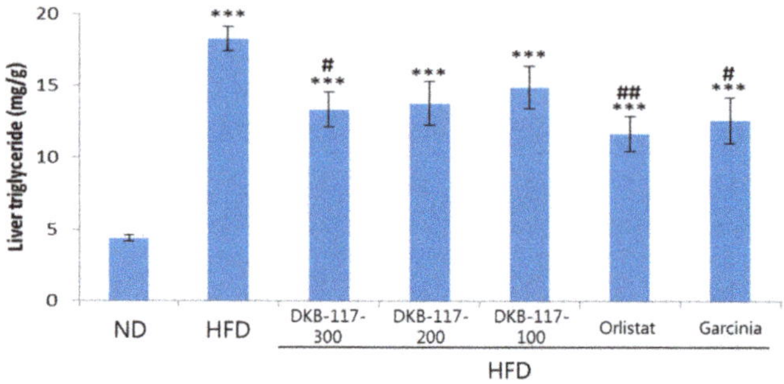

Figure 5. Effect of DKB-117 extracts on liver TG levels in HFD-induced obese mice. The values are presented as means ± SEMs. *** A significant difference at the $p < 0.001$ level versus the ND. ##/# A significant difference at the $p < 0.01/p < 0.05$ level versus the HFD. ND, normal diet control group; HFD, high-fat diet control group; DKB-117-300, DKB-117 300 mg/kg; DKB-117-200, DKB-117 200 mg/kg; DKB-117-100, DKB-117 100 mg/kg; Orlistat, orlistat 40 mg/kg; Garcinia, *Garcinia cambogia* 200 mg/kg.

As a result of the analysis of TG content in liver, the HFD-treated group showed a significant increase ($p < 0.001$) as compared with the ND-treated group, while the DKB-117-100-, orlistat- and *Garcinia cambogia*–treated groups showed a significant decrease as compared with the HFD-treated group ($p < 0.05$ or $p < 0.001$). Similar to the results of weight loss, the results here showed that DKB-117-300 intake reduced liver lipid accumulation through a mechanism of digestive enzyme inhibition (Figure 5).

The amount of carbohydrates in the feces was not different between the HFD and ND groups. The DKB-117-100 group showed a significant increase ($p < 0.01$) in carbohydrate content as compared with that in the HFD group, while the orlistat group showed a significant increase in TGs, TCHO, and carbohydrates at the eighth week (Figure 6).

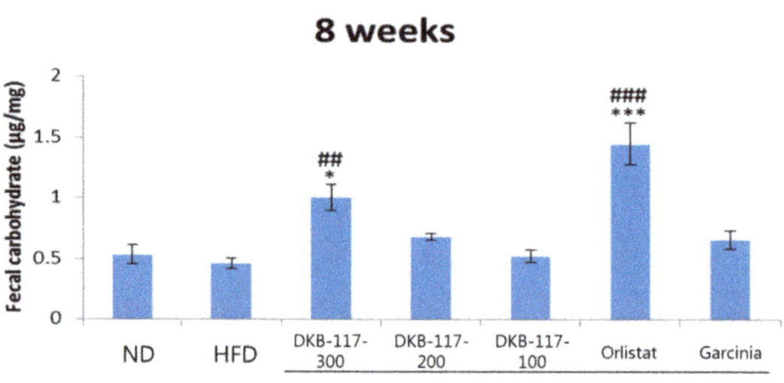

Figure 6. Effects of DKB-117 extracts on feces carbohydrate levels. The values are presented as means ± SEMs. ***/* A significant difference at the $p < 0.001/p < 0.05$ level versus the ND. ###/## A significant difference at the $p < 0.001/p < 0.01$ level versus the HFD. ND, normal diet control group; HFD, high-fat diet control group; DKB-117-300, DKB-117 300 mg/kg; DKB-117-200, DKB-117 200 mg/kg; DKB-117-100, DKB-117 100 mg/kg; Orlistat, orlistat 40 mg/kg; Garcinia, *Garcinia cambogia* 200 mg/kg.

4. Discussion

In the present study, the effects of PM and PF extracts on digestive enzymes were assessed. In in vitro studies, PM extract and PF extract have increased dose-dependent inhibitory activity on α-amylase (IC_{50} value: 6.13 mg/mL) and pancreatic lipase (IC_{50} value: 1.68 mg/mL), respectively. Notably, α-amylase, one of the digestive enzymes secreted from the pancreas and salivary glands, is involved in important biological processes such as the digestion of carbohydrates. α-Amylase inhibitors are well known to be effective in reducing postprandial hyperglycemia by slowing the digestion of carbohydrates and absorbing postprandial glucose. Reducing postprandial hyperglycemia prevents glucose uptake into adipose tissue, inhibiting the synthesis and accumulation of triacylglycerol [24,25].

On the other hand, pancreatic lipase is one of the most important enzymes for the digestion of dietary triacylglycerols. It is well known that dietary lipid is not directly absorbed from the intestines unless it has been subjected to the action of pancreatic lipase [24]. It has been clinically reported that a pancreatic lipase inhibitor, orlistat, prevented obesity and hyperlipidemia through the increment of fat excretion in the feces and the inhibition of pancreatic lipase. Based on this fact, the inhibition of these digestive enzymes is an important factor in the treatment of obesity [26,27].

We confirmed the anti-obesity effect of DKB-117 by way of digesting enzyme inhibition through in vivo testing. As a result of the test, DKB-117 extracts revealed effects of reducing weight and total fat in HFD-induced obese C57BL/6J mice. Micro-CT imaging was performed to quantify fat volume (total fat, abdominal fat, and subcutaneous fat). This method can readily discriminate between subcutaneous and abdominal fat [28]. Our results revealed that DKB-117 was able to reduce these fats. The in vivo test results demonstrate that DKB-117 administration regulates serum biochemical parameters (TCHO, TG, HDL, and LDL) in HFD-induced obese mice. These results are associated with an anti-hypercholesterolemic effect of PF [13,14]. In addition, the DKB-117 extracts increased TG and carbohydrate emissions in feces compared to those in the HFD group. It is expected that DKB-117 extracts promote the release of fat and carbohydrates due to the α-amylase inhibitory effect and the lipase inhibitory effect. *Garcinia cambogia,* which is currently used as a dietary supplement, is known to help reduce body fat by inhibiting the enzymes needed to synthesize carbohydrates into fat [29].

On the other hand, the DKB-117 extracts are expected to have an anti-obesity effect through a complex mechanism that promotes carbohydrate release through the inhibition of carbohydrate-degrading enzymes and which inhibits lipid absorption through lipase inhibition. It was observed that DKB-117 can effectively inhibit weight gain in animal experiments through a complex mechanism of inhibition of pancreatic lipase activity and inhibition of amylase activity, which is considered to be a suitable anti-obesity measure for Koreans who use carbohydrate as a staple food.

While the existing research on herbal extract (*Garcinia cambogia, Plantago psyllium, Morus alba*) single-action mechanisms is underway, it has been confirmed through this study that they are effective in anti-obesity in the battle against DKB-117 multifunction machines. Based on the results of in vitro and in vivo tests, the anti-obesity effects of DKB-117 in human application tests will be confirmed. Furthermore, it is determined that active ingredients should be investigated through the study on the separation of components of DKB-117.

In the future, the safety and efficacy of the DKB-117 extracts should be demonstrated through ongoing clinical trials; we plan to search for the activity compound in the DKB-117 extracts as well as perform additional testing of enzyme activity for lipid metabolism so that it can be widely used as a functional food material.

5. Conclusions

In summary, results of the present study revealed that DKB-117 extracts possess significant anti-obesity activities. These data clearly show that DKB-117 extracts are expected to have an anti-obesity effect through a complex mechanism that promotes carbohydrate release through the inhibition of carbohydrate-degrading enzymes while blocking lipid absorption through lipase inhibition.

Pancreatic lipase is an enzyme that is secreted from the pancreas and hydrolyzes the ester bonds of triglycerides to produce glycerol and fatty acids.

Decomposed glycerol and fatty acids are absorbed by the mucosal cells of the small intestine and are used as an energy source, but fat that has not been used as an energy source is synthesized into triacylglycerol again through the monoacylglycerol pathway and accumulated in the body.

Suppressing pancreatic lipase activity inhibits hydrolysis of triacylglycerol into glycerol and fatty acids, thus inhibiting liposuction through the small mucous membrane, reducing the amount accumulated in the body to prevent obesity.

By suppressing the a-amylase activity, the hydrolysis of ingested polysaccharides is inhibited, so that the absorption of excessively ingested carbohydrates in the body can be reduced and obesity can be prevented. It was confirmed that DKB-117 exhibits an anti-obesity effect by a method in which not only suppression of fat absorption but also suppression of excess carbohydrate absorption act together.

Based on the above results, our team plans to secure the anti-obesity effect and safety of DKB-117 extract through clinical trials.

Author Contributions: D.H.K. and Y.H.P. conceived and designed the experiments, prepared the plant extract, and carried out all the experimental works; formal analysis, J.S.L., H.I.J., and K.W.L.; writing—review and editing, D.H.K., Y.H.P., and T.H.K. All authors have read and agreed to the published version of the manuscript.

Funding: This research received no external funding.

Conflicts of Interest: The authors declare that they have no conflict of interest.

References

1. Wang, Y.C.; McPherson, K.; Marsh, T.; Gortmaker, S.L.; Brown, M. Health and economic burden of the projected obesity trends in the USA and the UK. *Lancet* **2011**, *378*, 815–825. [CrossRef]
2. WHO. Obesity and Overweight. 2016. Available online: http://www.who.int/mediacentre/factsheets/fs311/en/ (accessed on 1 April 2020).
3. The GBD 2015 Obesity Collaborators. Health Effects of Overweight and Obesity in 195 Countries over 25 Years. *N. Engl. J. Med.* **2017**, *377*, 13–27. [CrossRef]
4. Obata, A.; Okauchi, S.; Kimura, T. Advanced breast cancer in a relatively youngman with severe obesity and type 2 diabetes mellitus. *J. Diabetes Investig.* **2017**, *8*, 395–396. [CrossRef]
5. Tziomalos, K.; Krassas, G.E.; Tzotzas, T. The use of sibutramine in the management of obesity and related disorders: An update. *Vasc. Health Risk Manag.* **2009**, *5*, 441–452.
6. Hursel, R.; Westerterp-Plantenga, M.S. Thermogenic ingredients and body weight regulation. *Int. J. Obes.* **2010**, *34*, 659–669. [CrossRef] [PubMed]
7. Sun, N.N.; Wu, T.Y.; Chau, C.F. Natural dietary and herbal products in anti-obesity treatment. *Molecules* **2016**, *21*, 1351. [CrossRef] [PubMed]
8. Park, J.P.; Kim, J.H.; Park, M.K.; Yun, J.W. Potential agents for cancer and obesity treatment with herbal medicines from the green garden. *Biotechnol. Bioprocess Eng.* **2011**, *16*, 1065–1076. [CrossRef]
9. Amin, K.A.; Nagy, M.A. Effect of Carnitine and herbal mixture extract on obesity induced by high fat diet in rats. *Diabetol. Metab. Syndr.* **2009**, *1*, 17. [CrossRef] [PubMed]
10. Abdulwahid, S.J.; Asma, G.I. Studying the effect of anti-amylase inhibitor extracted from white kidney bean (*Phaseolus vulgaris*) in treat diabetes and obesity in an affected mice. *Int. J. Curr. Microbiol. Appl. Sci.* **2014**, *3*, 97–106.
11. Barrett, M.L.; Udani, J.K. A proprietary alpha-amylase inhibitor from white bean (*Phaseolus vulgaris*): A review of clinical studies on weight loss and glycemic control. *Nutr. J.* **2011**, *10*, 24. [CrossRef]
12. Zhao, M.; Huang, C.; Chen, Q.; Wu, X.; Qu, J.; Zhang, J. Genetic variability and population structure of the mushroom *Pleurotus eryngii* var. *tuoliensis*. *PLoS ONE* **2013**, *8*, e83253. [CrossRef] [PubMed]
13. Jo, K.J.; Ghim, J.; Kim, J.; Lee, H.; Lee, T.G.; Kim, J.I.; Kim, Y.; Byun, J.W.; Min, B.S.; Son, J.S.; et al. Water extract of *Pleurotus eryngii* var. *ferulae* prevents high-fat diet-induced obesity by inhibiting pancreatic Lipase. *J. Med. Food* **2019**, *22*, 178–185. [CrossRef] [PubMed]
14. Alam, N.; Lee, T.S. Antihyperlipidemic activities of *Pleurotus ferulae* on biochemical and histological function in hypercholesterolemic rats. *J. Res. Med. Sci.* **2011**, *16*, 776–786. [PubMed]

15. Maury, J.; Issad, T.; Perdereau, D.; Gouhot, B.; Ferré, P.; Girard, J. Effect of acarbose on glucose homeostasis, lipogenesis and lipogenic enzyme gene expression in adipose tissue of weaned rats. *Diabetologia* **1993**, *36*, 503–509. [CrossRef]
16. Wilson, J.J.; Ingledew, W.M. Isolation and characterization of *Schwanniomyces alluvius* amylolytic enzymes. *Appl. Environ. Microbiol.* **1982**, *44*, 301–307. [CrossRef]
17. Eom, S.H.; Lee, M.S.; Lee, E.W.; Kim, Y.M.; Kim, T.H. Pancreatic lipase inhibitory activity of phlorotannins isolated from *Eisenia bicyclis*. *Phytother. Res.* **2013**, *27*, 148–151. [CrossRef]
18. Birari, R.B.; Bhutani, K.K. Pancreatic lipase inhibitors from natural sources: Unexplored potential. *Drug Discov. Today* **2007**, *12*, 879–889. [CrossRef]
19. Sung, Y.Y.; Kim, D.S.; Choi, G.; Kim, S.H.; Kim, H.K. Dohaekseunggi-tang extract inhibits obesity, hyperlipidemia, and hypertension in high-fat diet-induced obese mice. *BMC Complement. Altern. Med.* **2014**, *4*, 372. [CrossRef]
20. Hong, S.J.; Lee, J.H.; Kim, E.J.; Yang, H.J.; Park, J.S.; Hong, S.K. Anti-obesity and anti-diabetic effect of Neoagarooligosaccharides on high-fat diet-induced obesity in mice. *Mar. Drugs* **2017**, *15*, 90. [CrossRef]
21. Heck, A.M.; Yanovski, J.A.; Calis, K.A. Orlistat, a new lipase inhibitor for the management of obesity. *Pharmacotherapy* **2000**, *20*, 270–279. [CrossRef]
22. Machado, M.V.; Cortez-pinto, H. Non-alcoholic fatty liver disease: What the clinician needs to know. *World J. Gastroenterol.* **2014**, *20*, 12956–12980. [CrossRef] [PubMed]
23. Donnelly, K.L.; Smith, C.I.; Schwarzenberg, S.J.; Jessurun, J.; Boldt, M.D.; Parks, E.J. Sources of fatty acids stored in liver and secreted via lipoproteins in patients with nonalcoholic fatty liver disease. *J. Clin. Investig.* **2005**, *115*, 1343–1351. [CrossRef] [PubMed]
24. Rani, N.; Vasudeva, N.; Sharma, S.K. Quality assessment and anti-obesity activity of Stellaria media (Linn.) Vill. *BMC Complement. Altern. Med.* **2012**, *3*, 145. [CrossRef] [PubMed]
25. Hernell, O.; Staggers, J.E.; Carey, M.C. Physical-chemical behavior of dietary and biliary lipids intestinal digestion and absorption 2 phase analysis and aggregation states of luminal lipids during duodenal fat digestion in healthy adult human beings. *Biochemistry* **1990**, *29*, 2041–2056. [CrossRef] [PubMed]
26. Han, L.K.; Kimura, Y.; Kawashima, M.; Takaku, T.; Taniyama, T.; Hayashi, T.; Zheng, Y.N.; Okuda, H. Anti-obesity effects in rodents of dietary teasaponin, a lipase inhibitor. *Int. J. Obes. Relat. Metab. Disord.* **2001**, *25*, 1459–1464. [CrossRef]
27. Drent, M.L.; Larsson, I.; William Olsson, T.; Quaade, F.; Czubayko, F.; Bergmann, K.; Strobel, W.; Sjostrom, L.; van der Veen, E.A. Orlistat (Ro 18-0647), a lipase inhibitor, in the treatment of human obesity: A multiple dose study. *Int. J. Obes. Relat. Metab. Disord.* **1995**, *19*, 221–226.
28. Judex, S.; Luu, Y.K.; Ozcivici, E.; Adler, B.; Lublinsky, S.; Rubin, C.T. Quantification of adiposity in small rodents using micro-CT. *Methods* **2010**, *50*, 14–19. [CrossRef]
29. Heymsfield, S.B.; Allison, D.B.; Vasselli, J.R.; Pietrobellj, A.; Greenfield, D.; Nunez, C. *Garcinia cambogia* (Hydroxycitric Acid) as a potential antiobesity agent. *JAMA* **1998**, *280*, 1596–1600. [CrossRef]

nutrients

MDPI

Review

Macroalgae—A Sustainable Source of Chemical Compounds with Biological Activities

Elena-Suzana Biris-Dorhoi [1], Delia Michiu [2], Carmen R. Pop [1], Ancuta M. Rotar [1], Maria Tofana [1], Oana L. Pop [1], Sonia A. Socaci [1,*] and Anca C. Farcas [1,*]

1 Department of Food Science, University of Agricultural Sciences and Veterinary Medicine Cluj-Napoca, 3-5 Calea Manastur, 400372 Cluj-Napoca, Romania; elena.biris@usamvcluj.ro (E.-S.B.-D.); carmen-rodica.pop@usamvcluj.ro (C.R.P.); anca.rotar@usamvcluj.ro (A.M.R.); maria.tofana@usamvcluj.ro (M.T.); oana.pop@usamvcluj.ro (O.L.P.)

2 Department of Food Technology, University of Agricultural Sciences and Veterinary Medicine Cluj-Napoca, 3-5 Calea Manastur, 400372 Cluj-Napoca, Romania; delia.michiu@usamvcluj.ro

* Correspondence: sonia.socaci@usamvcluj.ro (S.A.S.); anca.farcas@usamvcluj.ro (A.C.F.); Tel.: +40-745-472-589 (S.A.S.); +40-740-600-924 (A.C.F.)

Received: 17 September 2020; Accepted: 6 October 2020; Published: 11 October 2020

Abstract: Nowadays, one of the most important research directions that concerns the scientific world is to exploit the earth's resources in a sustainable way. Considering the increasing interest in finding new sources of bioactive molecules and functional products, many research studies focused their interest on demonstrating the sustainability of exploiting marine macroalgal biomass as feedstock for wastewater treatment and natural fertilizer, conversion into green biofuels, active ingredients in pharmaceutical and nutraceutical products, or even for the production of functional ingredients and integration in the human food chain. The objective of the present paper was to provide an overview on the recent progress in the exploitation of different macroalgae species as a source of bioactive compounds, mainly emphasizing the latter published data regarding their potential bioactivities, health benefits, and industrial applications.

Keywords: macroalgae; bioactive compounds; bioactivities; antimicrobial; antiproliferative activity; polysaccharides

1. Introduction

Algae are part of a heterogeneous group of photosynthetic organisms. The division includes multicellular organisms, macroalgae or seaweed (reaching sizes of up to 60 m in length), and unicellular organisms, also known as microalgae (measuring from 1 mm to several cm). One way to classify macroalgae is on the basis of their pigmentation: (i) brown seaweed (*Phaeophyceae*), (ii) red seaweed (*Rhodophyceae*), and (iii) green seaweed (*Chlorophyceae*) [1].

Algae are distributed in diverse and extreme environments. They are valuable due to their high content in compounds with different biological activities, including both complex organic compounds and primary and secondary metabolites. Worth mentioning, among them are phytopigments (xanthophylls and carotenoids), polyunsaturated fatty acids (PUFAs) comprising docosahexaenoic acid (DHA), phenolic compounds, tannins, peptides, lipids, enzymes, vitamins, carbohydrates, terpenoids, and others. Thus, algae are a viable and economical biomass source of valuable compounds with potential applications in the nutraceutical, pharmaceutical, chemical, food, and cosmetic industries due to their biologically active and regenerative properties [2–6].

In recent years, macroalgae have gained more and more interest owed to their various health promoting properties that can decrease the risks of many chronic diseases and even help to extend the lifespan [7,8]. Macroalgae can also be used for wastewater treatment or as a natural fertilizer in agriculture, therefore improving the quality of the products and minimizing the need for chemical fertilizers [9–11]. The potential of macroalgae as a source of renewable energy is also of considerable interest. These aquatic organisms have the ability to mitigate carbon dioxide emissions and nowadays are being used as feedstock to produce "clean" or so-called "third generation biofuels" [12].

The most important applications of algae are synthetized in Figure 1.

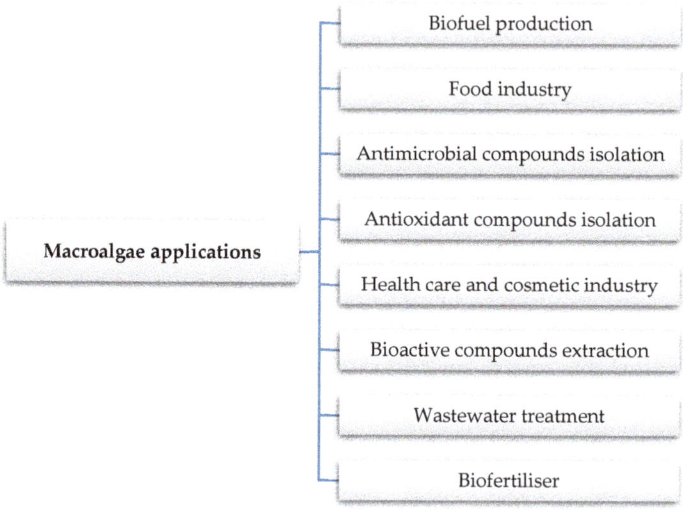

Figure 1. The main applications of macroalgae.

This review focuses on the recent progress in exploitation of different macroalgae species as a source of bioactive compounds, mainly emphasizing the latter published data (between 2010 and 2020) regarding the health benefits, their bioactivities, and potential applications.

2. Algae Chemical Composition

The use of different marine macroalgae (seaweed) as sources of bioactive compounds had the advantage to exploit an under-utilized renewable natural resource. It was demonstrated that this biomass produced a broad spectrum of nutrient and bioactive secondary metabolites. The chemical composition of macroalgae varies considerably due to both environmental conditions (light intensity, growth habitat, seawater salinity, temperature) and genetic differences among species [2,13].

Macroalgae have a protein content that can range from 7 to 31% of dry weight and a lipid content ranging from 2 to 13% of dry weight [14]. A considerable amount of carbohydrate can also be found in macroalgae (up to 32–60% of dry weight).

Regarding the macroalgae content in micronutrients, they are a good source of vitamins, especially of the B-group representatives (i.e., B_1, B_{12}), as well as the lipophilic vitamins A and E (tocopherol) [13,15–17]. The richness in vitamin B_{12} propels the macroalgae-based products as dietary supplements for a vegan lifestyle, considered to be at risk for vitamin B_{12} deficiency [18]. Within the mineral composition, the most significant microelements present in the seaweeds are usually potassium, sodium, magnesium, and calcium, accounting for more than 97% of the total mineral content. Other microelements such as copper, iron, manganese, and zinc are found in small amounts (ranging from 0.001 to 0.094% of seaweeds' dry weight) [19].

2.1. Protein and Amino Acid Composition

Proteins are a major class of compounds, essential for human nutrition. For food products, the amount of protein is considered a quality parameter, but of equal importance for human health is the protein quality (e.g., protein composition in amino acids, the ratio of essential amino acids, their digestibility, and bioavailability). It is well known that seaweeds can be used as a nutrient source, especially in developing countries. In this sense, macroalgae is considered a sustainable nutrient alternative source, mainly due to high-value proteins.

Nine of the 21 amino acids are considered essential for humans, namely: histidine, isoleucine, leucine, lysine, methionine, phenylalanine, threonine, tryptophan, and valine. Proteins of animal source have a chemical score of 1.0, meaning that animal proteins contain all the essential amino acids in a minimum proportion necessary for the human body. Instead, the chemical score for cereal proteins normally ranges from 0.4 to 0.6, while the one for algae proteins ranges from 0.75 to 1.0 indicating that the protein quality of algae is superior to most terrestrial plants [14]. Therefore, macroalgae are able to cover the human requirements for essential amino acids [13].

The protein content of marine algae differs according to species. Although the protein level is generally low in brown seaweeds (3–15% on dry weight basis (DW)), and moderate in green seaweeds (9–26% DW), in red seaweeds the content can reach 47% DW [20]. One gram of algae meal from algae with the highest protein levels (e.g., *Enteromorpha intestinalis*, *Palmaria palmata*, and *Vertebrata lanosa)* contains equal to or higher amounts of all of the essential amino acids compared to rice, corn, and wheat. In addition, the lysine content was reported to be three to nine times higher. The approximate amount of free amino acids can range from 2 to 14.5%, the lowest amount being reported in the green algae and highest in the red varieties [13]. If we consider nonessential amino acids, the green seaweed proteins contain high levels of glutamic and aspartic acids (that can have a concentration up to 26 and 32% of the total amino acids), but also alanine and glycine [20].

The seaweed varieties that have a high protein level can be used as ingredients in the manufacturing process of different foods. *Porphyra* species are known to be used in the famous sushi preparations. The same seaweeds are also processed into roasted products (such as yaki-nori) or they can be boiled in soy sauce (tsukudani-nori) [6]. For instance, species such as *Ulva pertusa*, *Enteromorpha* sp., and *Monostroma* sp. (protein levels of 26, 19, and 20% dw, respectively) are mixed together to create a food product called "aonori" (or green laver), a protein rich product very appreciated in Japan. In Europe and Canada, *Palmaria palmata* is often used as a food ingredient. Due to its high protein content (up to 35% dw), this specie of algae can be processed into dry flakes and used to obtain different functional products [21,22].

2.2. Lipid and Fatty Acid Composition

The lipid content is relatively low in macroalgae species, with values less than 5% w/dw. Variations in the quantity and in fatty acids profile can be attributed to both environmental (light intensity, seawater salinity, temperature) and genetic differences among species. In general, it has been observed that brown species have a higher lipid content compared to green varieties [23,24].

However, nearly half of lipids are polyunsaturated fatty acids such as eicosapentaenoic acid (EPA) and arachidonic acid (AA). Red and brown algae are rich in EPA and AA, while green seaweeds such as *Ulva pertusa* predominantly contain hexadecatetraenoic, oleic, and palmitic acids, and also significant levels of PUFAs, such as linoleic acid (18:2n-6) and α-linolenic acid (18:3n-3) [23,25]. Moreover, the ratio between ω-6 and ω-3 and the ratio between PUFAs and SFAs (saturated fatty acids) found in red and brown algae are more favorable for human health than those found in green algae [26].

Besides the fatty acids, the lipidic fraction of macroalgae contains glycolipids and phospholipids. Glycolipids are carbohydrates (mono- or oligosaccharide) that are linked to a lipid (through a glycosidic bound), being essential components of the cellular membrane. Several studies were conducted on different glycolipids from seaweed (e.g., monoglycosyl diacylglycerol subfraction from *Fucus distichus;* monogalactosyl diacylglycerols from *Sargassum horneri;* sulfoglycolipids from *Porphyra crispata),* showing their anti-inflammatory and antiproliferative effects, respectively [27–29]. Regarding lipid extraction, Ramola et al. [30] and Margareta et al. [31] found that the most efficient solvent was a mixture of chloroform: methanol (2:1) with an efficiency of 14% compared to hexane (2:1) with 12.5% efficiency.

2.3. Carbohydrates

Carbohydrates, which include mono-, oligo- and polysaccharides, are considered an important and irreplaceable source of energy necessary to support different functions of the human body and its physical activity [32]. Of these, algal polysaccharides such as alginates, carrageenan, fucoidan, and laminarin, were found to exhibit a wide spectrum of biological activities, amongst which the antioxidant, antithrombotic, anti-inflammatory, and neuroprotective activities are the most studied [33]. Additionally, the nondigestible oligosaccharides can act as prebiotic agents gaining increased attention due to their positive influence on the gut flora [34–36]. The qualitative differences of the sugar backbone, the molecular weight, and also the sulfation degree vary the pharmacological effects of algal carbohydrates [37].

In general, edible seaweeds present a variable carbohydrate content. Reported to fresh weight, carbohydrates levels vary from 4.1/100 g wet weight in *Ulva* sp., to 13.1/100 g wet weight in species *Ascophyllum nodosum*, with species *Laminaria digitata* having a content of 9.9/100 g wet weight [3]. Furthermore, the *Undaria pinnatifida* species contains carbohydrates in a content of 9.14% [38]. Related to dry matter, the total carbohydrate concentrations in the seaweed species represent up to 76% of their dry weight. In this sense, species such as *U. pinnatifida* contains between 45 and 52% total carbohydrates. A high content of carbohydrates can be also found in *Saccharina japonica* brown algae (51.9% of dry weight), *Gracilaria chilensis* red algae (66.1% of dry weight), and *Ulva compresa* green algae (48.2% of dry weight) [17,39,40].

According to literature data, the sulfated polysaccharides represents one of the main constituents [41,42]. The highest contents are found in genera such as *Ascophyllum*, *Porphyra*, and *Palmaria*. Additionally, species *Kappaphycus alvarezii* and *Eucheuma spinosum* present a content of polysaccharides up to 56 and 40%, respectively [43]. Carrageenans are one of the major constituents of red seaweed cell walls representing 30 to 75% of the algal dry weight. Ulvans are the major constituents of green seaweeds cell walls representing 8 to 29% of the algal dry weight. Alginates and fucans are the major constituents of brown seaweeds cell walls representing between 17 and 45%, respectively 5 to 20% of the algal dry weight. Furthermore, brown seaweeds contain laminarin up to 35% of the algal dry weight [44]. Porphyran, a linear sulfated polysaccharide, was found to be one of the main components isolated from *Pyropia yezoensis* (edible red alga cultivated and consumed mainly in East and Southeast Asia) [45]. Considering the applicability in the food industry, seaweed polysaccharides such as agar, alginates, and carrageenan are the most important and economically feasible obtained products due to their rheological gelling and thickening properties [17].

2.4. Pigments

Macroalgae use light as energy source and pigments play a key role in gathering solar energy. These pigments absorb light from the visible spectrum [46,47]. Carotenoids are orange/red pigments that absorb light energy and then pass it on to chlorophyll, therefore playing a secondary role in photosynthesis [48]. Carotenoids supplement the light gathering potential of the algae. There is an alteration for both chlorophyll and carotenoid content in seaweeds depending on the ultraviolet radiation (UV) levels throughout the year. Both chlorophyll and carotenoid pigments possess antioxidant and chemo preventive properties [37,49]. The main algal pigments in commercial use at the present time comprise: beta-carotene, astaxanthin, lutein, phycocyanin, chlorophyll, and fucoxanthin.

2.4.1. Carotenoids

Carotenoids are terpenoid pigments widely distributed that are divided into two main groups: carotenes (unsaturated hydrocarbons) and xanthophylls (carotenes' oxygenated derivatives—of which in algae the most representatives are fucoxanthin, astaxanthin, lutein, and zeaxanthin) [50,51]. β-carotene is the major unsaturated hydrocarbon in brown and green seaweeds [37,52].

Carotenoids are a well-known as antioxidant agents [53]. Fucoxanthin exerts strong 2,2-diphenyl-1-picrylhydrazyl (DPPH) radical scavenging activity, most of it under anoxic conditions, being also recognized to exhibit anti-inflammatory properties. Its inhibitory activity against proinflammatory agents, such as nitric oxide (NO), tumor necrosis factor alpha (TNF-α), interleukin-1β, prostaglandin E2 (PGE2), and interleukin-6 (IL-6) was reported in [54]. A number of studies suggest that fucoxanthin is a promising and upcoming anticancer and antitumor agent and can suppress metastatic potential [55,56]. It also exhibited favorable levels of chemopreventive and/or chemotherapeutic activities against different human colon cancer cell lines, in combination with troglitazone being able to induce cell apoptosis via DNA fragmentation [57]. *U. pinnatifida* ethanolic extract, rich in fucoxanthin, was found to improve the plasma and lipid profile in high-fat diet mice. Aki et al. [58] investigated the effects of the seaweed carotenoids on unsaturated fatty acid metabolism in a hepatocyte culture (BRL-3A). The molecular mechanism revealed that fucoxanthin and its physiological metabolite, fucoxanthinol, caused alterations in fatty acid composition, leading to a decrease in EPA and the accumulation of docosahexaenoic acid.

2.4.2. Chlorophylls

Chlorophylls are pigments which contain in their structure a central magnesium ion, playing a functional role in the algae photosynthesis process but also a protective role ensuring the algal tissue integrity against oxidative stress that may be excessive UV radiation [37]. Terrestrial plants and brown algae are dominated by chl a, while chl b is mainly related to green algae. Additionally, brown algae are considered the main source of chl c, while chl d is specific to red algae [59]. Chlorophyll is known to be converted into pheophytin, pyropheophytin, and pheo-phorbide in processed vegetable food and following ingestion by humans. These derivates show antioxidant and antimutagenic effect and may play a significant role in cancer prevention [17]. Beside the biological activities and health effects of the different chlorophyll catabolites, the seaweeds can also be considered as an alternative to replace the synthetic pigments used in the food industry.

2.5. Phenolic Compounds

Among the current interests of the scientific community is to find a sustainable source of bioactive molecules in order to reduce the use of synthetic compounds. In this sense, macroalgae phenolic compounds have gained particular attention due to their specific bioactivities and health-promoting benefits, including antioxidant, antiproliferative, antimicrobial, antiallergic, antidiabetic, and neuroprotective properties [60–63]. Similar to terrestrial plants, these secondary metabolites are essential to the normal growth and development of macroalgae, supporting the natural defense system against various disturbing factors such as diseases, injuries, and environmental aggression [64]. Structurally, phenolic compounds from terrestrial plants are derived from gallic and ellagic acid, while the algal compounds are derived from polymerised phloroglucinol units (1,3,5-trihydroxybenzene) [17].

The phenolic compounds present in macroalgae vary from simple molecules, such as phenolic and cinnamic acids or flavonoids, to the more complex phlorotannin polymeric structures, their concentration being closely dependent on a number of intrinsic and extrinsic factors, such as species, seasonal variations, and environmental conditions [64]. Of all the seaweed phenolic metabolites, the main attention has been focused on phlorotannins (phloroglucinol, eckol, 7-phloroeckol, 6,6-bieckol, phlorofucofuroeckol A, fucodiphloroethol), identified in considerable quantities in brown *Ecklonia* species [40]. Other compounds such as hydroxybenzoic acid derivatives (gallic, p-hydroxybenzoic, vanillic, and syringic acids), hydroxycinnamic acids (caffeic, ferulic, sinapic, and p-coumaric acids), flavonoids (epicatechin, epigallocatechin, rutin, quercitrin, hesperidin, myricetin, and kaempferol), and bromophenols were identified in variable concentrations in all green, red, and brown species [17,40,64].

3. Algae Applications in the Food Industry

Algae species have been used as plain food since ancient times. In Asia and in the East, the tradition of eating algae is a long-standing one, while in the Western countries, the interest in consuming algae-based products is quite recent but gaining increasing terrain [65]. Algae have manifold uses in different industry fields as a result of their rich chemical composition and content of bioactive substances. Moreover, their gelling, thickening, and stabilizing properties have driven the isolation and development of products such as agar, alginate, and carrageenan [66,67]. Due to these properties, algae have a main use in the food industry as hydrocolloids or as functional ingredients in different fish and meat products (steaks, frankfurters, or sausages), milk-based and fermented products [68–70], or cereal-based products (flour, pasta, bread, and biscuits) [4,21]. Moreover, these algae-based hydrocolloids are of utmost importance for food industry innovative fields such as molecular gastronomy.

In the dairy industry, algae were added in order to improve the nutritional value of cheese and other milk-based products [68]. *Laminaria* was added to smoked cheese, yoghurt, and milk deserts, giving them not only improved nutritional properties but also good sensory characteristics. *Laminaria saccharina* algae from the North Sea can be also introduced into cottage cheese or fresh cheese composition in order to improve their iodine content [69]. In addition to their nutritional properties, algae have been shown to have the ability to increase product stability during shelf-life due to the presence of compounds with antibacterial and antioxidant potential. In this sense, it was demonstrated that when algae Wakame (*U. pinnatifida*) and Kombu (*Laminaria japonica*) were added into the cheese composition, the product quality was maintained for a longer storage period [70]. In the meat industry, algae were added in the diet of lambs and chickens in order to improve the content of DHA, EPA [71,72], and antioxidants [73,74].

Recent studies on the bioactivity of some common species, such as *H. elongata* and *U. pinnatifida*, recommend their use in the composition of functional foods, due to the high content of antioxidants and the potential to alleviate the metabolic syndrome [75]. A wide range of studies reported the high potential of using algae as a source of prebiotics [76]. Wang et al. [77] proved that rats which had 2.5% alginate supplementation exhibited an increase in *Bifidobacterium* and *Lactobacillus*. A dietary supplementation of 1% laminarin was proved to result in an increase in *Bifidobacterium* number in rats [78]. *U. pinnatifida* and *Porphyra ternera* extracts fed to rats resulted in lower bacterial enzyme activity in the cecum, and also, the enzymatic activities that were reduced are implicated in the conversion of procarcinogens to carcinogens, therefore implying a possible link between seaweed extract intake and the reduced risk of colon cancer [79].

The most use species of algae in the food industry are summarized in Table 1 below.

Table 1. The most used species of algae in food industry and their general characteristics.

Laminaria digitata	Dark brown, to 2 m in length; presents a claw-like holdfast, a smooth, flexible stipe, and also a laminate blade to 1.5 m long split into finger-like segments [80]
	The species is rich in alginates, mannitol, and amino acids. [20,81]
	Very rich in iodine; therefore, this seaweed promotes slimming and has antiseptic properties [82]
Dictyota dichotoma	Thallus is flat and leaf-like, up to 30 mm long and 5–30 mm broad. The fronds are thin and translucent; colors differ from olive to golden-brown [80]
	Produces large amounts of heterogeneous polysaccharides when submitted to the extraction procedures used to obtain fucoidans [83]
Undaria pinnatifida	Thallus fixed by a ramified holdfast [84]
	Rich source of eicosapentaenoic acid, an omega-3 fatty acid, and presents high levels of sodium, calcium, iodine, thiamine, and niacin [16,38]

Table 1. *Cont.*

Enteromorpha linza	Can be distinguished by its smooth thallus, most with a ruffled margin [85]
	Rich in essential amino acids, fatty acids, vitamins, dietary fiber, and resistant protein [2,3,86]
Ecklonia cava	A perennial brown alga and exists mainly in subtidal areas off the coast of Japan and Korea [87]
	Eckol isolated from *E. cava* attenuates oxidative stress-induced cell damage in lung fibroblast cells [88]
	Exhibits radical scavenging activity, but also antiplasmin inhibiting activity, antimutagenic activity, bactericidal activity, human immunodeficiency virus type 1 (HIV-1) reverse transcriptase, and protease inhibition [89]

4. Health Effects

Seaweeds contain a large variety of bioactive compounds that may be involved in the prevention and treatment of many diseases. They have several mechanisms for disease prevention and/or treatment. In this regard, some epidemiological, clinical, and meta-analysis studies associate the lower incidence of different chronic diseases, such as cancer, cardiovascular deficiency, diabetes, Parkinson disease, obesity related disorders, and metabolic syndrome, with a diet profile that includes seaweed consumption [49,90–93]. The main benefits on human health are presented in Figure 2.

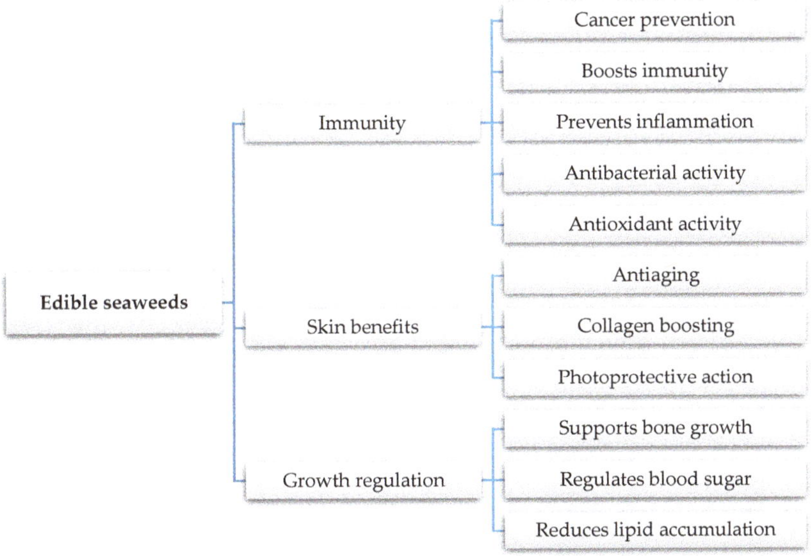

Figure 2. Health effects of macroalgae on human health and wellbeing.

4.1. Blood Pressure, Sugar, and Fat Reduction

Seaweeds are known to be rich in linolenic acid and its derivatives. These compounds can reduce blood viscosity and also smooth the interaction between blood vessels and vasoconstrictor substances. It was shown that when linolenic acid concentration increased by 1%, the blood pressure can diminish by 5 mmHg [90,94–96]. In this sense, Ryan et al. [86] analyzed the possible reduction in blood pressure using DHA algae oil and found that blood pressure reduction and heart rate were significantly reduced.

Alginate was shown to reduce blood sugar level, with sodium alginate supplementation of patients with diabetes type II leading to a decrease in the blood peak glucose level [96–99]. Porphyran and peptides were also proved to reduce blood sugar and blood pressure in rats and rabbits (*Porphyra yezoensis* in 1.6 g/L and 0.47 mg/mL) [100,101]. Fucoidan can reduce blood fat and sugar levels by disrupting fat absorption. Fucoidan is known to improve endoplasmic reticulum stress-reduced insulin sensitivity through adenosine monophosphate-activated protein kinase activation, and it can restore lipid homeostasis in mice with type II diabetes [102,103]. Linolenic acid aids the transformation of low-density lipoprotein (LDL) cholesterol to high-density lipoprotein (HDL) cholesterol and, therefore, can regulate fat metabolism [90,96,104].

4.2. Anticoagulant and Antithrombotic Properties

According to WHO, cardiovascular and cerebrovascular diseases have become the main cause of population mortality. Sulfated polysaccharides extracted from algae possess anticoagulant and antithrombotic properties [90]. In this regard, Ustyuzhanina et al. [105] showed that chemical transformation of branched xylofucans isolated from the brown algae *Punctaria plantaginea* into highly sulfated linear fucans effectively inhibited clot formation, having similar antithrombotic and anticoagulant effects to that of the heparinoid Clexane (enoxaparin) and the native fucoidan from *S. latissima*. *E. cava* was proved to be a great source of bioactive marine polyphenols, with antihyperglycaemic, antihyperlipidaemic, anti-inflammatory, and antioxidant effects, supported by evidence from in vitro studies as well as from those from human and animal trials already completed [60].

4.3. Antiaging, Antidepression, and Antifatigue Properties

A number of different physical and physiological factors are relevant when it comes to aging. The healthy function of the kidney and spleen plays an important role in human health. Seaweeds and seaweed-derived bioactive substances regulate the nervous system function, repairing DNA, promoting immunity, removing free radicals, regulating endocrine function, promoting healthy metabolism, and enhancing the kidney and spleen function [106]. *Fucus vesiculosus* aqueous extract increases the expression of integrin molecules. Topical application of the extract had a positive effect on the thickness and mechanical properties of human skin [90,96,106,107]. Polysaccharides have a large number of applications in the cosmetic industry. They act as rheology modifiers, suspending agents or wound-healing agents [108]. Carotenoids are powerful antioxidants possessing anti-inflammatory and antiaging properties. Several studies reported that astaxanthin, a xanthophyll carotenoid found also in macroalgae, can lower the oxidative stress protecting the mitochondria from the cumulative reactive oxygen species damage. Furthermore, astaxanthin was show to exhibit neuroprotective effects suggesting its possible use in the therapeutic treatment or prevention of neurodegenerative diseases such as Alzheimer's or Parkinson's disease [109,110].

Miyake et al. [111] performed a study on seaweed consumption and depressive symptoms during pregnancy concluding that a rich seaweed diet can be associated with a lower prevalence of depressive symptoms during pregnancy. Seaweed polysaccharides also possess antifatigue properties [112]. Higher hemoglobin, more oxyhemoglobin dissociation, and enhanced release of oxygen are responsible for the antifatigue property.

4.4. Antimicrobial and Antioxidant Potential

The antimicrobial compounds in algae are from several chemical classes, their level varying during algal growth and during seasons. In this sense, it was demonstrated that the *Polysiphonia* type produces antibiotic compounds constantly throughout the year, the *Laminaria* type has the maximum production during the winter, the *Dictyota* type during the summer, while *Codium* type has the best efficiency during the spring [113].

Extracts obtained with different solvents from a wide range of algae species, including *Ulva fasciata, Bryopsis plumosa, Chaetomorpha antennina, Acrosiphonia orientalis, Sargassum wightii, Grateloupia filicina, Hypnea pannosa, Gracilaria corticate Portieria hornemannii, Cheilosporum spectabile, Centroceras clavulatum, Chnoospora bicanaliculata, and Padina tetrastromatica*, were tested for their antimicrobial activity against *E. coli, S. aureus*, and *S. pyogenes*. From the tested solvents, the mixture between methanol and toluene (3:1 v/v) had the highest efficiency in extracting the compounds with antimicrobial potential from fresh biomass [114,115]. In another study, Tuney et al. [116] used methanol, acetone, diethyl ether, and ethanol to extract the bioactive compounds from 11 seaweed species. Diethyl ether extracts of fresh *C. mediterranea*, *E. linza*, *U. rigida*, *G. gracilis*, and *E. siliculosus* exerted high antimicrobial effects (10–15-mm halo) against several organisms (including *Enterococcus faecalis*, *Staphylococcus aureus*, *Pseudomonas aeruginosa*, and *Escherichia coli*). Instead, Bhuyar et al. [117] tested the ethanolic extract of the red alga *Kappaphycus alvarezii* against *Bacillus cereus*, the results indicating an inhibition zone with less than 10 mm of diameter.

The *Phylum Rhodophyta* (red algae) is recognized as one of the oldest groups of algae, characterized by the presence of phycoerythrin (a red protein-pigment complex), carrageenan (a sulfated polysaccharide), and phlorotannins. All of these compounds having strong antimicrobial activity. Another red alga extracts, *Symphyocladia latiuscula*, were proved to exhibit antimicrobial activity against a broad spectrum of microorganisms, the strongest antimicrobial effect being observed against *Vibrio mimicus* (50 μg/mL) and *Vibrio vulnificus* (50 μg/mL) [113,118].

Species such as *Laminaria saccharina*, *Laminaria digitata*, *Himanthalia elongata*, *Palmaria palmata*, and *Enteromorpha spirulina* are recognized as edible algae. Among these, *H. elongata* contains considerable amount of phenolics, tannins, and flavonoids. These antioxidant compounds that have a significant DPPH scavenging activity (50% inhibition (EC_{50}) level at 0.125 μg/mL extract) can promote *H. elongata* as a natural alternative for food preservation. Moreover, the *H. elongata* methanolic extract at a concentration of 6% inhibited the growth of food spoilage (*Pseudomonas aeruginosa* and *Enterococcus faecalis*) and food pathogenic microorganisms (*Listeria monocytogenes* and *Salmonella abony*). Lower concentrations of the same brown seaweed extract (3%) extended the lag phase and decreased the exponential growth rate and final population densities of microorganisms in the culture [61,119].

The antimicrobial activity of other bioactive compounds extracted from marine algae was assessed against various microorganisms such as *Staphylococcus aureus*, *Salmonella choleraesuis*, *Mycobacterium smegmatis*, *Candida albicans*, and *Escherichia coli*. From the isolated compounds, three of them (namely cycloeudesmol (10–50 μg/mL), laurinterol (1–5 μg/mL), and debromolaurinterol (10–50 μg/mL)) exhibited antimicrobial activity at concentrations close to that of streptomycin (complete inhibition after 48 h) [120,121].

Studies of Al-Saif et al. [122] revealed the high antimicrobial potential of several algae strains (*Ulva reticulate*, *Caulerpa occidentalis*, *Cladophora socialis*, *Dictyota ciliolate*, and *Gracilaria dendroides*) against *Escherichia coli* (ATCC 25322), *Enterococcus faecalis* (ATCC 29212), *Pseudomonas aeruginosa* (ATCC 27853), and *Staphylococcus aureus* (ATCC 29213). The chloroform extract of *Gracilaria dendroides* had the highest antimicrobial activity against *E. coli* (32.6 mm inhibition zone). Wahidi et al. [123] tested the antimicrobial activity of extracts of macroalgae from Marrocan Atlantic coast against Gram-positive (*Bacillus subtilis* and *Staphylococcus aureus*) and Gram-negative (*Escherichia coli* and *Pseudomonas aeruginosa*) bacteria. Their results showed that the ethanolic extract of *Cystoseira brachycarpa* (500 μg/disc) had the highest inhibition diameter (>20 mm) for all tested bacteria, similar to that of control (rifampicine 30 μg).

In general, the microbial species on which the algae extracts have the strongest inhibitory activity are *Staphylococcus aureus* [9], *Escherichia coli* [124], *Salmonella* spp [62,125,126], *Bacillus cereus* [127,128], and *Listeria monocytogenes* [129,130]. For example, 100% ethanolic extracts of *Pithophora oedogonium* and *Botrydiopsis arhiza*, at concentrations of 2, 4, 6, and 8 mg/mL, were investigated for their antimicrobial activity against *Salmonella* and *Staphylococcus* sp. While *B. arhiza* extracts showed no inhibition capacity, the *P. oedogonium* extract (4 mg/mL) inhibits the growth of the above-mentioned strains [126]. Jang and Lee [128] evaluated the antibacterial potential of 51 Korean domestic algae methanolic extracts against foodborne pathogens, such as *B. cereus*, *S. aureus*, and *L. monocytogenes*. From the tested extracts, microorganisms were specifically sensitive to *Laurencia okamurae* Yamada and *Dictyopteris undulata* Holmes extracts which exerted antibacterial potential comparable with that of streptomycin [128]. *C. linum* methanolic extract at a concentration of 500 μg/mL was most effective against *B. cereus*, with a 27 mm inhibition zone, comparable with that of the standard antibiotic (chloramphenicol, 100 μg/mL).

The high antimicrobial activity of the *C. linum* methanolic extract may be associated with its significant phenolic content (672.3 mg/g gallic acid equivalent), and high scavenging activity (IC_{50} 9.8 μg/mL) [62].

In the effort of finding new natural antimicrobials, the algae represent a rich source of bioactive compounds with manifold activities. In this direction, several studies were conducted assessing different fractions of methanolic or ethanolic seaweed extracts. For example, the ethyl acetate soluble fraction of *E. cava* methanolic extract exhibits high antibacterial activity against *L. monocytogenes* having an minimum inhibitory concentration (MIC) value of 256 μg/mL and an minimum bactericidal concentration (MBC) value of 512 μg/mL. Instead, the chloroform fraction of the ethanolic extract of *Myagropsis myagroides* was even more efficient in inhibiting the *L. monocytogenes* growth, with an MIC value of 63 μg/mL [129,130].

One worthy nutritional property of algae is linked to their high content of polyphenols, flavonoids, and carotenoids [131]. The major phenolic compounds isolated from the marine algae included anthraquinones, coumarins, and flavonoids, with rutin, quercetin, and kaempferol flavonoids being identified in all the algal species. According to Al-Saif et al. [122], the highest concentration of these three flavonoids was found in alga *Gracilaria dendroides* (rutin, 10.5 mg/kg; quercetin 7.5 mg/kg; kaempferol 15.2 mg/kg). These compounds were proved to be the most effective flavonoids in inhibiting bacterial growth (*E. coli, P. aeruginosa, S. aureus, E. faecalis*). The eckol (phlorotannin compound) isolated from the ethyl acetate extracts of *E. cava* species showed potential antimicrobial activity against methicillin resistant *S. aureus*, the MIC values ranging from 125 to 250 μg/mL [132]. In the case of phlorotanins isolated from *E. bicyclis*, namely eckol, dieckol, dioxinodehydroeckol, fucofuroeckol-A, 7-phloroeckol, and phlorofucofuroeckol-A, the MIC values for the inhibition of *S. aureus* and methicillin-resistant *S. aureus* ranged between 32 and 64 μg/mL [133]. The antioxidant activity of three representative Black Sea macroalgae, *Ulva lactuca* (green algae), *Cystoseira barbata* (brown algae), and *Ceramium rubrum* (red algae), was assessed according to the antioxidative capacity in lipid soluble substances procedure (ACL method). Of these, *C. barbata* showed the highest antioxidant activity (141.5 Trolox equivalent units, nmols/g dry weight) [134].

4.5. Antiallergic Effect

The worldwide trend is to use natural substances to cure allergies, and this has led to an increased interest in algal bioactive compounds, particularly in seaweed phenols. From the phenolic compounds, curcumin, epigallocatechin gallate, flavonoids, and quercetin were proved to have a significant antiallergic activity [20,21]. Additionally, fucoidan extracted from *U. pinnatifida* was proved to reduce the chemical and immunological responses in an animal model [21,135,136].

The porphyran, a sulfate polysaccharide isolated from *Porphyra tenera* and *Porphyra yezoensis*, is also known to possess antiallergic properties. The oral administration of porphyran (obtained from dried nori, 2% in drinking water) to mice with ear edema suppressed the evolution of the disease [137]. Aside from the antiallergic potential, the porphyran was also found to exert anti-inflammatory activity, their reactive oxygen species scavenging potential being considered the main mechanism responsible for this action [21]. Phlorotannins from *E. arborea* have been used since ancient times as folk medicine due to their antiallergic properties as reported by literature data [138]. Phlorotannins, carotenoids, polysaccharides, PUFAs, and phycocyanins were all found to exhibit antiallergic properties [139].

4.6. Anticancer Properties

For several decades, macroalgae have been promoted for their potential role in preventing cancer occurrence, tumor progression, and even health recovery after radio- or chemotherapy treatments [78,136]. Iodine may also exert an anticancer effect, due to the ability to cause apoptosis in cancer cells. The same property can be attributed to the omega-3 fatty acids such as stearidonic acid and hexadecatetraenoic acid found in edible marine algae such as *Undaria* and *Ulva* up to 40% of total fatty acids [140].

Alginate, laminaran, fucoidan, and many other seaweed polysaccharides were proven to have antitumor activities. A high amount of polysaccharide (~65% of polysaccharide in total dry weight) can be found in the many seaweeds such as *Ulva, Ascophyllum, Porphyra*, and *Palmaria* [141]. Alginate is able to clean up the intestinal tract, therefore improving immunity and intestinal tract health levels, reducing the risk of cancer. Laminaran and fucoidan are able to induce apoptosis in order to prevent cancer, but some unidentified seaweed polysaccharides can also exhibit direct or indirect antitumor effects. *Sargassum latifolium* inhibited cytochrome P450 1A and glutathione S-transferases and reduced 1301 cell viability, inducing apoptosis [142]. *Ulva fasciata* extract inhibits the growth of tumor cells in human colon cancer by 50% at a concentration of 200 μg/mL [143]. *Gracilariopsis lemaneiformis* was shown to have antitumor activity by inducing apoptosis in several cancer cell lines (e.g., human lung cancer cell line A549, the gastric cancer cell line MKN28, and the mouse melanoma cell line B16) attributed to neutral polysaccharide with a linear structure of repeated disaccharide agarobiose units [144]. Furthermore, the extracts of *Hydroclathrus clathratus* and its purified polysaccharide fractions were able to suppress the ascitic Sarcoma 180 tumor growth and prolonged the life span of the tumor-bearing mice with 30–40%, while presenting low toxicity to the normal cells [145].

Fucoidans extracted from *Dictyota ciliolata, Padina sanctae-crucis*, and *Sargassum fluitans* brown algae were reported to possess a high antioxidant activity being able to protect HepG2 cells from oxidative stress. Moreover, at a concentration of 2mg/mL, none of the fucoidan extracts had cytotoxic effects [146]. The isolated fucoidan from *Sargassum polycystum* exhibited potent antioxidant, anticancer, and antiproliferative properties against human breast cancer cell line MCF 7 at 150 g/mL and an IC50 of 50 g/mL [147]. Fucoidans from *Fucus evanescens* at a concentration of 800 μg/mL were reported to have anticancer properties by inhibiting the proliferation of melanoma SK-MEL-28 cell [148]. These findings underline once more that algae fucoidans are worth considering for the development of functional food, supplements, or drugs that can be used in the prevention of oxidative stress induced diseases.

The administration of extracts of the red alga *Eucheuma cottonii* to rats significantly improved the oxidative state of cells, and contributed to the tumor suppression response against MB-MDA-431 cell lines [149].

Meroditerpenoids are reported to be present mostly in brown algae. Metabolites such as fallahydroquinone, fallaquinone, fallachromenoic acid, sargaquinone, sargaquinoic acid, sargahydroquinoic acid, and sargachromenol were identified from the *Sargassum fallax* brown algae and reported to possess lower to moderate antitumor activity against a P388 Murine Leukaemia cell line [150].

4.7. Anti-Inflammatory Property

Inflammation is one of the most complex medical problems. It can be initiated by several factors, namely environmental pollution, chemical intoxication, and bacterial infection, which lead to injury or death of cells. Approximately 20% of all human cancers are caused due to chronic inflammation [151]. Several studies on the anti-inflammatory effect of different species of algae are listed in Table 2 below.

Table 2. The anti-inflammatory effect of different species of algae.

Algae Species	Active Extract/Compound	Biological Activity	Reference
Phorphyra dentate	Methanolic extracts	Anti-inflammatory effect in lipopolysaccharide (LPS) induced mouse RAW 264.7 macrophages cell line.	[152]
Caulerpa mexicana	Methanolic extracts	Decrease the xylene-induced ear edema and reduce cell migration to different sites.	[153]
Myagropsis myagroides	Fucoxanthin	LPS-stimulated RAW 264.7 macrophages.	[154]
Ulva reticulate	Methanolic extracts	Carrageenan-induced hind paw edema in rats and peritonitis in acute and chronic inflammatory models.	[155]
Laminaria saccharina	Sulfated polysaccharides	Inhibits leukocyte recruitment in rat and the neutrophil adhesion to platelets.	[156]

Table 2. *Cont.*

Algae Species	Active Extract/Compound	Biological Activity	Reference
Porphyra dioica *Palmaria palmate* *Chondrus crispus*	β-carotene fucoxanthin PUFA	Able to inhibit LPS-induced inflammatory pathways in human macrophages.	[157]
Caulerpa cupressoides	Sulfated polysaccharides	Decrease neutrophils migration. Strongly reduced the carrageenan-induced rat paw edema.	[158]
Dictyota menstrualis	Heterofucan	Binds to the surface of leucocytes and inhibits migration of leucocytes to the site of injury. Inhibit the chemical-induced leukocyte migration into the peritoneal cavity.	[159]
Dictyopteris prolifera *Grateloupia lanceolata* *Grateloupia filicina*	Ethanolic extract	Concentration-dependent reduction of LPS-induced prostaglandin E2 production. Suppresses the expression of inducible nitric oxide synthase (iNOS) and cyclooxygenase-2 (COX-2) at the protein level in RAW 264.7 cells. Reduced the release of tumor necrosis factor alpha (TNF-α) and interleukin 6 (IL-6) into the medium.	[160]
Gracilaria cornea	Sulfated polysaccharide fraction	Significantly inhibits rat paw edema induced by different inflammatory agents (carrageenan and dextran, histamine and L-arginin). Downregulates interleukin-1β (IL-1β), TNF-α, and COX-2 mRNA and protein levels.	[161]
Porphyra yezoensis	Dc-porphyran	Inhibits nitric oxide (NO) production in LPS-stimulated RAW 264.7 cells.	[162]
Pyropia yezoensis	Astaxanthin Xanthophyl	Anti-inflammatory action.	[163]
Lobophora variegata	Fucans	Inhibits the paw edema, plasma exudation, nitrite content, and leukocyte migration.	[164]
Red algae	Carrageenan, Fucoidan, Chondroitin	Lowered the expression of inducible nitric oxide synthase (iNOS). Inhibited the expressions of TNF-α, IL-1β, and interferon-c (IFN-c). Repressed pro-inflammatory cytokines and suppressed the activity of COX-2.	[41,165–167]
Fucus vesiculosus	Fucoidan	Anti-atopic dermatitis	[103,107]

4.8. Antifungal Effects

The antifungal effects of macroalgae were also assessed by many studies. Species such as *E. cava* were found to have potential as a novel antifungal agent against *T. rubrum* at an MIC of dieckol of 200 μM [168], while *Halimeda tuna* methanolic extracts were very effective against *Aspergillus niger*, *Aspergillus flavus*, *Aspergillus alternaria*, *Candida albicans*, and *Epidermophyton floccosum* [7]. Studies of Ertürk et al. [169] evaluated the antifungal activity of *Enteromorpha linza* and *Padina pavonica*, and it was found to be stronger than the standard antifungal activity (100 units of nystatin) against *Aspergillus niger* and *Candida albicans*. Instead, the acetone extract of *T. conoides* exerted a mild antifungal capability against *Aspergillus niger*, with an inhibition zone diameter close to 3 mm [170]. In a study performed on 45 chronic asthmatic patients with acute respiratory distress, Mickymaray et al. [171] observed the highest antifungal activity against *C. albicans* in the case of *L. paniculata*, followed by *U. prolifera*, *Cladophoropsis* sp., *A. specifera*, and *Tydemania* sp. ethanolic fraction extracts. The minimum fungicidal concentration and MIC values of the above algal ethanolic fractions ranged between 125 and 1000 μg/mL and 125 and 500 μg/mL, respectively [171].

Like most bioactivities of algae, the antifungal activity of seaweed extracts can be related to the presence of phenolic compounds, polyunsaturated fatty acids, and various terpenoids [172,173]. For example, bromophenols were successfully isolated from the red alga *Odonthalia corymbifera*, and others obtained by chemical transformation of bis(hydroxyphenyl)methanes with bromine were tested for antimicrobial and antifungal activities by Oh et al. [174]. Among the natural bromophenols, 2,2′,3,3′-tetrabromo-4,4′,5,5′-tetrahydroxydiphenylmethane isolated from *Odonthalia corymbifera* showed good antifungal activity against *C. albicans*, *A. fumigatus*, *T. rubrum*, and *T. mentagrophytes* while two of the synthesized compounds exhibited effective antibacterial activity against *S. aureus*, *B. subtilis*, *M. luteus*, *P. vulgaris*, and *S. typhimurium* [174].

4.9. Antiviral Effects

Although the research regarding the antiviral potential of algae against food-borne viruses is gaining interest in recent years, currently, the available data are still scarce. The main compounds from algae that have been proved to have antiviral potential are sulphated polysaccharides, including fucoidan, sulphoglycolipids, carrageenan, and sesquiterpene hydroquinone. Marine-derived polysaccharides and their lower molecular weight oligosaccharide derivatives have been shown to possess a variety of antiviral activities and also exert antioxidant and antimicrobial effects. In general, algal polysaccharides can suppress the DNA replication and inhibit the host cell colonization by the virus. For example, the antiviral potential of polysaccharides from brown seaweeds revealed a significant inhibiting activity against hepatitis B virus (HBV) DNA polymerase, therefore affecting its replication [175]. The antiviral activity of these polysaccharides is exerted through suppression of virus adhesion to the host cells (*U. pinnatifida*, *Cystoseira indica*, *Ascophylum nodosum*) [176]. Fucoindan extract from *Cladosiphon okamurans* was used to inhibit New Castle Disease Virus in vitro in early stages of viral infection (0–60 min post-infection), the compound displaying high selectivity index (IS50 > 2000) for inhibiting syncytia formation [177]. *Grateloupia indica*, *Scinaia hatei*, *Gracilaria* corticata, *Stoechospermum marginatum*, *Cystoseira indica*, and *Caulerpa racemosa* sulfated polysaccharide extracts were screened for antiviral activity against the four serotypes of dengue virus (DENV). DENV-2 was the most susceptible serotype to all polysulfates, with inhibitory concentration 50% values in the range 0.12–20 μg/mL [178]. Krylova et al. [179] found that modified and native fucoidans from marine macroalgae *Fucus evanescens* presented antiviral properties against herpes virus or human immunodeficiency virus. Eom et al. [180] studied the antiviral activity of phlorotannin from *Eisenia bicyclis*. The results showed a strong antiviral potential against norovirus (murine norovirus, MNV) with EC_{50} of 0.9 μM [180]. Serkedjieva [181] analyzed the influence of *Ceramium rubrum* water extract on the reproduction of a range of influenza viruses in vitro and in ovo. The results showed that the virus-inhibitory effect was selective, dose-dependent, and strain-specific. At a concentration over 0.5 mg/mL, the extract also inhibited the reproduction of herpes simplex virus (HSV) type 1 with MIC_{90} of 1.4 mg/mL.

However, there is still the need for more research to comprehensively understand the antiviral action mechanisms of algae compounds and to benefit from their use as functional ingredients in pharmaceutical and food industries [124].

5. Conclusions and Future Trends

Nowadays, macroalgae are gaining more interest due to their demonstrated health promoting properties. They can be seen as a valuable source of bioactive compounds that can sustain the human health, preventing or reducing the convalescence period for various diseases, due to their antioxidant, anti-inflammatory, antiproliferative, antiviral, and antibacterial activities.

Although algae have been extensively analyzed regarding their content in biologically active compounds, the potential beneficial and toxicological effects on the human body are still of major interest. The identification of compounds directly responsible for the antimicrobial, antiviral, and anticancer activities of algae is still a relatively incipient domain that must be elucidated. Furthermore, the research must be oriented on their use as a substitute for antibiotics through a viable and sustainable approach, this strategy representing progress in solving the major emerging problems related to antibiotic resistance.

A future valorization strategy can be sustained trough an integrated biorefinery concept developed based on cost-effective and environmentally-friendly extraction methods. In this context, more research is needed to evaluate nutritional properties and mechanisms underlying the health benefits of a wide variety of macroalgal products.

Author Contributions: Conceptualization, E.-S.B.-D. and A.C.F.; methodology, C.R.P. and A.M.R.; investigation, D.M.; writing—original draft preparation, E.-S.B.-D.; writing—review and editing, A.C.F., S.A.S., and O.L.P.; supervision, M.T.; project administration, A.C.F. and S.A.S. All authors have read and agreed to the published version of the manuscript.

Funding: The publication was supported by funds from the National Research Development Projects to finance excellence (PFE)-37/2018–2020 granted by the Romanian Ministry of Research and Innovation.

Acknowledgments: The work was supported by two grants of the Ministry of Research and Innovation, CNCS-UEFISCDI, project number PN-III-P1–1.1-PD-2016–0869 and PN-III-P1–1.1-TE-2016–0973, within PNCDI III.

Conflicts of Interest: The authors declare no conflict of interest.

References

1. Brodie, J.; Lewis, J. Introduction in unravelling the algae: The past, present, and future of algal systematics. In *Unravelling the Algae: The Past, Present, and Future of Algal Systematics*, 1st ed.; Brodie, J., Lewis, J., Eds.; CRC Press: Boca Raton, FL, USA, 2007.
2. Skrovankova, S. Seaweed vitamins as nutraceuticals, Toldra, F. Ed. *Adv. Food Nutr. Res.* **2011**, *64*, 357–369. [PubMed]
3. Mahadevan, K. Seaweeds: A sustainable food source. In *Seaweed Sustainability*, 1st ed.; Tiwari, B.K., Troy, D.J., Eds.; Academic Press: London, UK, 2015; pp. 347–364.
4. Bueno, A.M.; Marcílio Candido, T.; Morocho-Jácome, A.L.; Rolim Baby, A.; Robles Velasco, M.V.; M de Carvalho, J.C. Cosmetic attributes of algae–A review. *Algal Res.* **2017**, *25*, 483–487.
5. Garcia-Vaquero, M.; Ummat, V.; Tiwari, B.; Rajauria, G. Exploring ultrasound, microwave and ultrasound–microwave assisted extraction technologies to increase the extraction of bioactive compounds and antioxidants from brown macroalgae. *Mar. Drugs* **2020**, *18*, 172. [CrossRef]
6. Hayes, M. Seaweeds: A nutraceutical and health food. In *Seaweed Sustainability*, 1st ed.; Tiwari, B.K., Troy, D.J., Eds.; Academic Press: London, UK, 2015; pp. 365–387.
7. Indira, K.; Balakrishnan, S.; Srinivasan, M.; Bragadeeswaran, S.; Balasubramanian, T. Evaluation of in vitro antimicrobial property of seaweed (Halimeda tuna) from Tuticorin coast, Tamil Nadu, Southeast coast of India. *Afr. J. Biotechnol.* **2013**, *12*, 284–289.
8. Ganesan, A.; Tiwari, U.; Rajauria, G. Seaweed nutraceuticals and their role in disease prevention. *Food Sci. Hum. Wellness.* **2019**, *8*, 252–263. [CrossRef]
9. Biris-Dorhoi, E.S. *Cercetari Privind Utilizarea Algelor Marine in Bioremedierea Apelor Uzate Industrial–o Tehnica Eco-Friendly in Rezolvarea Poluarii Mediului*; Mega: Cluj-Napoca, Romania, 2018; pp. 66–117.
10. Negreanu-Pîrjol, B.; Negreanu-Pîrjol, T.; Paraschiv, G.-M.; Bratu, M.M.; Sirbu, R.; Roncea, F.; Meghea, A. Physical-chemical characterization of some green and red macrophyte algae from the Romanian Black Sea littoral. *Sci. Study Res. Chem. Chem. Eng.* **2011**, *12*, 173–184.
11. Boukhari, E.L.; Barakate, M.; Bouhia, Y.; Lyamlouli, K. Trends in seaweed extract based biostimulants: Manufacturing process and beneficial effect on soil-plant systems. *Plants* **2020**, *9*, 359. [CrossRef] [PubMed]
12. Barbot, Y.N.; Hashem, A.; Benz, R. A Review on the valorization of macroalgal wastes for biomethane production. *Mar. Drugs* **2016**, *14*, 120. [CrossRef]
13. Mæhre, K.H.; Malde, M.K.; Eilertsen, K.-E.; Elvevoll, E.O. Characterization of protein, lipid and mineral contents in common Norwegian seaweeds and evaluation of their potential as food and feed. *J. Sci. Food Agric.* **2014**, *94*, 3281–3290. [CrossRef]
14. Kazir, M.; Abuhassira, Y.; Robin, A.; Nahor, O.; Luo, J.; Israel, A.; Golberg, A.; Livney, Y.D. Extraction of proteins from two marine macroalgae, *Ulva* sp. and *Gracilaria* sp., for food application, and evaluating digestibility, amino acid composition and antioxidant properties of the protein concentrates. *Food Hydrocol.* **2019**, *87*, 197–203. [CrossRef]
15. Wells, M.L.; Potin, P.; Craigie, J.S.; Raven, J.A.; Merchant, S.S.; Helliwell, K.E.; Smith, A.G.; Camire, M.E.; Brawley, S.H. Algae as nutritional and functional food sources: Revisiting our understanding. *J. Appl. Phycol.* **2017**, *29*, 949–982. [CrossRef]
16. Hamid, N.; Ma, Q.; Boulom, S.; Liu, T.; Zheng, Z.; Balbas, J.; Robertson, J. Seaweed minor constituents. In *Seaweed Sustainability*, 1st ed.; Tiwari, B.K., Troy, D.J., Eds.; Academic Press: London, UK, 2015; pp. 193–242.

17. Holdt, S.L.; Kraan, S. Bioactive compounds in seaweed: Functional food applications and legislation. *J. Appl. Phycol.* **2011**, *23*, 543–597. [CrossRef]
18. Rizzo, G.; Laganà, A.S.; Chiara Rapisarda, A.M.; Grazia La Ferrera, G.M.; Buscema, M.; Rossetti, P.; Nigro, A.; Muscia, V.; Valenti, G.; Sapia, F.; et al. Vitamin B12 among vegetarians: Status, assessment and supplementation. *Nutrients* **2016**, *8*, 767. [CrossRef] [PubMed]
19. Rodrigues, D.; Freitas, A.C.; Pereira, L.; Rocha-Santos TA, P.; Vasconcelos, M.W.; Roriz, M.; Rodriguez-Alcala, L.M.; Gomes AM, P.; Duarte, A.C. Chemical composition of red, brown and green macroalgae from Buarcos bay in Central West Coast of Portugal. *Food Chem.* **2015**, *183*, 197–207. [CrossRef] [PubMed]
20. Fleurence, J.; Morançais, M.; Dumay, J. Seaweed proteins. In *Proteins in Food Processing*, 2nd ed.; Yada, R.Y., Ed.; Woodhead Publishing: Cambridge, UK, 2018; pp. 245–262.
21. Fleurence, J.; Gall, E.A. Antiallergic Properties. In *Seaweed in Health and Disease Prevention*, 1st ed.; Fleurence, J., Levine, I., Eds.; Academic Press: London, UK, 2016; pp. 389–420.
22. Dawczynski, C.; Schubert, R.; Jahreis, G. Amino acids, fatty acids, and dietary fibre in edible seaweed products. *Food Chem.* **2007**, *103*, 891–899. [CrossRef]
23. Biancarosa, I.; Belghit, I.; Bruckner, C.G.; Liland, N.S.; Waagbo, R.; Amlund, H.; Heesch, S.; Lock, E.J. Chemical characterization of 21 species of marine macroalgae common in Norwegian waters: Benefits of and limitations to their potential use in food and feed. *J. Sci. Food Agric.* **2018**, *98*, 2035–2042. [CrossRef] [PubMed]
24. Jeon, S.-M.; Kim, H.-J.; Woo, M.-N.; Lee, M.-K.; Chul Shin, Y.; Bok Park, Y.; Choi, M.-S. Fucoxanthin-rich seaweed extract suppresses body weight gain and improves lipid metabolism in high-fat-fed C57BL/6J mice. *Biotechnol. J.* **2010**, *5*, 961–969. [CrossRef]
25. Peng, Y.; Hu, J.; Yang, B.; Lin, X.-P.; Zhou, X.-F.; Yang, X.-W.; Liu, Y. Chemical composition of seaweeds. In *Seaweed Sustainability*; Tiwari, B.K., Troy, D.J., Eds.; Academic Press: London, UK, 2015; pp. 79–124.
26. Jayasinghe, G.D.T.M.; Jinadasa, B.K.K.K.; Chinthaka, S.D.M. Study on lipid content and fatty acid profile of four marine macro algae (Seaweeds) collected from South East Coast of Sri Lanka. *Asian J. Chem. Pharm. Sci.* **2018**, *1*, 1–6. [CrossRef]
27. Kellogg, J.; Esposito, D.; Grace, M.H.; Komarnytsky, S.; Lila, M.A. Alaskan seaweeds lower inflammation in RAW 264.7 macrophages and decrease lipid accumulation in 3T3-L1 adipocytes. *J. Func. Foods* **2015**, *15*, 396–407. [CrossRef]
28. Ma, A.-C.; Chen, Z.; Wang, T.; Song, N.; Yan, Q.; Fang, Y.-C.; Guan, H.-S.; Liu, H.-B. Isolation of the molecular species of monogalactosyldiacylglycerols from brown edible seaweed *Sargassum horneri* and their inhibitory effects on triglyceride accumulation in 3T3-L1 adipocytes. *J. Agric. Food Chem.* **2014**, *62*, 11157–11162. [CrossRef]
29. Tsai, C.-J.; Sun Pan, B.A. Identification of sulfoglycolipid bioactivities and characteristic fatty acids of marine macroalgae. *J. Agric. Food Chem.* **2012**, *60*, 8404–8410. [CrossRef] [PubMed]
30. Ramola, B.; Kumar, V.; Nanda, M.; Mishra, Y.; Tyagi, T.; Gupta, A.; Sharma, N. Evaluation, comparison of different solvent extraction, cell disruption methods and hydrothermal liquefaction of *Oedogonium macroalgae* for biofuel production. *Biotech. Rep.* **2019**, *22*, e00304. [CrossRef] [PubMed]
31. Margareta, W.; Nagarajan, D.; Changa, J.S.; Lee, D.J. Dark fermentative hydrogen production using macroalgae (*Ulva* sp.) as the renewable feedstock. *Appl. Energy* **2020**, *262*, 114547. [CrossRef]
32. Hofheins, J. An overview of macronutrients. In *Essentials of Sport Nutrition and Supplements*; Antonio, J., Kalman, D., Stout, J.R., Greenwood, M., Willoughby, D.S., Haff, G.G., Eds.; Humana Press: Totowa, NJ, USA, 2008; pp. 237–249.
33. Praveen, M.A.; Karthika Parvathy, K.R.; Balasubramanian, P.; Jayabalan, R. An overview of extraction and purification techniques of seaweed dietary fibers for immunomodulation on gut microbiota. *Trends Food Sci. Technol.* **2019**, *92*, 46–64. [CrossRef]
34. Patel, S.; Goyal, A. Functional oligosaccharides: Production, properties and applications. *World J. Microbiol. Biotechnol.* **2011**, *27*, 1119–1128. [CrossRef]
35. Fleita, D.; El-Sayed, M.; Rifaat, D. Evaluation of the antioxidant activity of enzymatically-hydrolyzed sulfated polysaccharides extracted from red algae *Pterocladia capillacea*. *LWT–Food Sci. Technol.* **2015**, *63*, 1236–1244. [CrossRef]

36. Xu, J.; Xu, L.-L.; Zhou, Q.-W.; Hao, S.-X.; Zhou, T.; Xie, H.-J. Isolation, purification, and antioxidant activities of degraded polysaccharides from *Enteromorpha prolifera*. *Int. J. Biol. Macromol.* **2015**, *81*, 1026–1030. [CrossRef]
37. Koutsaviti, A.; Ioannou, E.; Roussis, V. Bioactive seaweed substances. In *Bioactive Seaweeds for Food Applications*, 1st ed.; Quin, Y., Ed.; Academic Press: London, UK, 2018; pp. 25–52.
38. Wang, L.; Park, Y.; Jeon, Y.; Ryu, B. Bioactivities of the edible brown seaweed, *Undaria pinnatifida*: A review. *Aquaculture* **2018**, *495*, 873–880. [CrossRef]
39. Lafarga, T.; Acién-Fernándeza, F.G.; Garcia-Vaquero, M. Bioactive peptides and carbohydrates from seaweed for food applications: Natural occurrence, isolation, purification, and identification. *Algal. Res.* **2020**, *48*, 101909. [CrossRef]
40. Leandro, A.; Pacheco, D.; Cotas, J.; Marques, J.C.; Pereira, L.; Gonçalves, A. Seaweed's bioactive candidate compounds to food industry and global food security. *Life (Basel)* **2020**, *10*, 140. [CrossRef]
41. Ullah, S.; Khalil, A.A.; Shaukat, F.; Song, Y. Sources, extraction and biomedical properties of polysaccharides. *Foods* **2019**, *8*, 304. [CrossRef] [PubMed]
42. Venkatesan, J.; Sukumaran, A.; Kim, S.-K. Introduction to seaweed polysaccharides. In *Seaweed Polysaccharides*, 1st ed.; Venkatesan, J., Anil, S., Kim, S.K., Eds.; Elsevier: Amsterdam, The Netherlands, 2017; pp. 1–9.
43. Bouanati, T.; Colson, E.; Moins, S.; Cabrera, J.C.; Eeckhaut, I.; Raquez, J.M.; Gerbaux, P. Microwave-assisted depolymerization of carrageenans from *Kappaphycus alvarezii* and *Eucheuma spinosum*: Controlled and green production of oligosaccharides from the algae biomass. *Algal. Res.* **2020**, *51*, 102054. [CrossRef]
44. Vera, J.; Castro, J.; Gonzalez, A.; Moenne, A. Seaweed polysaccharides and derived oligosaccharides stimulate defense responses and protection against pathogens in plants. *Mar. Drugs* **2011**, *9*, 2514–2525. [CrossRef] [PubMed]
45. Ueno, M.M.N.; Siddiqui, R.; Watashi, K.; Wakita, T.; Kato, N.; Ikeda, M.; Okimura, T.; Isaka, S.; Oda, T.; Ariumi, Y. Acidic polysaccharides isolated from marine algae inhibit the early step of viral infection. *Int. J. Biol. Macromol.* **2019**, *124*, 282–290. [CrossRef]
46. Dumay, J.; Morançais, M. Proteins and Pigments. In *Seaweed in Health and Disease Prevention*, 1st ed.; Fleurence, J., Levine, I., Eds.; Academic Press: London, UK, 2016; pp. 275–318.
47. Koyande, A.K.; Chewa, K.W.; Rambabub, K.; Taoc, Y.; Chud, D.T.; Show, P.L. Microalgae: A potential alternative to health supplementation for humans. *Food Sci. Hum. Well.* **2019**, *8*, 16–24. [CrossRef]
48. Aryee AN, A.; Agyei, D.; Akanbi, T.O. Recovery and Utilization of Seaweed Pigments in Food Processing. *Curr. Opin. Food Sci.* **2018**, *19*, 113–119. [CrossRef]
49. Lordan, S.; Ross, R.P.; Stanton, C. Marine bioactives as functional food ingredients: Potential to reduce the incidence of chronic diseases. *Mar. Drugs* **2011**, *9*, 1056–1100. [CrossRef]
50. Pangestuti, R.; Kim, S.K. Carotenoids, bioactive metabolites derived from seaweeds. In *Springer Handbook of Marine Biotechnology*; Kim, S.K., Ed.; Springer: Berlin/Heidelberg, Germany, 2015; pp. 815–821.
51. Prasanna, K.I.; King, P.; Prasad, V.S.R.K. Adsorption of zinc from aqueous solution using marine green algae—*Ulva fasciata* sp. *Chem. Eng. J.* **2007**, *129*, 161–166. [CrossRef]
52. Pádua, D.; Rocha, E.; Gargiulo, D.; Ramos, A.A. Bioactive compounds from brown seaweeds: Phloroglucinol, fucoxanthin and fucoidan as promising therapeutic agents against breast cancer. *Phytochem. Lett.* **2015**, *14*, 91–98. [CrossRef]
53. Erdogan, A.; Demirel, Z.; Dalay, M.; Eroglu, A.E. Fucoxanthin content of *Cylindrotheca closterium* and its oxidative stress mediated enhancement. *Turk. J. Fish. Aquat. Sci.* **2016**, *16*, 491–498. [CrossRef]
54. Kim, K.N.; Heo, S.J.; Yoon, W.J.; Kang, S.M.; Ahn, G.; Yi, T.H.; Jeon, Y.J. Fucoxanthin inhibits the inflammatory response by suppressing the activation of NF-κB and MAPKs in lipopolysaccharide-induced RAW 264.7 macrophages. *Eur. J. Pharmacol.* **2010**, *649*, 369–375. [CrossRef] [PubMed]
55. Chung, T.W.; Choi, H.J.; Lee, J.Y.; Jeong, H.S.; Kim, C.H.; Joo, M.; Choi, J.Y.; Han, C.W.; Kim, S.Y.; Choi, J.S.; et al. Marine algal fucoxanthin inhibits the metastatic potential of cancer cells. *Biochem. Biophys. Res. Commun.* **2013**, *439*, 580–585. [CrossRef] [PubMed]
56. Martin, L.J. Fucoxanthin and its metabolite fucoxanthinol in cancer prevention and treatment. *Mar. Drugs* **2015**, *13*, 4784–4798. [CrossRef] [PubMed]
57. Hussain, E.; Wang, L.J.; Jiang, B.; Riaz, S.; Butt, G.Y.; Shi, D.Y. A review of the components of brown seaweed as potential candidates in cancer therapy. *RSC Adv.* **2016**, *6*, 12592–12610. [CrossRef]

58. Aki, T.; Yamamoto, M.; Takahashi, K.; Tomita, T.; Rieko, T.; Kazuhiro, I.; Seji, K.; Masashi, H.; Kazuo, M.; Kazuhisa, O. Regulation of polyunsaturated fatty acid biosynthesis by seaweed fucoxanthin and its metabolite in cultured hepatocytes. *Lipids* **2014**, *49*, 133–141. [CrossRef]
59. Pereira, D.M.; Valentão, P.; Andrade, P.B. Marine natural pigments: Chemistry, distribution and analysis. *Dye Pigment.* **2014**, *111*, 124–134. [CrossRef]
60. Murray, M.; Dordevic, A.L.; Ryan, L.; Bonham, M.P. An emerging trend in functional foods for the prevention of cardiovascular disease and diabetes: Marine algal polyphenols. *Crit. Rev. Food Sci. Nutr.* **2018**, *58*, 1342–1358. [CrossRef]
61. Gupta, S.; Cox, S.; Rajauria, G.; Abu-Ghannam, N.; Jaiswal, A.K. Growth inhibition of common food spoilage and pathogenic microorganisms in the presence of brown seaweed extracts. *Food Bioproc. Tech.* **2010**, *5*, 1–10. [CrossRef]
62. Senthilkumar, P.; Sudha, S. Antioxidant and antibacterial properties of methanolic extract of green seaweed *Chaetomorpha linum* from Gulf of Mannar: Southeast Coast of India. *Jundishapur J. Microbiol.* **2012**, *5*, 411–415. [CrossRef]
63. Cotas, J.; Leandro, A.; Monteiro, P.; Pacheco, D.; Figueirinha, A.; Gonçalves AM, M.; da Silva, G.J.; Pereira, L. Seaweed phenolics: From extraction to applications. *Mar. Drugs* **2020**, *18*, 384. [CrossRef]
64. Santos, S.; Félix, R.; Pais, A.; Rocha, S.M.; Silvestre, A. The quest for phenolic compounds from macroalgae: A review of extraction and identification methodologies. *Biomolecules* **2019**, *9*, 847. [CrossRef] [PubMed]
65. Ścieszka, S.; Klewicka, E. Algae in food: A general review, Critical Reviews. *Food Sci. Nutr.* **2019**, *59*, 3538–3547.
66. Cofrades, S.; Benedí, J.; Garcimartin, A.; Sánchez-Muniz, F.J.; Jimenez-Colmenero, F. A comprehensive approach to formulation of seaweed-enriched meat products: From technological development to assessment of healthy properties. *Food Res. Internat.* **2017**, *99*, 1084–1094. [CrossRef] [PubMed]
67. Garicano Vilar, E.; Ouyang, H.; O'Sullivan, M.G.; Kerry, J.P.; Hamill, R.M.; O'Grady, M.N.; Mohammed, H.O.; Kilcawley, K.N. Effect of salt reduction and inclusion of 1% edible seaweeds on the chemical, sensory and volatile component profile of reformulated frankfurter. *Meat Sci.* **2020**, *161*, 108001. [CrossRef] [PubMed]
68. Roohinejad, S.; Kouba, M.; Barba, F.J.; Saljoughian, S.; Amid, M.; Greiner, R. Application of seaweeds to develop new food products with enhanced shelf-life, quality and health-related beneficial properties. *Food Res. Internat.* **2017**, *99*, 1066–1083. [CrossRef]
69. Cofrades, S.; Serdaroglu, M.; Jiménez-Colmenero, F. Design of healthier foods and beverages containing whole algae. In *Functional Ingredients from Algae for Foods and Nutraceuticals*, 1st ed.; Dominguez, H., Ed.; Woodhead Publishing: Cambridge, UK, 2013; pp. 609–633.
70. Lalic, L.M.; Berkovic, K. The influence of algae addition on physicochemical properties of cottage cheese. *Milchwissenschaft* **2005**, *60*, 151–154.
71. Hopkins, D.L.; Clayton, E.H.; Lamb, T.A.; van de Ven, R.J.; Refshauge, G.; Kerr, M.J.; Bailes, K.; Lewandowski, P.; Ponnampalam, E.N. The impact of supplementing lambs with algae on growth, meat traits and oxidative status. *Meat Sci.* **2014**, *98*, 135–141. [CrossRef]
72. Ponnampalam, E.N.; Burnett, F.V.; Norng, S.; Hopkins, D.L.; Plozza, T.; Jacobs, J.L. Muscle antioxidant (vitamin E) and major fatty acid groups, lipid oxidation and retail colour of meat from lambs fed a roughage-based diet with flaxseed or algae. *Meat Sci.* **2015**, *111*, 154–160. [CrossRef]
73. Bortoluzzi, C.; Schmidt, J.M.; Bordignon, H.L.F.; Fülber, L.M.; Layter, J.R.; Fernandes, J.I.M. Efficacy of yeast derived glucomannan or algae-based antioxidant or both as feed additives to ameliorate mycotoxicosis in heat stressed and unstressed broiler chickens. *Livest. Sci.* **2016**, *193*, 20–25. [CrossRef]
74. Xiao, R.; Power, R.F.; Mallonee, D.; Crowdus, C.; Brennan, K.M.; Ao, T.; Pierce, J.L.; Dawson, K.A. A comparative transcriptomic study of vitamin E and an algae-based antioxidant as antioxidative agents: Investigation of replacing vitamin E with the algae-based antioxidant in broiler diets. *Poult. Sci.* **2011**, *90*, 136–146. [CrossRef]
75. Rico, D.; Martín Diana, A.B.; Milton-Laskibar, I.; Fernández-Quintela, A.; Silván, J.M.; Rai, D.K.; Choudhary, A.; Peñas, E.; de Luis, D.A.; Martínez-Villalueng, C. Characterization and in vitro evaluation of seaweed species as potential functional ingredients to ameliorate metabolic syndrome. *J. Funct. Foods* **2018**, *46*, 185–194. [CrossRef]
76. Charoensiddhi, S.; Abraham, R.E.; Su, P.; Zhang, W. Seaweed and seaweed-derived metabolites as prebiotics. *Adv. Food Nutr. Res.* **2020**, *91*, 97–156. [PubMed]

77. Wang, Y.; Han, F.; Hu, B.; Li, J.B.; Yu, W.G. In vivo prebiotic properties of alginate oligosaccharides prepared through enzymatic hydrolysis of alginate. *Nut. Res.* **2006**, *26*, 597–603. [CrossRef]
78. Kuda, T.; Yano, T.; Matsuda, N.; Nishizawa, M. Inhibitory effects of laminaran and low molecular alginate against the putrefactive compounds produced by intestinal microflora in vitro and in rats. *Food Chem.* **2005**, *91*, 745–749. [CrossRef]
79. Gudiel-Urbano, M.; Goni, I. Effect of edible seaweeds (*Undaria pinnatifida* and *Porphyra ternera*) on the metabolic activities of intestinal microflora in rats. *Nut. Res.* **2002**, *22*, 323–331. [CrossRef]
80. Baweja, P.; Kumar, S.; Sahoo, D.; Levine, I.A. Biology of Seaweeds. In *Seaweed in Health and Disease Prevention*, 1st ed.; Fleurence, J., Levine, I., Eds.; Academic Press: London, UK, 2016; pp. 41–106.
81. Zaharudin, N.; Salmeán, A.A.; Dragsted, L.O. Inhibitory effects of edible seaweeds, polyphenolics and alginates on the activities of porcine pancreatic α-amylase. *Food Chem.* **2018**, *245*, 1196–1203. [CrossRef]
82. Reboleira, J.; Freitas, R.; Pinteus, S.; Silva, J.; Alves, C.; Pedrosa, R.; Bernardino, S. Brown Seaweeds. In *Nonvitamin and Nonmineral Nutritional Supplements*; Nabavi, S.M., Silva, A.S., Eds.; Academic Press: London, UK, 2019; pp. 171–176.
83. Rabanal, M.; Ponce NM, A.; Navarro, D.A.; Gómez, R.M.; Stortz, C.A. The system of fucoidans from the brown seaweed *Dictyota dichotoma*: Chemical analysis and antiviral activity. *Carbohydr. Polym.* **2014**, *101*, 804–811. [CrossRef]
84. Shibneva, S.Y.; Skriptsova, A.V. Morphological variability of *Undaria pinnatifida* (Harvey) Suringar, 1873 (*Phaeophyceae, Laminariales*) in Peter the Great Bay, Sea of Japan. *Russ. J. Mar. Biol.* **2012**, *38*, 381–391. [CrossRef]
85. Guidone, M.; Thornber, C.; Wysor, B.; O'Kelly, C.J. Molecular and morphological diversity of Narragansett Bay (RI, USA) Ulva (Ulvales, Chlorophyta) populations. *J. Phycol.* **2013**, *49*, 979–995. [CrossRef]
86. Zhong, R.; Wan, X.; Wang, D.; Zhao, C.; Liu, D.; Gao, L.; Wang, M.; Wu, C.; Nabavid, S.M.; Daglia, M.; et al. Polysaccharides from marine *Enteromorph*a: Structure and function. *Trends Food Sci. Technol.* **2020**, *99*, 11–20. [CrossRef]
87. Wijesinghe, W.A.J.P.; Jeon, Y.J. Exploiting biological activities of brown seaweed *Ecklonia cava* for potential industrial applications: A review. *Int. J. Food Sci. Nutr.* **2012**, *63*, 225–235. [CrossRef] [PubMed]
88. Kang, K.A.; Lee, K.H.; Chae, S.; Zhang, R.; Jung, M.S.; Lee, Y.; Kim, S.Y.; Kim, H.S.; Joo, H.G.; Park, J.W.; et al. Eckol isolated from *Ecklonia cava* attenuates oxidative stress induced cell damage in lung fibroblast cells. *FEBS Lett.* **2005**, *579*, 6295–6304. [CrossRef] [PubMed]
89. Ahn, M.J.; Yoon, K.D.; Min, S.Y.; Lee, J.S.; Kim, J.H.; Kim, T.G.; Kim, S.H.; Kim, N.G.; Huh, H.; Kim, J. Inhibition of HIV-1 reverse transcriptase and protease by phlorotannins from the brown alga *Ecklonia cava*. *Biol. Pharm. Bull.* **2004**, *27*, 544–547. [CrossRef] [PubMed]
90. Déléris, P.; Nazih, H.; Bard, J.-M. Seaweeds in Human Health. In *Seaweed in Health and Disease Prevention*, 1st ed.; Fleurence, J., Levine, I., Eds.; Academic Press: London, UK, 2016; pp. 319–367.
91. Guh, D.P.; Zhang, W.; Bansback, N.; Amarsi, Z.; Birmingham, C.L.; Anis, A.H. The incidence of co-morbidities related to obesity and overweight: A systematic review and meta-analysis. *BMC Public Health* **2009**, *9*, 88. [CrossRef]
92. Unnikrishnan, M.K.; Veerapur, V.; Nayak, Y.; Paul, P.; Mathew, G. Antidiabetic, antihyperlipidemic and antioxidant effects of the flavonoids. In *Polyphenols in Human Death and Disease*, 1st ed.; Watson, R., Preedy, V., Zibadi, S., Eds.; Elsevier: Amsterdam, The Netherlands, 2014; pp. 143–161.
93. Souza, R.B.; Frota, A.F.; Sousa, R.S.; Cezario, N.A.; Santos, T.B.; Souza LM, F.; da Cunha, R.M.S. Neuroprotective effects of sulphated agaran from marine alga *Gracilaria cornea* in rat 6-hydroxydopamine Parkinson's disease model: Behavioural, neurochemical and transcriptional alterations. *Basic Clin. Pharmacol. Toxicol.* **2017**, *120*, 159–170. [CrossRef]
94. Djousse´, L.; Arnett, D.K.; Pankow, J.S.; Hopkins, P.N.; Province, M.A.; Ellison, R.C. Dietary linolenic acid is associated with a lower prevalence of hypertension in the NHLBI family heart study. *Hypertension* **2005**, *45*, 368–373. [CrossRef]
95. Cardoso, S.M.; Pereira, O.R.; Seca, A.M.L.; Pinto, D.C.G.A.; Silva, A.M.S. Seaweeds as preventive agents for cardiovascular diseases: From nutrients to functional foods. *Mar. Drugs* **2015**, *13*, 6838–6865. [CrossRef]
96. Yimin, Q. Health benefits of bioactive seaweed substances. In *Bioactive Seaweeds for Food Applications*, 1st ed.; Quin, Y., Ed.; Academic Press: London, UK, 2018; pp. 180–199.

97. Hall, A.C.; Fairclough, A.C.; Mahadevan, K.; Paxman, J.R. *Ascophyllum nodosum* enriched bread reduces subsequent energy intake with no effect on post-prandial glucose and cholesterol in healthy, overweight males. A pilot study. *Appetite* **2012**, *58*, 379–438. [CrossRef]
98. Odunsi, S.T.; Vázquez-Roque, M.I.; Camilleri, M.; Papathanasopoulos, A.; Clark, M.M.; Wodrich, L.; Lempke, M.; McKinzie, S.; Ryks, M.; Burton, D.; et al. Effect of alginate on satiation, appetite, gastric function, and selected gut satiety hormones in overweight and obesity. *Obesity (Silver Spring)* **2010**, *18*, 1579–1584. [CrossRef]
99. Sharifuddin, Y.; Chin, Y.-X.; Lim, P.-E.; Phang, S.-M. Potential bioactive compounds from seaweed for diabetes management. *Mar. Drugs* **2015**, *13*, 5447–5491. [CrossRef]
100. Umekawa, H.; Takuya, T.; Midori Hirayama, Y.W.; Hattori, M.; Araki, T. Effect of *Porphyra* extract on blood pressure in spontaneously hypertensive rats. *Nippon Shokuhin Kagaku Kogaku Kaishi* **2008**, *55*, 502–505. [CrossRef]
101. Qu, W.; Ma, H.; Pan, Z.; Luo, L.; Wang, Z.; He, R. Preparation and antihypertensive activity of peptides from *Porphyra yezoensis*. *Food Chem.* **2010**, *123*, 14–20. [CrossRef]
102. Jeong, Y.T.; Kim, Y.D.; Jung, Y.M.; Park, D.C.; Lee, D.S.; Ku, S.K.; Li, X.; Lu, Y.; Chao, G.H.; Kim, K.J.; et al. Low molecular weight fucoidan improves endoplasmic reticulum stress-reduced insulin sensitivity through AMP-activated protein kinase activation in L6 myotubes and restores lipid homeostasis in a mouse model of type 2 diabetes. *Mol. Pharmacol.* **2013**, *84*, 147–157. [CrossRef] [PubMed]
103. Lin, H.V.; Tsou, Y.C.; Chen, Y.T.; Lu, W.J.; Hwang, P.A. Effects of low-molecular-weight fucoidan and high stability fucoxanthin on glucose homeostasis, lipid metabolism, and liver function in a mouse model of type II diabetes. *Mar. Drugs* **2017**, *15*, 113. [CrossRef]
104. Kim, J.H.; Kim, J.; Park, Y. Trans-10, cis-12 conjugated linoleic acid enhances endurance capacity by increasing fatty acid oxidation and reducing glycogen utilization in mice. *Lipids* **2012**, *47*, 855–863. [CrossRef]
105. Ustyuzhanina, N.E.; Bilan, M.I.; Gerbst, A.G.; Ushakova, N.A.; Tsvetkova, E.A.; Dmitrenok, A.S.; Usov, A.I.; Nifantiev, N.E. Anticoagulant and antithrombotic activities of modified xylofucan sulfate from the brown alga *Punctaria plantaginea*. *Carbohydr. Polym.* **2016**, *136*, 826–833. [CrossRef]
106. Fujimura, T.; Tsukahara, K.; Moriwaki, S.; Kitahara, T.; Sano, T.; Takema, Y. Treatment of human skin with an extract of *Fucus vesiculosus* changes its thickness and mechanical properties. *J. Cosmet. Sci.* **2002**, *53*, 1–9.
107. Tian, T.; Chang, H.; He, K.; Ni, Y.; Li, C.; Hou, M.; Chen, L.; Xu, Z.; Chen, B.; Ji, M. Fucoidan from seaweed *Fucus vesiculosus* inhibits 2, 4-dinitrochlorobenzene-induced atopic dermatitis. *Int. Immunopharmacol.* **2019**, *75*, 105823. [CrossRef]
108. Wang HM, D.; Chen, C.C.; Huynh, P.; Chang, J.S. Exploring the potential of using algae in cosmetics. *Biores. Tech.* **2015**, *184*, 355–362. [CrossRef]
109. Kidd, P. Astaxanthin, cell membrane nutrient with diverse clinical benefits and anti-aging potential. *Altern. Med. Rev.* **2011**, *16*, 355–364.
110. Barbosa, M.; Valentão, P.; Andrade, P.B. Bioactive compounds from macroalgae in the new millennium: Implications for neurodegenerative diseases. *Mar. Drugs* **2014**, *12*, 4934–4972. [CrossRef] [PubMed]
111. Miyake, Y.; Tanaka, K.; Okubo, H.; Sasaki, S.; Arakawa, M. Seaweed consumption and prevalence of depressive symptoms during pregnancy in Japan: Baseline data from the Kyushu Okinawa Maternal and Child Health Study. *BMC Pregnancy Childbirth* **2014**, *14*, 301. [CrossRef] [PubMed]
112. Shao, J.T.; Wang, M.Y.; Zheng, L.B. Antifatigue effect of *Gracilaria eucheumoides* in mice. *Exp. Ther. Med.* **2013**, *6*, 1512–1516. [CrossRef] [PubMed]
113. Parsaeimehr, A.; Lutzu, G.A. Algae as a Novel Source of Antimicrobial Compounds: Current and Future Perspectives. In *Antibiotic Resistance*, 1st ed.; Kon, K., Rai, M., Eds.; Academic Press: London, UK, 2016; pp. 377–413.
114. Shanmughapriya, S.; Manilal, A.; Sujith, S.; Selvin, J.; Kiran, G.S.; Natarajaseenivasan, K. Antimicrobial activity of seaweeds extracts against multiresistant pathogens. *Ann. Microbiol.* **2008**, *58*, 535–541. [CrossRef]
115. Prarthana, J.; Maruthi, K.R. Fresh water algae as a potential source of bioactive compounds for aquaculture and significance of solvent system in extraction of antimicrobials. *Asian J. Sci. Res.* **2019**, *12*, 18–28. [CrossRef]
116. Tüney, I.; Çadirci, B.H.; Ünal, D.; Sukatar, A. Antimicrobial activities of the extracts of marine algae from the Coast of Urla (Izmir, Turkey). *Turk. J. Biol.* **2006**, *30*, 171–175.
117. Bhuyar, P.; Rahim, M.H.; Sundararaju, S.; Maniam, G.P.; Govindan, N. Antioxidant and antibacterial activity of red seaweed; *Kappaphycus alvarezii* against pathogenic bacteria. *Glob. J. Environ. Sci. Manag.* **2020**, *6*, 47–58.

118. Lim, C.W.; Leem J., S.; Cho, Y.J. Structures and some properties of the antimicrobial compounds in the red alga, *Symphyocladia latiuscula*. *Kor. J. Fish Aquatic. Sci.* **2000**, *33*, 280–287.
119. Cox, S.; Abu-Ghannam, N.; Gupta, S. An assessment of the antioxidant and antimicrobial activity of six species of edible Irish seaweeds. *Int. Food Res. J.* **2010**, *17*, 205–220.
120. Sims, J.J.; Donnell, M.S.; Leary, J.V.; Lacy, G.H. Antimicrobial agents from marine algae. *Antimicrob. Agents Chemother.* **1975**, *7*, 320–321. [CrossRef]
121. Gamal, A.A.E. Biological importance of marine algae. *Saudi Pharm. J.* **2010**, *18*, 1–25. [CrossRef] [PubMed]
122. Al-Saif, S.S.; Abdel-Raouf, N.; El-Wazanani, H.A.; Aref, I.A. Antibacterial substances from marine algae isolated from Jeddah coast of Red sea, Saudi Arabia. *Saudi J. Biol. Sci.* **2014**, *21*, 57–64. [CrossRef] [PubMed]
123. Wahidi, M.E.; Amraoui, B.E.; Amraoui, M.E.; Bamhaoud, T. Screening of antimicrobial activity of macroalgae extracts from the Moroccan Atlantic coast. *Ann. Pharm. Fr.* **2015**, *73*, 190–196. [CrossRef]
124. Pina-Pérez, M.C.; Rivas, A.; Martínez, A.; Rodrigo, D. Antimicrobial potential of macro and microalgae against pathogenic and spoilage microorganisms in food. *Food Chem.* **2017**, *235*, 34–44. [CrossRef] [PubMed]
125. Sanmukh, S.G.; Benedict, B.; Udhaya, R.; Krishna, K.; Sandhya, S.; Waman, N.P. Bioactive compounds derived from microalgae showing antimicrobial activities. *Aquaculture* **2014**, *5*, 100224. [CrossRef]
126. Danyal, A.; Mubeen, U.; Malik, K.A. Investigating two native algal species to determine antibiotic susceptibility against some pathogens. *Curr. Res. J. Biol. Sci.* **2013**, *5*, 70–74. [CrossRef]
127. Lee, J.H. Antibacterial effect of marine algae against oral-borne patogens. *Res. J. Med. Plants* **2014**, *8*, 196–203.
128. Jang, K.; Lee, J. Investigation of anti-bacterial activity against food-borne pathogens among korean domestic algae. *Adv. J. Food Sci. Technol.* **2015**, *7*, 490–495. [CrossRef]
129. Nshimiyumukiza, O.; Kang, S.K.; Kim, H.J.; Mog, Y.; Lee, E.H.; Han, H.N.; Kim, Y.; Kim, D.H.; Kim, J.H.; Eom, S.H. Synergistic Antibacterial activity of *Ecklonia cava* (Phaeophyceae: Laminariales) against *Listeria monocytogenes* (Bacillales: Listeriaceae). *Fish Aquac. Sci.* **2015**, *18*, 1–6. [CrossRef]
130. Lee, S.Y.; Kim BW, R.; Lim, S.I.; Ahn, D.H. Antibacterial mechanism of *Myagropsis myagroides* extract on *Listeria monocytogenes*. *Food Control* **2014**, *42*, 23–28. [CrossRef]
131. Rajauria, G. Optimization and validation of reverse phase HPLC method forqualitative and quantitative assessment of polyphenols in seaweed. *J. Pharmaceut. Biomed. Anal.* **2018**, *148*, 230–237. [CrossRef] [PubMed]
132. Choi, J.G.; Kang, O.H.; Brice, O.O.; Lee, Y.S.; Chae, H.S.; Oh, Y.C.; Sohn, D.H.; Park, H.; Choi, H.G.; Kim, S.G.; et al. Antibacterial activity of *Ecklonia cava* against methicillin-resistant *Staphylococcus aureus* and *Salmonella* spp. *Foodborne Pathog. Dis.* **2010**, *7*, 435–441. [CrossRef] [PubMed]
133. Eom, S.H.; Kim, Y.M.; Kim, S.-K. Antimicrobial effect of phlorotannins from marine brown algae. *Food Chem. Toxicol.* **2012**, *50*, 3251–3255. [CrossRef] [PubMed]
134. Negreanu -Pirjol, T.; Negreanu -Pirjol, B.; Sirbu, R.; Paraschiva, G.M.; Meghea, A. Comparative studies regarding the antioxidative activity of some therapeutic marine algae species along the Romanian Black Sea Coast. *J. Environ. Prot. Ecol.* **2012**, *13*, 1744–1750.
135. Ngo, D.H.; Kim, S.K. Marine bioactive peptides as potential antioxidantsm. *Curr. Protein Pept. Sci.* **2013**, *14*, 189–198. [CrossRef]
136. Pimentel, F.B.; Alves, R.C.; Harnedy, P.A.; FitzGerald, R.J.; Oliveira, M.B.P.P. Macroalgal-derived protein hydrolysates and bioactive peptides: Enzymatic release and potential health enhancing properties. *Trends Food Sci. Technol.* **2019**, *93*, 106–124. [CrossRef]
137. Ishihara, K.; Oyamada, C.; Matsushima, R.; Murata, M.; Muraoka, T. Inhibitory effect of porphyran, prepared from dried "Nori", on contact hypersensitivity in mice. *Biosci. Biotechnol. Biochem.* **2005**, *69*, 1824–1830. [CrossRef]
138. Sugiura, Y.; Matusda, K.; Yamada, Y. Anti-allergic phlorotannins from the edible alga, *Eisenia arborea*. *Food Sci. Technol. Res.* **2007**, *13*, 54–60. [CrossRef]
139. Vo, T.; Ngo, D.; Kim, S. Marine algae as a potential pharmaceutical source for anti-allergic therapeutics. *Proc. Biochem.* **2012**, *47*, 386–394. [CrossRef]
140. Van Ginneken, V.J.; Helsper, J.P.; de Visser, W.; van Keulen, H.; Brandenburg, W.A. Polyunsaturated fatty acids in various macroalgal species from North Atlantic and tropical seas. *Lipids Health Dis.* **2011**, *10*, 104. [CrossRef]
141. Sakthivel, R.; Pandima Devi, K. Antioxidant, anti-inflammatory and anticancer potential of natural bioactive compounds from seaweeds. In *Studies in Natural Products Chemistry*, 1st ed.; Rahman, A., Ed.; Elsevier: Amsterdam, The Netherlands, 2019; Volume 63, pp. 113–154.

142. Gamal-Eldeen, A.M.; Ahmed, E.F.; Abo-Zeid, M.A. In vitro cancer chemopreventive properties of polysaccharide extract from the brown alga, *Sargassum latifolium*. *Food Chem. Toxicol.* **2009**, *47*, 378–1384. [CrossRef] [PubMed]
143. Ryu, M.J.; Kim, A.D.; Kang, K.A.; Chung, H.S.; Kim, H.S.; Suh, I.S.; Chang, W.Y.; Hyun, J.W. The green algae *Ulva fasciata* Delile extract induces apoptotic cell death in human colon cancer cells. *In Vitro Cell Dev. Biol. Anim.* **2013**, *49*, 74–81. [CrossRef] [PubMed]
144. Kang, Y.; Wang, Z.J.; Xie, D.; Sun, X.; Yang, W.; Zhao, X.; Xu, N. Characterization and potential antitumor activity of polysaccharide from *Gracilariopsis lemaneiformis*. *Mar. Drugs* **2017**, *15*, 100. [CrossRef] [PubMed]
145. Wang, H.; Chiu, L.; Ooi, V.; Ang, P. A potent antitumor polysaccharide from the edible brown seaweed *Hydroclathrus clathratus*. *Bot. Mar.* **2010**, *53*, 265–274. [CrossRef]
146. Chale-Dzul, J.; Freile-Pelegrín, Y.; Robledo, D.; Moo-Puc, R. Protective effect of fucoidans from tropical seaweeds against oxidative stress in HepG2 cells. *J. Appl. Phycol.* **2017**, *29*, 2229–2238. [CrossRef]
147. Palanisamy, S.; Vinosha, M.; Marudhupandi, T.; Rajasekar, P.; Prabhu, N.M. Isolation of fucoidan from *Sargassum polycystum* brown algae: Structural characterization, in vitro antioxidant and anticancer activity. *Int. J. Biol. Macromol.* **2017**, *102*, 405–412. [CrossRef]
148. Menshova, R.V.; Shevchenko, N.M.; Imbs, T.I.; Zvyagintseva, T.N.; Malyarenko, O.S.; Zaporoshets, T.S.; Ermakova, S.P. Fucoidans from brown alga *Fucus evanescens*: Structure and biological activity. *Front. Mar. Sci.* **2016**, *3*, 129. [CrossRef]
149. Namvar, F.; Mohamed, S.; Fard, S.G.; Behravan, J.; Mustapha, N.M.; Alitheen NB, M.; Othman, F. Polyphenol-rich seaweed (*Eucheuma cottonii*) extract suppresses breast tumour via hormone modulation and apoptosis induction. *Food Chem.* **2012**, *130*, 376–382. [CrossRef]
150. Reddy, P.; Urban, S. Meroditerpenoids from the southern australian marine brown alga *Sargassum fallax*. *Phytochem.* **2009**, *70*, 250–255. [CrossRef]
151. Sornsiri, J.; Srisook, K.; Pornngam, P.; Sootanan, P. Prediction of biochemical mechanism of anti-inflammation explained from two marine-derived bioactive compounds. *Agric. Nat. Res.* **2018**, *52*, 588–595. [CrossRef]
152. Kazlowska, K.; Hsu, T.; Hou, C.C.; Yang, W.C.; Tsai, G.J. Antiinflammatory properties of phenolic compounds and crude extract from *Porphyra* dentate. *J. Ethnopharmacol.* **2010**, *128*, 123–130. [CrossRef] [PubMed]
153. Bitencourt MA, O.; Dantas, G.R.; Lira, D.P.; Filho JM, B.; Miranda GE, C.; Santos BV, O.; Souto, J.T. Aqueous and methanolic extracts of *Caulerpa mexicana* suppress cell migration and ear edema induced by inflammatory agents. *Mar. Drugs* **2011**, *9*, 1332–1345. [CrossRef] [PubMed]
154. Heo, S.J.; Yoon, W.Y.; Kim, K.N.; Ahn, G.N.; Kang, S.M.; Kang, D.H.; Affan, A.; Oh, C.; KyoJung, W.; JinJeon, Y. Evaluation of anti-inflammatory effect of fucoxanthin isolated from brown algae in lipopolysaccharide-stimulated RAW 264.7 macrophages. *Food Chem. Toxicol.* **2010**, *48*, 2045–2051. [CrossRef] [PubMed]
155. Hong, D.D.; Hien, H.M.; Anh, H.T. Studies on the analgesic and anti-inflammatory activities of *Sargassum swartzii* (Turner), *C. Agardh* (Phaeophyta) and *Ulva reticulata Forsskal* (Chlorophyta) in experiment animal models. *Afr. J. Biotechnol.* **2011**, *10*, 2308–2314.
156. Croci, D.O.; Cumashi, A.; Ushakova, N.A.; Preobrazhenskaya, M.E.; Piccoli, A.; Totani, L.; Ustyuzhanina, N.E.; Bilan, M.I.; Usov, A.I.; Grachev, A.A. Fucans, but not fucomannoglucuronas, determine the biological activities of sulfated polysaccharides from *Laminaria saccharina* brown seaweed. *PLoS ONE* **2011**, *6*, e17283. [CrossRef]
157. Robertson, R.C.; Guihéneuf, F.; Bahar, B.; Schmid, M.; Stengel, D.B.; Fitzgerald, G.F.; Ross, R.P.; Stanton, C. The anti-inflammatory effect of algae-derived lipid extracts on lipopolysaccharide (LPS)-stimulated human THP-1 macrophages. *Mar. Drugs* **2015**, *13*, 5402–5424. [CrossRef]
158. Rodrigues JA, G.; Oliveira Vanderlei ED, S.; Silva, L.M.; de Araujo, I.W.; de Queiroz, I.N.; de Paula, G.A.; Abreu, T.M.; Ribeiro, N.A.; Bezerra, M.M.; Chaves, H.V. Antinociceptive and anti-inflammatory activities of a sulfated polysaccharide isolated from the green seaweed *Caulerpa cupressoides*. *Pharmacol. Rep.* **2012**, *64*, 282–292. [CrossRef]
159. Albuquerque, I.R.L.; Cordeiro, S.L.; Gomes, D.L.; Dreyfuss, J.L.; Filgueira, L.G.A.; Leite, E.L.; Nader, H.B.; Rocha, H.A.O. Evaluation of anti-nociceptive and anti-inflammatory activities of a heterofucan from *Dyctiota menstrualis*. *Mar. Drugs* **2013**, *11*, 2722–2740. [CrossRef]
160. Yang, Y.I.; Jung, S.H.; Lee, K.T.; Choi, J.H. Dieckol isolated from edible brown algae Ecklonia cava induces apoptosis of ovarian cancer cells and inhibits tumor xenografr growth. *J. Cancer Res. Clin. Oncol.* **2015**, *68*, 141–255.

161. Coura, C.O.; Souza, R.B.; Rodrigues, J.A.; Vanderlei Ede, S.; de Arau´ jo, I.W.; Ribeiro, N.A.; Frota, A.F.; Ribeiro, K.A.; Chaves, H.V. Mechanisms involved in the anti-inflammatory action of a poly sulfated fraction from Gracilaria cornea in rats. *PLoS ONE* **2015**, *10*, e0119319. [CrossRef]
162. Isaka, S.; Cho, K.; Nakazono, S.; Abu, R.; Ueno, M.; Kim, D.; Oda, T. Antioxidant and anti-inflammatory activities of porphyran isolated from discolored nori (*Porphyra yezoensis*). *Int. J. Biol. Macromol.* **2015**, *74*, 68–75. [CrossRef] [PubMed]
163. Ha, Y.; Lee, W.H.; Jeong, J.; Park, M.; Ko, J.Y.; Kwon, O.W.; Lee, J.; Kim, Y.J. *Pyropia yezoensis* extract suppresses IFN-Gamma- and TNF-Alpha-induced proinflammatory chemokine production in HaCaT cells via the down-regulation of NF-κB. *Nutrients* **2020**, *12*, 1238. [CrossRef] [PubMed]
164. Siqueira, R.C.L.; da Silva, M.S.J.; de Alencar, D.B.; de Pires, F.A.; de Alencar, N.M.; Pereira, M.G.; Cavada, B.S.; Sampaio, A.H.; Farias, W.R.L.; Assreuy, A.M.S. In vivo anti-inflammatory effect of a sulfated polysaccharide isolated from the marine brown algae *Lobophora variegate*. *Pharm. Biol.* **2011**, *49*, 167–174. [CrossRef] [PubMed]
165. Agyemang, K.; Han, L.; Liu, E.; Zhang, Y.; Wang, T.; Gao, X. Recent advances in *Astragalus membranaceus* anti-diabetic research: Pharmacological effects of its phytochemical constituents. *Evid. Based Complement. Altern. Med.* **2013**, *654643*, 1–9. [CrossRef]
166. Volpi, N. Anti-inflammatory activity of chondroitin sulphate: New functions from an old natural macromolecule. *Inflammopharmacology* **2011**, *19*, 299–306. [CrossRef]
167. Wijesekara, I.; Pangestuti, R.; Kim, S.K. Biological activities and potential health benefits of sulfated polysaccharides derived from marine algae. *Carbohydr. Polym.* **2011**, *84*, 14–21. [CrossRef]
168. Lee, M.H.; Lee, K.B.; Oh, S.M. Antifungal activities of dieckol isolated from the marine brown alga *Ecklonia cava* against *Trichophyton rubrum*. *J Korean Soc. Appl. Biol. Chem.* **2010**, *53*, 504–507. [CrossRef]
169. Ertürk, O.; Taş, B. Antibacterial and antifungal effects of some marine algae. *Kafkas Univ. Vet. Fak. Derg.* **2011**, *17*, 121–124.
170. Manivannan, K.; Karthikai, D.G.; Anantharaman, P.; Balasubramanian, T. Antimicrobial potential of selected brown seaweeds from Vedalai coastal waters, Gulf of Mannar. *Asian Pac. J. Trop. Biomed.* **2011**, *1*, 114–120. [CrossRef]
171. Mickymaray, S.; Alturaiki, W. Antifungal efficacy of marine macroalgae against fungal isolates from bronchial asthmatic cases. *Molecules* **2018**, *23*, 3032. [CrossRef]
172. Hamed, S.M.; Abd El-Rhman, A.A.; Abdel-Raouf, N.; Ibraheem, I.B.M. Role of marine macroalgae in plant protection & improvement for sustainable agriculture technology review. *Beni-Suef Univ. J. Basic Appl. Sci.* **2018**, *7*, 104–110.
173. Shobier, A.H.; Abdel Ghani, S.A.; Barakat, K.M. GC/MS spectroscopic approach and antifungal potential of bioactive extracts produced by marine macroalgae. *Egypt J. Aqua. Res.* **2016**, *42*, 289–299. [CrossRef]
174. Oh, K.; Lee, J.H.; Chung, S.C.; Shin, J.; Shin, H.J.; Kimd, H.K.; Lee, H.S. Antimicrobial activities of the bromophenols from the red alga *Odonthalia corymbifera* and some synthetic derivatives. *Bioorg. Med. Chem. Lett.* **2008**, *18*, 104–108. [CrossRef]
175. Wang, W.; Wang, S.X.; Guan, H.-S. The antiviral activities and mechanisms of marine polysaccharides: An overview. *Mar. Drugs* **2012**, *10*, 2795–2816. [CrossRef] [PubMed]
176. Ahmadi, A.; Zorofchian, M.S.; Zandi, A.Z. Antiviral potential of algae polysaccharides isolated from marine sources: A review. *Biomed. Res. Int.* **2015**, 825203. [CrossRef]
177. Elizondo-Gonzalez, R.; Cruz-Suarez, E.; Ricque-Marie, D.; Mendoza-Gamboa, E.; Rodriguez-Padilla, C.; Trejo-Avila, L.M. In vitro characterization of the antiviral activity of fucoidan from *Cladosiphon okamuranus* against Newcastle Disease Virus. *Virol. J.* **2012**, *9*, 307. [CrossRef]
178. Pujol, C.A.; Ray, S.; Ray, B.; Damonte, E.B. Antiviral activity against dengue virus of diverse classes of algal sulfated polysaccharides. *Int. J. Biol. Macromol.* **2012**, *51*, 412–416. [CrossRef]
179. Krylova, N.V.; Ermakova, S.P.; Lavrov, V.F.; Leneva, I.A.; Kompanets, G.G.; Iunikhina, O.V.; Nosik, M.N.; Ebralidze, L.K.; Falynskova, I.N.; Silchenko, A.S.; et al. The comparative analysis of antiviral activity of native and modified fucoidans from brown algae *Fucus evanescens* in vitro and in vivo. *Mar. Drugs* **2020**, *18*, 224. [CrossRef]

180. Eom, S.H.; Moon, S.Y.; Lee, D.S.; Kim, H.J.; Park, K.; Lee, E.W.; Kim, Y.M. *In vitro* antiviral activity of dieckol and phlorofucofuroeckol-A isolated from edible brown alga *Eisenia bicyclis* against murine norovirus. *Algae* **2015**, *30*, 241–246. [CrossRef]
181. Serkedjieva, J. Antiviral activity of the red marine alga *Ceramium rubrum*. *Phytother. Res.* **2004**, *18*, 480–483. [CrossRef]

nutrients

MDPI

Article

Effects of *Euglena gracilis* Intake on Mood and Autonomic Activity under Mental Workload, and Subjective Sleep Quality: A Randomized, Double-Blind, Placebo-Controlled Trial

Ayaka Nakashima [1,*], Kosuke Yasuda [1], Ako Murata [1], Kengo Suzuki [1] and Naoki Miura [2]

1 Euglena Co. Ltd., Tokyo 108-0014, Japan; kosuke.yasuda@euglena.jp (K.Y.); ako.murata@euglena.jp (A.M.); suzuki@euglena.jp (K.S.)

2 Miura Clinic, Medical Corporation Kanonkai, Osaka 530-0044, Japan; miura@miura-cl.jp

* Correspondence: nakashima@euglena.jp; Tel.: +81-3-5442-4907

Received: 23 August 2020; Accepted: 20 October 2020; Published: 23 October 2020

Abstract: While the human body maintains homeostasis by altering the balance in the autonomic nervous, endocrine, and immune systems, a prolonged imbalance in these systems can result in physical and mental symptoms, including a decline in sleep quality and work efficiency. *Euglena gracilis* (*Euglena*) is a single-celled microalga with the properties of both plants and animals and contains abundant nutrients, such as vitamins, minerals, amino acids, and fatty acids, which have various beneficial health effects. This study evaluated the effects of *Euglena* intake on the mood states and stress coping under mental workload tasks, and subjective sleep quality. We assigned men and women aged 20 to 64 years to *Euglena* and placebo intake groups, and measured indices related to the autonomic nervous system, psychological states, and sleep quality together with the application of workload stress before food intake, and 4, 8, and 12 weeks after commencing intake. *Euglena* intake regulated the autonomic nervous system under a workload and improved psychological parameters and sleep conditions. These results indicate that the consumption of *Euglena* may regulate the balance of the autonomic nervous system during stress and may have a favorable effect on psychological status and sleep quality.

Keywords: *Euglena gracilis*; paramylon; autonomic nervous system

1. Introduction

Stress is a general term for the defense response to external stimuli, such as pain, cold, and infection, and can include mental tension and worry [1]. Recently, there has been considerable interest in the effects of psychological stress on the body, particularly in light of the discovery of a neural pathway that transmits stress signals from specific areas of the cerebral cortex, which processes psychological stress and emotions, to the hypothalamus, which controls the sympathetic nervous system [2]. The human body is constantly exposed to stress and works to maintain a constant internal environment via the control of the autonomic nervous, endocrine, and immune systems for protection [3]. However, excessive stress weakens the body's stress resistance, resulting in physical and psychological symptoms, including a deterioration in the quality of sleep and reduced work efficiency [4].

The autonomic nervous system, which includes the sympathetic and parasympathetic nervous systems, plays several important roles in the regulation of biological functions, such as respiration/circulation, digestion/absorption, secretion, and metabolism in a coordinated manner to maintain homeostasis [5]. For example, an imbalance in the autonomic nervous system may prevent the activation of the sympathetic nervous system, even during activities when it is typically activated, or the activation of the parasympathetic nervous system, even during recovery and rest. These conditions lead to an inability to concentrate and various ailments, such as poor sleep quality, reduced immunity, poor bowel movements, as well as various diseases [5]. Eliminating the underlying stress is an effective approach to balancing the autonomic nervous system. However, adjustments in lifestyle and nutrition are also effective.

Euglena gracilis (*Euglena*) is a single-celled microalga with plant and animal properties. *Euglena* consists of various nutrients, such as vitamins, minerals, amino acids, and fatty acids, and are used as nutritional and general supplements [6]. Previous studies have confirmed that *Euglena* intake effectively suppresses elevated blood glucose levels [7], suppresses fat accumulation [8,9], attenuates lifestyle-related disease symptoms [10], promotes immune function [11,12], and improves bowel movements [13]. Autonomic nervous system activity plays an important role in maintaining homeostasis and is involved in biological processes such as lipolysis, blood glucose regulation, immune function, and digestion. For example, the autonomic nervous system of the pancreas contributes to the regulation of blood glucose levels by regulating insulin secretion [14]; in white adipose tissue the autonomic nervous system is involved in lipolysis [14,15]; in the spleen, it is involved in immunity [16,17], and in the stomach and intestines it is involved in the promotion of peristaltic movements [18]. These findings led us to hypothesize that various beneficial health effects of *Euglena* intake are mediated by its regulation of the autonomic nervous system balance.

In this study, we investigated the effects of the ingestion of *Euglena*-containing food for 12 weeks on the autonomic nervous system, psychological factors, and quality of sleep in men and women aged 20 to 64 years suffering from decreased motivation and a decline in sleep quality.

2. Materials and Methods

2.1. Materials

Euglena gracilis powder was obtained from euglena Co. Ltd. (Tokyo, Japan). The placebo food was the same formula but without *Euglena* powder.

2.2. Ingestion Study

2.2.1. Target Population

The study was conducted between July to November 2019. Subjects were healthy men and women between the ages of 20 and 64 years who were presented with reduced motivation based on the vitality (VT) measured by 36-Item Short-Form Health Survey (SF-36) and poor quality of sleep based on the Pittsburgh Sleep Quality Index (PSQI) total score. The purpose and content of the study were thoroughly explained to the subjects both verbally and in writing, and the subjects consented to participation by providing written informed consent. This study was conducted in accordance with the tenets of the Declaration of Helsinki and the Ethical Guidelines for Medical and Health Research Involving Human Subjects. Protection of the human rights of the subjects was always ensured, and the study was implemented under the supervision of a doctor with the approval of the ethics review committee of the Miura Clinic (approval number: R1816, approval date: 27 June 2019). This trial is registered with UMIN ID: 000037726.

2.2.2. Test Method

This study was designed and implemented as a double-blind placebo-controlled trial. Subjects were selected after screening based on the level of motivation and quality of sleep. After baseline (week 0) measurements were obtained before the first dose, and the subjects received 500, 1000, or 3000 mg of *Euglena* powder or placebo powder (starch), divided into two doses per day (after breakfast and after dinner) for 12 weeks. The same measurements were taken at 4, 8, and 12 weeks after commencing intake. Subjects were instructed not to eat or drink after 10:00 p.m. on the day prior to the test and not to consume anything other than water until the designated meal was consumed on test day. On the examination day, subjects consumed the designated meal (200 g of rice) at least 4 h before they visited the hospital and had only water until the end of the examination. On the day measurements were taken, the subjects underwent a subjective index survey using Visual Analog Scale (VAS), Profile of Mood States 2nd Edition (POMS2), SF-36, PSQI, Ogri-Shirakawa-Azumi Sleep Inventory MA Version (OSA-MS), autonomic system response indicator measurements, saliva collection, blood collection, and physical measurements such as height, weight and body mass index (BMI). After resting for 30 min, the subjects were subjected to the Uchida–Kraepelin [19–22] test for workload stress. The test took 30 min, including 15 min for the first half and 15 min for the second half with a break in the middle, while changing each line of simple single-digit addition once a minute. The correct answer rate for 15 min was calculated. The autonomic system response indicator measurements, saliva collection, and subjective index survey were repeated immediately after the Uchida–Kraepelin test. After resting for 60 min, the autonomic system response indicator measurements, saliva collection, and subjective index survey were performed again. During the test, the subjects drank water (about 100 mL) between the second subjective index survey and before starting the 60 min rest period.

Autonomic System Response Indicator Measurements

The balance of the autonomic nervous system, an indicator of fatigue and stress, was evaluated at three points for each test, i.e., before Uchida–Kraepelin stress loading, immediately after Uchida–Kraepelin stress loading, and 60 min after Uchida–Kraepelin stress loading. The tests were conducted before intake and at 4, 8, and 12 weeks after starting intake. The vital monitor VM-302 (Fatigue Science Laboratory Inc., Osaka, Japan) was used to simultaneously measure the pulse wave (PPG) and cardiac wave (ECG) signals, and to analyze high-frequency (HF) and low-frequency (LF) components of heart rate fluctuation [21]. HF reflects the parasympathetic response and LF reflects the sympathetic response.

Measurements of Health-Related Quality of Life through the 36-Item Short-Form Health Survey (SF-36)

The health of the subjects was surveyed with a quality of life questionnaire before intake and at 4, 8, and 12 weeks after starting intake using an SF-36 v2 [23–25], based on universally applicable health-related concepts. A five-point scale was used for responses to each of the 36 questions, and the questions included eight items; physical functioning (PF), role physical (RP), bodily pain (BP), general health (GH), vitality (VT), social functioning (SF), role emotional (RE), and mental health (MH).

Measurements of Mood States Using the Profile of Mood States 2nd Edition (POMS2)

Six items, anxiety and tension, hostility and anger, depression and discouragement, vitality and activity, fatigue and lethargy, and confusion and perplexity, were evaluated based on the T score using the Japanese version of the POMS2 [22,26,27] at three points for each test, before Uchida–Kraepelin stress loading, after Uchida–Kraepelin stress loading, and 60 min after Uchida–Kraepelin stress loading. Measurements were obtained before intake and at 4, 8, and 12 weeks after starting intake.

Visual Analog Scale (VAS)

The 100-mm VAS was used to quantify fatigue, irritability, concentration, mood, motivation, tension, and satisfaction with sleep for 1 week at three points for each test, before Uchida–Kraepelin stress loading, after Uchida–Kraepelin stress loading, and 60 min after Uchida–Kraepelin stress loading, before intake and at 4, 8, and 12 weeks after starting intake. Each VAS question had an endpoint of 0 indicating a "positive situation" and an endpoint of 100 indicating a "negative situation." Subjects were asked to mark the position on the line that represented their mood at the time. The distance from the endpoint of 0 to the mark was measured in mm.

Measurements of Quality of Sleep Using the Ogri-Shirakawa-Azumi Sleep Inventory MA Version (OSA-MA)

The sleep profile of the subjects was confirmed using OSA-MA [28,29], which includes five items related to sleep: sleepiness on rising, initiation and maintenance of sleep, frequent dreaming, refreshing, and sleep length. Items were assessed before intake and at 4, 8, and 12 weeks after starting intake. As for the score polarity, the direction of the good feeling of sleep had the highest score.

Pittsburgh Sleep Quality Index (PSQI)

The Japanese versionof the PSQI (PSQI-J) [29,30] was used to evaluate sleep quality before intake and at 4, 8, and 12 weeks after starting intake. The questions refer to the subject's sleep profile in the past month and can be assigned to seven components: subjective sleep quality, sleep latency, sleep duration, sleep efficiency, sleep disturbance, use of sleep medication, and daytime dysfunction. These seven categories are assigned scores from 0 to 3, and a higher total score (0 to 21) indicates a worse sleep quality.

Saliva and Blood Samples

Saliva samples were collected to measure cortisol, a stress hormone, at three points for each test, before Uchida–Kraepelin stress loading, after Uchida–Kraepelin stress loading, and 60 min after Uchida–Kraepelin stress loading, before intake and at 4, 8, and 12 weeks after starting intake. A blood test was also performed at these times to confirm the safety of ingesting the test substance before intake and at 12 weeks.

2.2.3. Statistical Analyses

The baseline characteristics of the participants were analyzed using one-way analysis of variance (ANOVA). After accounting for normality, two-way repeated-measures ANOVAs were performed since the autonomic nervous system response indices, salivary cortisol, POMS2, SF-36, OSA, and Uchida–Kraepelin test (correct answer rate) were normally distributed. Sessions (weeks 0, 4, 8, and 12) were within-subject independent variables and the study group (placebo and *Euglena* intake groups) were between-group independent variables. If a significant main effect of group, or a session x group, the interaction was found, we then followed up with post hoc comparisons using Dunnett's test. Kruskal–Wallis test was performed for VAS and PSQI since no normal distribution was found. If the Kruskal–Wallis test was significant, the Steel's test was then performed. Dunnett's test or Steel's test was used to compare the measured values at each measurement time point and/or the degree of change from week 0 to after test among the four groups. Numerical values are shown as means ± standard deviation (SD). Significance levels were set as follows: <5% in two-sided tests indicated a significant difference. In addition, < 10% in the two-tailed test showed a significant trend. R version 3.5.1 (R development Core Team), and JMP (ver. 12) were used for statistical analyses.

3. Results

3.1. Subjects

In total, 122 participants were enrolled as the analysis set. After screening, 80 individuals were selected for the study. As three subjects withdrew during the trial, data for 77 subjects were analyzed (Figure 1).

Baseline data for subjects are shown in Table 1. There were no significant differences in parameters measured at baseline between the placebo and *Euglena* intake groups.

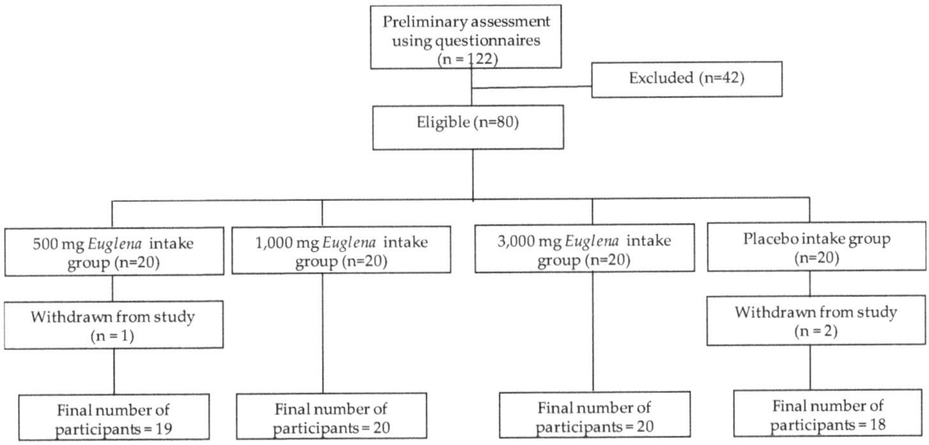

Figure 1. Flow chart for subject participation.

3.2. *Safety Evaluation*

Four complaints of symptoms were received during the study period, including fatigue, constipation, headache, and stiffness in the shoulder, although the symptoms were present before study participation and, although the cause could not be identified, the symptoms were mild and thus were judged by the investigator to be unproblematic. No abnormal values were found for vital signs such as blood pressure and pulse, nor in the blood test results. Subjects' blood pressure, pulse rate, and blood test results are summarized in Supplementary Tables S1 and S2. Although some items significantly differed between weeks, most average values remained within the range of the reference value. No adverse events or other associated issues were observed.

3.3. *Efficacy Evaluation*

3.3.1. Autonomic System Response Indicator Measurements

The low-frequency/high-frequency (LF/HF) ratio is summarized in Table 2. LF and HF, which represent the balance of the sympathetic nervous system and the parasympathetic nervous system in the autonomic nervous system, were measured at three points for each test (before Uchida–Kraepelin stress loading, after Uchida–Kraepelin stress loading, and 60 min after Uchida–Kraepelin stress loading) before intake and at 4, 8, and 12 weeks after starting intake. The LF/HF ratio was 3.7 in the placebo intake group, 3.3 in the *Euglena* 500 mg intake group, 4.1 in the 1000 mg intake group, and 4.7 in the *Euglena* 3000 mg intake group after Uchida–Kraepelin stress loading at week 0. The LF/HF ratio was 6.3 in the placebo intake group, 4.7 in the *Euglena* 500 mg intake group, 2.9 in the 1000 mg intake group, and 5.1 in the *Euglena* 3000 mg intake group after Uchida–Kraepelin stress loading at week 4 of intake. These findings suggest that the continuous intake of *Euglena* reduced disruptions in the balance of the autonomic nervous system. However, there were no significant differences.

The mean change amount of the LF/HF ratio before and after Uchida–Kraepelin loading at week 0 was 0.7. Thus, we defined subjects with a change amount greater than 0.7 as being more prone to workload stress based on Uchida–Kraepelin (Table 3). The LF/HF ratio in the *Euglena* 1000 mg intake group after Uchida–Kraepelin stress loading at week 4 of intake was significantly lower than that in the placebo intake group ($p = 0.024$, Dunnett's test). This suggests that the intake of *Euglena* 1000 mg regulates the balance of the autonomic nervous system during work stress. No difference was found in the balance of the autonomic nervous system in subjects without Uchida–Kraepelin stress. After week 8 of intake, these differences were difficult to discern.

Table 1. Baseline data for subjects. After accounting for normality, the baseline characteristics of the participants were analyzed by one-way ANOVA or Kruskal–Wallis test.

									SF-36	LH/HF			PSQI
Group	***n***	**Male (*n*)**	**Female (*n*)**		**Age (Years)**	**Height**	**Weight**	**BMI**	**VT**	**Before**	**After**	**Δ After**	**Total Score**
Placebo	18	8	10	Mean	47.3	164.2	58.9	21.7	35.8	3.1	3.7	0.5	7.7
				SD	9.5	8.0	8.5	1.7	12.1	2.5	3.5	2.9	2.6
Euglena 500 mg	19	9	10	Mean	47.8	165.2	60.1	22.0	39.8	3.1	3.3	0.3	8.1
				SD	11.5	6.4	9.8	2.8	14.6	2.2	2.5	2.2	3.1
Euglena 1000 mg	20	9	11	Mean	46.0	164.6	59.9	22.0	33.2	3.1	4.1	1.0	7.4
				SD	10.7	10.8	10.9	2.3	12.4	2.4	6.6	5.8	2.4
Euglena 3000 mg	20	9	11	Mean	48.0	162.8	59.9	22.5	34.1	3.9	4.7	0.8	6.8
				SD	9.7	8.4	10.6	2.8	11.7	5.0	3.5	3.8	2.2
p-value (one-way ANOVA)					0.931	0.838	0.980	0.804	0.381	0.822	0.775	0.931	-
p-value (Kruskal–Wallis test)					-	-	-	-	-	-	-	-	0.556

BMI: body mass index, SF-36: 36-Item Short-Form Health Survey, VT: vitality, LF/HF: low-frequency/high-frequency, PSQI: Pittsburgh Sleep Quality Index.

Table 2. Summary of LF/HF ratios, an indicator of autonomic nervous system balance. Low frequency (LF) and high frequency (HF) indicate sympathetic and parasympathetic nervous, respectively. After accounting for normality, a two-way repeated measure ANOVA was performed. The results were not significantly different.

			LF/HF 0 Weeks			LF/HF 4 Weeks			LF/HF 8 Weeks			LF/HF 12 Weeks		
Group	n		**Before**	**After**	**60 min after**	**Before**	**After**	**60 min after**	**Before**	**After**	**60 min after**	**Before**	**After**	**60 min after**
Placebo	18	Mean	3.1	3.7	2.7	4.0	6.3	4.8	3.7	4.3	4.3	3.1	3.3	2.7
		SD	2.5	3.5	2.6	5.2	6.3	6.8	3.7	3.4	3.3	2.3	2.7	2.6
Euglena 500 mg	19	Mean	3.1	3.3	2.8	4.4	4.7	3.9	3.2	3.4	2.5	3.6	3.9	2.9
		SD	2.2	2.5	2.0	5.2	4.3	4.1	3.3	2.8	1.9	3.0	3.4	2.7
Euglena 1000 mg	20	Mean	3.1	4.1	3.4	3.5	2.9	2.1	2.6	3.3	3.1	2.5	3.1	2.7
		SD	2.4	6.6	5.2	3.3	2.2	1.2	3.3	2.7	4.1	2.7	2.4	2.8
Euglena 3000 mg	20	Mean	3.9	4.7	4.5	3.7	5.1	3.5	3.7	3.9	3.9	4.0	4.1	2.7
		SD	5.0	3.5	5.3	3.7	5.3	2.7	3.2	3.2	4.5	3.4	2.9	1.8

LF/HF: low-frequency/high-frequency.

Table 3. Summary of LF/HF ratios, an indicator of autonomic nervous system balance (Subjects defined as susceptible to stress). After accounting for normality, a two-way repeated measure ANOVA was performed. The results showed a session x group interaction in "After" ($p = 0.036$) and were followed up with a post-hoc comparison using Dunnett's test. Dunnett's test was used for intergroup comparisons of the means of the placebo group and *Euglena* intake groups before and after Uchida–Kraepelin stress loading, and 60 min after loading at 0, 4, 8, and 12 weeks.

			LF/HF 0 Weeks			LF/HF 4 Weeks			LF/HF 8 Weeks			LF/HF 12 Weeks		
Group	n		**Before**	**After**	**60 min after**	**Before**	**After**	**60 min after**	**Before**	**After**	**60 min after**	**Before**	**After**	**60 min after**
Placebo	8	Mean	3.1	6.0	4.2	4.0	10.3	7.0	5.4	5.4	5.5	3.1	4.2	3.2
		SD	2.5	4.1	3.3	3.4	7.8	9.7	4.9	4.3	3.1	2.9	3.6	3.8
Euglena 500 mg	8	Mean	2.5	4.6	2.7	4.5	5.0	4.4	2.7	4.9	2.3	3.4	4.0	3.5
		SD	1.8	2.4	2.0	5.2	4.8	4.8	3.3	3.5	1.1	2.6	3.3	3.1
Euglena 1000 mg	8	Mean	3.1	8.1	5.9	4.8	2.6 *	2.6	3.8	4.9	5.7	2.8	2.5	3.5
		SD	2.1	9.2	7.8	4.4	1.1	1.2	4.7	3.4	5.6	3.3	1.6	3.9
Euglena 3000 mg	10	Mean	3.3	6.5	6.6	5.2	7.4	4.9	4.7	5.4	5.9	3.8	4.6	3.3
		SD	2.3	3.1	6.2	4.6	6.2	3.1	3.5	3.7	5.7	2.4	2.8	2.0

* $p < 0.05$ vs. the placebo intake group. LF/HF: low-frequency/high-frequency.

3.3.2. Analysis of SF-36

Table 4 summarizes the SF-36 results. The role physical (RP) values were significantly higher at weeks 8 and 12 of treatment with *Euglena* 3000 mg intake group (53.5 ± 5.2 and 55.4 ± 3.5) than in the placebo intake group (47.3 ± 8.4 and 49.2 ± 8.2), suggesting that intake of *Euglena* improves RP (p = 0.036, p = 0.021, Dunnett's test). At week 12 of intake, vitality (VT) tended to be higher in the 3000 mg *Euglena* intake group (51.2 ± 7.2) than in the placebo group (45.3 ± 5.5), indicating that the continuous intake of *Euglena* may also improve vitality (p = 0.065, Dunnett's test).

Table 4. Analyses of role physical (RP) and vitality (VT) based on SF-36. After accounting for normality, a two-way repeated measure ANOVA was performed. The results showed a main effect of group in RP (p = 0.023) and a session x group interaction in VT (p = 0.052). Therefore, we followed up with post-hoc comparisons using Dunnett's test. Dunnett's test was used for intergroup comparisons of mean values in the placebo intake group and *Euglena* intake group at 0, 4, 8, and 12 weeks.

	Group	n		0 Weeks	4 Weeks	8 Weeks	12 Weeks
RP	Placebo	18	Mean	44.2	48.5	47.3	49.2
			SD	10.0	8.7	8.4	8.2
	Euglena 500 mg	19	Mean	43.9	48.7	48.7	50.2
			SD	10.3	8.8	6.7	7.8
	Euglena 1000 mg	20	Mean	44.3	44.1	47.5	50.2
			SD	9.6	10.4	9.5	7.6
	Euglena 3000 mg	20	Mean	46.9	54.1	53.5 *	55.4 *
			SD	8.3	3.9	5.2	3.5
VT	Placebo	18	Mean	39.9	46.8	44.3	45.3
			SD	5.9	6.5	8.7	5.4
	Euglena 500 mg	19	Mean	41.8	46.2	47.8	49.0
			SD	7.1	9.4	10.9	9.4
	Euglena 1000 mg	20	Mean	38.6	42.7	43.9	45.1
			SD	6.0	8.7	10.1	9.1
	Euglena 3000 mg	20	Mean	39.0	48.6	49.7	51.2 †
			SD	5.7	7.3	6.2	7.2

* $p < 0.05$, † $p < 0.1$ vs. the placebo intake group. SF-36: 36-Item Short-Form Health Survey, RP: role physical, VT: vitality.

3.3.3. POMS Analysis

Table 5 shows the results of a POMS analysis. Results from the intake of 500 mg of *Euglena* indicate that the scores for friendliness were significantly higher than those of the placebo intake group before, immediately after, and 60 min after Uchida–Kraepelin stress loading at week 4 (p = 0.009, p = 0.007, and p = 0.0006, respectively, Dunnett's test); before, immediately after, and 60 min after Uchida–Kraepelin stress loading at week 8 (p = 0.004, p = 0.062, and p = 0.040, respectively, Dunnett's test); and immediately after and 60 min after Uchida–Kraepelin stress loading at week 12 (p = 0.007, and p = 0.003, respectively, Dunnett's test), respectively.

3.3.4. VAS Analysis

Tables 6 and 7 show the VAS results. The amount of change in tension from before Uchida–Kraepelin stress loading to 60 min after Uchida–Kraepelin stress loading in week 8 was 0.86 ± 1.67 in the placebo intake group, −0.41 ± 1.61 in the 500 mg *Euglena* intake group, and −0.29 ± 1.43 in the 3000 mg *Euglena* intake group, demonstrating a reduction in tension (p = 0.097, p = 0.099, Steel's test). The degree of change in tension from before Uchida–Kraepelin stress loading to 60 min after Uchida–Kraepelin stress loading in week 12 was 0.91 ± 1.77 in the placebo intake group and −0.74 ± 1.51 in the 1000 mg *Euglena* intake group, indicating a significant easing of tension (p = 0.023, Steel's test).

Table 5. Analyses of friendliness as determined by POMS2. After accounting for normality, a two-way repeated measure ANOVA was performed. The results showed a session x group interaction in "Before," "After," and "60 min after" ($p = 0.075$, $p = 0.009$, and $p = 0.012$, respectively) and a main effect of group in "Before," "After," and "60 min after" ($p = 0.058$, $p = 0.055$, and $p = 0.018$, respectively). Therefore, we followed up with post-hoc comparisons using Dunnett's test. Dunnett's test was used for intergroup comparisons of the mean values in the placebo intake group and *Euglena* intake groups before and after Uchida-Kraepelin stress loading and 60 min after loading at 0, 4, 8, and 12 weeks.

			0 Weeks			4 Weeks			8 Weeks			12 Weeks		
Group	***n***		**Before**	**After**	**60 min after**	**Before**	**After**	**60 min after**	**Before**	**After**	**60 min after**	**Before**	**After**	**60 min after**
Placebo	18	Mean	44.6	41.4	39.3	44.0	39.3	36.9	41.1	39.4	39.2	44.1	37.7	37.2
		SD	11.4	9.9	10.9	9.6	10.3	9.5	11.1	11.1	11.0	10.1	9.7	8.9
Euglena 500 mg	19	Mean	48.5	44.4	45.3	54.2 **	51.8 **	52.5 ***	52.9 **	48.2 †	48.9 *	50.8	49.7 **	50.6 **
		SD	11.7	11.1	13.3	12.1	13.0	13.6	13.0	14.5	15.0	12.9	15.1	15.9
Euglena 1000 mg	20	Mean	45.7	41.4	41.9	45.5	42.3	42.4	45.6	40.8	40.3	44.9	42.1	41.0
		SD	8.0	9.3	8.6	8.9	11.9	10.3	6.1	8.1	8.2	8.3	9.1	9.8
Euglena 3000 mg	20	Mean	48.1	44.3	43.3	46.9	42.4	42.9	47.6	43.5	41.9	44.9	42.9	41.8
		SD	10.4	11.0	12.2	9.8	13.2	14.3	11.7	11.6	12.1	12.5	12.0	11.8

*** $p < 0.001$, ** $p < 0.01$, * $p < 0.05$, † $p < 0.1$ vs. the placebo group. POMS2: Profile of Mood States 2nd Edition.

Table 6. Analyses of tension and irritability measured by the VAS. After accounting for normality, Kruskal–Wallis test was performed. If the Kruskal–Wallis test was significant, the Steel's test was then performed. Steel's test was used for intergroup comparison of the mean change from before Uchida–Kraepelin stress loading to immediately after, and 60 min after stress loading at 0, 4, 8, and 12 weeks in the placebo and *Euglena* intake groups.

				0 Weeks			4 Weeks			8 Weeks			12 Weeks		
	Group	*n*		Before	After	60 min after	Before	After	60 min after	Before	After	60 min after	Before	After	60 min after
Tension	Placebo	18	Mean	5.14	4.84	4.82	4.21	4.29	4.18	3.88	4.77	4.74	4.11	5.58	5.02
			SD	1.54	2.26	2.33	1.80	2.09	2.10	1.61	1.92	1.94	1.42	1.46	1.89
			Mean (Amount of change)	-	−0.30	−0.33	-	0.09	−0.02	-	0.89	0.86	-	1.47	0.91
			SD (Amount of change)	-	1.78	1.66	-	1.55	1.99	-	1.47	1.67	-	1.86	1.77
	Euglena 500 mg	19	Mean	3.74	3.39	2.68 *	3.30	3.09	2.77	2.63	2.59 **	2.22 **	2.81	3.32 *	2.71 **
			SD	1.79	2.12	1.80	2.27	2.36	1.96	1.95	1.85	1.77	1.95	2.16	2.11
			Mean (Amount of change)	-	−0.35	−1.05	-	−0.21	−0.53	-	−0.04	−0.41 †	-	0.51	−0.09
			SD (Amount of change)	-	1.66	1.94	-	1.56	1.31	-	1.72	1.61	-	1.67	1.94
	Euglena 1000 mg	20	Mean	4.84	4.58	3.84	3.74	4.21	3.56	3.20	3.92	3.33 †	3.94	3.69 *	3.20 *
			SD	1.50	1.89	1.78	1.51	1.41	1.49	1.50	1.95	1.70	1.65	1.81	1.65
			Mean (Amount of change)	-	−0.26	−1.00	-	0.47	−0.18	-	0.72	0.13	-	−0.25	−0.74 *
			SD (Amount of change)	-	1.96	1.87	-	1.65	1.38	-	1.29	1.08	-	1.92	1.52

Table 6. *Cont.*

Group	*n*		0 Weeks Before	0 Weeks After	0 Weeks 60 min after	4 Weeks Before	4 Weeks After	4 Weeks 60 min after	8 Weeks Before	8 Weeks After	8 Weeks 60 min after	12 Weeks Before	12 Weeks After	12 Weeks 60 min after
Euglena 3000 mg	20	Mean	4.61	4.29	3.59	3.23	3.60	2.82 †	3.06	3.37 †	2.78 **	3.15	3.38 **	2.72 **
		SD	1.33	2.21	2.42	1.66	1.89	1.75	1.85	1.90	1.54	1.81	2.14	2.17
		Mean (Amount of change)	-	−0.32	−1.02	-	0.37	−0.41	-	0.31	−0.29 †	-	0.23	−0.43
		SD (Amount of change)	-	1.42	2.13	-	2.08	2.50	-	1.57	1.43	-	2.12	2.14
p-value (Kruskal–Wallis test)			0.0663 †	0.1442	0.032 *	0.4204	0.2222	0.0975 †	0.1562	0.0102 *	0.0011 **	0.0988 †	0.0049 **	0.0038 **
p-value (Kruskal–Wallis test/Amount of change)				0.9468	0.6333		0.4346	0.6664		0.1564	0.0906 †		0.14	0.0682 †
Irritability Placebo	18	Mean	6.09	5.04	5.41	4.64	4.62	4.63	4.23	5.33	5.32	4.08	5.88	5.38
		SD	1.49	2.64	2.49	2.33	2.36	2.33	2.02	1.93	1.89	1.57	1.62	2.09
		Mean (Amount of change)	-	−1.06	−0.69	-	−0.02	−0.01	-	1.11	1.09	-	1.79	1.29
		SD (Amount of change)	-	2.99	2.91	-	2.67	2.39	-	1.99	1.92	-	2.00	2.02
Euglena 500 mg	19	Mean	4.60	3.36	3.23 †	2.85 †	3.05	2.72 *	2.61	2.97 *	2.29 **	2.56 †	3.33 *	2.97 *
		SD	2.45	2.56	3.02	2.50	2.56	2.46	2.15	2.56	2.30	2.03	2.87	2.55
		Mean (Amount of change)	-	−1.24	−1.37	-	0.21	−0.13	-	0.36	−0.32	-	0.76	0.41
		SD (Amount of change)	-	3.13	3.49	-	2.46	2.08	-	2.19	2.25	-	1.78	1.99
Euglena 1000 mg	20	Mean	5.54	5.70	4.69	3.61	5.04	4.82	3.58	4.65	4.11	4.21	4.48	3.67 †
		SD	1.97	2.52	2.52	2.42	2.53	2.63	2.37	2.52	2.51	2.49	2.65	2.40
		Mean (Amount of change)	-	0.16	−0.86	-	1.43	1.21	-	1.07	0.54	-	0.27 †	−0.54 **
		SD (Amount of change)	-	1.68	1.53	-	2.15	2.39	-	2.16	2.07	-	1.59	1.30
Euglena 3000 mg	20	Mean	4.67	4.49	3.62 †	3.21	4.03	3.25	3.44	4.11	3.12 **	2.83 †	4.35	3.42 *
		SD	2.68	2.53	2.46	2.25	2.40	2.02	2.75	2.47	1.98	2.14	2.45	2.45
		Mean (Amount of change)	-	−0.18	−1.05	-	0.82	0.04	-	0.68	−0.32	-	1.52	0.59
		SD (Amount of change)	-	3.27	2.61	-	2.62	2.47	-	2.92	2.34	-	2.87	2.67
p-value (Kruskal–Wallis test)			0.2245	0.0641 †	0.0649 †	0.1172	0.0827 †	0.0155 *	0.2593	0.0335 *	0.0007 ***	0.0317 *	0.0352 *	0.0199 *
p-value (Kruskal–Wallis test/Amount of change)			-	0.3628	0.8083	-	0.3075	0.3036	-	0.6082	0.3015	-	0.0927 †	0.0323 *

*** $p < 0.001$, ** $p < 0.01$, * $p < 0.05$, † $p < 0.1$ vs. the placebo intake group. VAS: visual Analog Scale.

Table 7. Analyses of the level of subjective satisfaction with sleep measured by the VAS. After accounting for normality, Kruskal–Wallis test was performed. If the Kruskal–Wallis test was significant, the Steel's test was then performed. Steel's test was used for intergroup comparison of the mean change at 0, 4, 8, and 12 weeks in the placebo and *Euglena* intake groups.

	Group	*n*		0 Weeks	4 Weeks	8 Weeks	12 Weeks	Δ 4 Weeks	Δ 8 Weeks	Δ 12 Weeks
Sleep	Placebo	18	Mean	6.93	6.04	5.84	6.28	−0.89	−1.09	−0.65
			SD	2.02	2.38	2.22	1.95	1.44	1.51	1.22
	Euglena 500 mg	19	Mean	6.77	5.73	4.74	4.55 †	−1.04	−1.87	−2.22 *
			SD	2.49	2.89	2.62	2.66	1.43	1.69	3.08
	Euglena 1000 mg	20	Mean	7.35	5.50	5.47	4.66 †	−1.85	−1.88	−2.69 *
			SD	1.19	2.34	2.22	2.47	1.85	2.29	2.31
	Euglena 3000 mg	20	Mean	7.65	5.04	4.98	4.24 *	−2.61 **	−2.67	−3.41 ***
			SD	1.50	2.52	2.50	2.37	1.75	1.93	2.26
p-value (Kruskal–Wallis test)				0.6878	0.5877	0.5464	0.0492 *	0.0040 **	0.1027	0.0014 **

*** $p < 0.001$, ** $p < 0.01$, * $p < 0.05$, † $p < 0.1$ vs. the placebo intake group. VAS: visual Analog Scale.

The amount of change in irritability from before Uchida–Kraepelin stress loading to immediately after Uchida–Kraepelin stress loading in week 12 was 1.79 ± 2.00 in the placebo intake group and 0.27 ± 1.59 in the 1000 mg *Euglena* intake group (p = 0.057 Steel's test). The amount of change in irritability from before Uchida–Kraepelin stress loading to 60 min after Uchida–Kraepelin stress loading in week 12 was 1.29 ± 2.02 in the placebo intake group and −0.54 ± 1.30 in the 1000 mg *Euglena* intake group, showing a significant easing of irritability caused by Uchida–Kraepelin stress (p = 0.008, Steel's test).

The amount of change in the level of subjective satisfaction with sleep from week 0 to 4 was -0.89 ± 1.44 in the placebo intake group and −2.61 ± 1.75 in the 3000 mg *Euglena* intake group, showing a significant increase in the level of subjective satisfaction with sleep (p = 0.007, Steel's test). The amount of change in the level of subjective satisfaction with sleep from week 0 to 12 was −0.65 ± 1.22 in the placebo intake group, −2.22 ± 3.08 in the 500 mg *Euglena* intake group, −2.69 ± 2.31 in the 1000 mg *Euglena* intake group, and −3.41 ± 2.26 in the 3000 mg *Euglena* intake group, showing a significant difference increase in the level of subjective satisfaction with sleep (p = 0.028, p = 0.012, p = 0.0006, respectively, Steel's test).

3.3.5. OSA Analysis

Table 8 shows the OSA results. Sleepiness on rising was significantly higher after 8 weeks of the continuous intake of *Euglena* 500 mg (47.7 ± 9.0) than in the placebo intake group (41.5 ± 5.3) (p = 0.027, Dunnett's test); at week 12 after the continuous intake of *Euglena* 500 mg (48.9 ± 10.8) than in the placebo intake group (41.4 ± 7.2) (p = 0.027, Dunnett's test); and at week 12 after the continuous intake of *Euglena* 3000 mg (48.3 ± 8.9) than in the placebo intake group (41.4 ± 7.2) (p = 0.040, Dunnett's test). These results suggest that the quality of sleep may improve with the intake of *Euglena*.

The "refreshing" index tended to be higher at week 8 after the continuous intake of *Euglena* 500 mg (47.0 ± 8.5) than in the placebo intake group (41.1 ± 4.7) (p = 0.055, Dunnett's test). This index was significantly higher in week 12 after the continuous intake of *Euglena* 500 mg (48.4 ± 8.0) than in the placebo intake group (41.6 ± 7.0) (p = 0.032, Dunnett's test). This index was significantly higher at week 12 after the continuous intake of *Euglena* 3000 mg (49.0 ± 8.2) than in the placebo intake group (41.6 ± 7.0) (p = 0.016, Dunnett's test). These results suggest that the quality of sleep may improve with the intake of *Euglena*.

Sleep length tended to be longer in week 12 after the continuous intake of *Euglena* 500 mg (51.3 ± 9.3) than in the placebo intake group (44.5 ± 7.1) (p = 0.060, Dunnett's test). Sleep length tended to be longer in week 12 after the continuous intake of *Euglena* 3000 mg (50.7 ± 10.9) than in the placebo intake group (44.5 ± 7.1) (p = 0.089, Dunnett's test). These findings suggest that the quality of sleep may improve with the intake of *Euglena*.

3.3.6. PSQI Analysis

Table 9 shows the results of a PSQI analysis. The total PSQI score was significantly lower at week 4 after the continuous intake of *Euglena* 3000 mg (5.1 ± 2.0) than in the placebo intake group (7.1 ± 2.6) (p = 0.039, Steel's test); at week 12 after the continuous intake of *Euglena* 3000 mg (4.2 ± 2.1) than in the placebo intake group (6.3 ± 2.3) (p = 0.026, Steel's test); and after the continuous intake of *Euglena* 1000 mg (4.7 ± 2.1) compared with the placebo intake group (6.3 ± 2.3) (p = 0.070, Steel's test). These results suggest that the quality of sleep may improve with the intake of *Euglena*.

Table 8. Summary of OSA analysis on Sleepiness on rising, Refreshing, and Sleep length. After accounting for normality, a two-way repeated measure ANOVA was performed. The results showed a session x group interaction ($p = 0.054$, $p = 0.016$, $p = 0.003$) and a main effect of group ($p = 0.052$, $p = 0.089$, $p = 0.780$), therefore, we followed up with post-hoc comparisons using Dunnett's test. Dunnett's test was used for intergroup comparisons of mean values in the placebo group and *Euglena* intake group at 0, 4, 8, and 12 weeks.

	Group	*n*		0 Weeks	4 Weeks	8 Weeks	12 Weeks	Δ 4 Weeks	Δ 8 Weeks	Δ 12 Weeks
Sleepiness on rising	Placebo	18	Mean	41.8	40.6	41.5	41.4	−1.2	−0.3	−0.4
			SD	7.0	5.7	5.3	7.2	6.0	3.4	5.9
	Euglena 500 mg	19	Mean	42.1	43.4	47.7 *	48.9 *	1.3	5.3 *	6.7 *
			SD	9.1	6.9	9.0	10.8	6.6	7.2	8.6
	Euglena 1000 mg	20	Mean	38.1	41.2	43.6	43.5	3.1	5.5 *	5.4 †
			SD	3.3	5.7	6.0	6.5	5.7	7.4	8.2
	Euglena 3000 mg	20	Mean	42.7	44.5	44.7	48.3 *	1.9	2.1	5.7 †
			SD	6.5	6.2	7.5	8.9	7.0	7.9	10.2
Refreshing	Placebo	18	Mean	41.2	41.5	41.1	41.6	0.3	−0.2	0.3
			SD	7.6	6.7	4.7	7.0	5.8	5.6	6.4
	Euglena 500 mg	19	Mean	43.2	42.0	47.0 †	48.4 *	−1.2	3.6	5.2
			SD	5.7	9.0	8.5	8.0	9.2	9.4	9.7
	Euglena 1000 mg	20	Mean	37.3	42.0	43.9	44.1	4.8	6.7 *	6.8 †
			SD	5.4	7.8	7.6	8.9	7.9	10.0	9.6
	Euglena 3000 mg	20	Mean	38.9	44.5	45.4	49.0 *	5.5	6.5 †	10.1 *
			SD	6.9	6.6	8.5	8.2	8.9	8.4	8.8
Sleep length	Placebo	18	Mean	43.5	47.7	44.1	44.5	4.2	0.6	1.0
			SD	8.2	8.4	8.9	7.1	7.0	5.8	6.5
	Euglena 500 mg	19	Mean	46.4	42.5	46.8	51.3 †	−3.8 *	0.2	4.9
			SD	6.9	13.9	8.0	9.3	12.2	7.3	9.9
	Euglena 1000 mg	20	Mean	42.7	43.9	47.2	46.3	1.2	4.4	3.6
			SD	6.4	8.1	10.9	8.1	8.3	11.2	9.5
	Euglena 3000 mg	20	Mean	40.8	45.2	49.1	50.7 †	4.4	8.3 *	9.9 **
			SD	9.4	6.9	10.0	10.9	8.0	10.7	7.7

** $p < 0.01$, * $p < 0.05$, † $p < 0.1$ vs. the placebo intake group. OSA: Ogri-Shirakawa-Azumi Sleep Inventory.

Table 9. Analysis of sleep based on PSQI. After accounting for normality, Kruskal–Wallis test was performed. If the Kruskal–Wallis test was significant, the Steel's test was then performed. Steel's test was used for intergroup comparisons of mean values in the placebo group and *Euglena* intake groups at 0, 4, 8, and 12 weeks.

	Group	*n*		0 Weeks	4 Weeks	8 Weeks	12 Weeks
Total score	Placebo	18	Mean	7.7	7.1	6.2	6.3
			SD	2.6	2.6	2.4	2.3
	Euglena 500 mg	19	Mean	8.1	5.9	5.5	5.7
			SD	3.1	2.2	1.5	2.3
	Euglena 1000 mg	20	Mean	7.4	5.8	5.4	4.7 †
			SD	2.4	2.3	2.6	2.1
	Euglena 3000 mg	20	Mean	6.8	5.1 *	4.9	4.2 *
			SD	2.2	2.0	2.4	2.1
p-value (Kruskal–Wallis test)				0.5557	0.0803 †	0.2304	0.0243 *

* $p < 0.05$, † $p < 0.1$ vs. the placebo intake group. PSQI: Pittsburgh Sleep Quality Index.

3.3.7. Analysis of Saliva Samples

No significant differences were observed in the salivary cortisol levels.

3.3.8. Uchida–Kraepelin Results

Table 10 shows the correct answer rates in the Uchida–Kraepelin test. The correct answer rate in the first half increased significantly in weeks 8 and 12 of intake of *Euglena* 1000 mg (p = 0.009, and p = 0.020, respectively, Dunnett's test).

Table 10. Correct answer rates (%) for the Uchida–Kraepelin test. After accounting for normality, a two-way repeated measure ANOVA was performed. The results showed a main effect of group (p = 0.098) in "First half;" therefore, we followed up with post-hoc comparisons using Dunnett's test. Dunnett's test was used for intergroup comparisons of mean values in the placebo group and *Euglena* intake groups at 0, 4, 8, and 12 weeks.

			0 Weeks		4 Weeks		8 Weeks		12 Weeks	
Group	*n*		**First Half**	**Latter Half**	**First Half**	**Latter Half**	**First Half**	**Latter Half**	**First Half**	**Latter Half**
Placebo	18	Mean	98.7	98.8	98.9	98.7	98.7	99.0	98.6	98.7
		SD	1.4	1.2	1.3	1.2	1.4	1.0	1.4	1.2
Euglena 500 mg	19	Mean	98.7	99.1	99.1	99.2	99.3 †	99.2	99.3	99.2
		SD	1.5	0.9	0.9	0.8	0.8	1.0	1.0	1.1
Euglena 1000 mg	20	Mean	99.2	99.2	99.3	99.2	99.6 **	99.5	99.5 *	99.4
		SD	0.5	0.5	0.7	0.6	0.5	0.5	0.6	0.6
Euglena 3000 mg	20	Mean	99.1	99.0	99.0	99.2	99.2	99.2	99.4 †	99.4
		SD	0.7	1.2	0.8	0.7	0.6	0.7	0.8	0.5

** $p < 0.01$, * $p < 0.05$, † $p < 0.1$ vs. the placebo intake group.

The correct answer rate in the first half tended to increase in week 8 with the continuous intake of *Euglena* 500 mg (p = 0.083, Dunnett's test), in week 12 with the continuous intake of *Euglena* 3000 mg (p = 0.067, Dunnett's test).

4. Discussion

We conducted a double-blind placebo-controlled trial to evaluate the effects of *Euglena* intake on various properties, such as the autonomic nervous system, psychological parameters, and quality of sleep. Our results indicated that the intake of *Euglena* regulates the autonomic nervous system balance and improves various symptoms caused by stress loading. In addition, we observed no adverse events with a maximum intake of 3000 mg of *Euglena* per day, confirming that this dose is safe.

The subjects included in the study were concerned about a lack of motivation and a decline in the quality of sleep. However, the sympathetic nervous system was in a dominant state in these individuals, with a high LF/HF ratio, which is an indicator of the autonomic nervous system balance, exceeding the normal value of 2.0. Previous studies have shown that Uchida–Kraepelin stress affects heart rate variability [31–34]. We selected subjects who felt that taking the Uchida–Kraepelin test was stressful had a substantial increase in their LF/HF ratios. We found that the intake of *Euglena* at 1000 mg suppressed the LF/HF ratio compared with that in the placebo intake group. When the LF/HF ratio was broken down into the components LF and HF, a decrease in LF had a relatively large impact upon 500 mg and 1000 mg Euglena intake. Therefore, *Euglena* may have suppressed the excitation of the sympathetic nervous system. The intake of *Euglena* significantly suppressed the LF/HF ratios in individuals subjected to work stress and appeared to regulate the autonomic nervous system balance. The most effective dose was 1000 mg.

Irritability and tension during work stress were maintained at high levels even 60 min after exposure to work stress in the placebo intake group; however, these parameters were suppressed in the *Euglena* intake groups of 500 mg or more and were significantly suppressed in the 1000 mg intake group, indicating an improvement in the psychological state. In other words, the intake of *Euglena* may have alleviated the tension caused by Uchida–Kraepelin stress loading.

In a POMS2 analysis, which evaluates emotions, the score for friendship (F) was significantly higher in the group taking 500 mg of *Euglena* per day at weeks 4, 8, and 12 after starting intake, both before and after Uchida–Kraepelin stress loading. This suggests that the continuous intake of 500 mg *Euglena* can enhance friendliness. However, this did not occur upon taking *Euglena* at doses other than 500 mg. In SF-36, which evaluates health-related quality of life, a significant difference was observed in daily physical function (role physical), which represents physical fatigue.

Improvement of sleep quality with the intake of *Euglena* was confirmed based on three sleep tests: VAS, OSA, and PSQI. These effects increased in a dose- and time-dependent manners. The parasympathetic nervous system is enhanced at the onset of sleep [35]; accordingly, the improvement in sleep quality by *Euglena* intake was likely due to an improvement in the balance of the autonomic nervous system.

Euglena intake also increased the correct answer rate on the Uchida–Kraepelin test. These findings suggest that the continuous intake of *Euglena* may improve concentration and work efficiency.

The autonomic nervous system includes the sympathetic nervous system and the parasympathetic nervous system. It regulates various functions, such as respiration/circulation, digestion/absorption, secretion, and metabolism, and serves an important role in the maintenance of homeostasis in the body. However, when stress signals are transmitted to the hypothalamus in the brain, the autonomic nervous system is excited, depending on the type of stress. An imbalance in the autonomic nervous system due to excessive stress may prevent the proper activation of the sympathetic or parasympathetic nervous system. These conditions may cause excessive irritability, an inability to concentrate, deterioration of sleep quality, weakened immunity, and various ailments, including problems with bowel movements. For example, patients with chronic fatigue syndrome caused by stress experience a decrease in parasympathetic nervous system function and a relative increase in sympathetic nervous system function, depending on the degree of subjective fatigue evaluated by the VAS. This shows that there is a relationship between fatigue and the autonomic nervous system [36]. Therefore, it is important to maintain a well-balanced autonomic nervous system. In this study, the intake of *Euglena* improved the autonomic nervous system balance, thereby improving irritability and tension as well as the quality of daily sleep. Previous studies have evaluated improvements in autonomic nervous system activity after nutritional supplementation in healthy subjects [37–40]. When stress is transmitted to the hypothalamus in the brain, a signal is sent to excite the autonomic nervous system, depending on the type of stress, and a command is simultaneously sent to the pituitary gland to stimulate the thyroid, adrenal medulla, and adrenal cortex via the release of hormones. In addition, signals from the hypothalamus stimulate the adrenal cortex in response to stress; the pituitary gland secretes the adrenocorticotrophic hormone, adrenocorticotropic hormone (ACTH), and the adrenal cortex secretes the hormone cortisol. In this way, the hypothalamus, which controls the autonomic nervous system, also controls the pituitary gland, which secretes various hormones and neurotransmitters. Therefore, a mutual relationship exists whereby an imbalance in the autonomic nervous system affects the secretion of hormones and neurotransmitters, thereby affecting the balance of the autonomic nervous system. Proteins, B vitamins, and minerals, such as magnesium, are required for hormones and neurotransmitters. A lack of vitamin B6 makes it difficult for the body to produce serotonin and melatonin, leading to insomnia and autonomic nervous system ataxia. The overconsumption of zinc is related to stress, and a shortage of vitamin B makes it difficult to produce energy in the mitochondria, which increases fatigue and malaise. *Euglena* consist of a wide variety of nutrients, such as vitamins, minerals, amino acids, and unsaturated fatty acids. The ingestion of *Euglena* assists in replenishing these nutrients in the body, which not only improves autonomic nervous system activity but also normalizes the secretion of hormones and neurotransmitters, contributing to the regulation of the autonomic nervous system balance. In this study, we found no changes in cortisol levels, which suggests that hormone levels are being affected rather than cortisol levels and that the stress of Uchida–Kraepelin may not affect cortisol production [41]; this warrants further investigation.

Meanwhile, previous studies have used in vitro cell-type-specific calcium imaging to prove that *Euglena* directly induces Ca^{2+} signaling in dorsal root ganglia (DRG) neurons, suggesting that *Euglena* excites visceral afferents [42]. Although the mechanism underlying the beneficial effects of *Euglena* in our study remains unclear, its ability to excite neurons may contribute to improvements in health-related quality of life [12]. Further studies are needed to better clarify the physiological role of visceral afferents in response to food constituents.

The storage polysaccharide paramylon in *Euglena* regulates the immune balance. Previous studies of mouse models have confirmed that the ingestion of *Euglena* or paramylon alleviates the symptoms of influenza infection [11], rheumatoid arthritis [43], and atopic dermatitis [44]. The autonomic nervous system is involved in the regulation of immunity, and immunity is generally enhanced in situations where the parasympathetic nervous system is dominant, contributing to defense against infection [16,17]. Cortisol also suppresses the immune system [45]. Previous studies have investigated the ability of paramylon to regulate immunity by binding to Dectin-1 expressed on intestinal immune cells [46,47]; however, it could also mediate immunity by the regulation of the autonomic nervous system and endocrine system.

The results of this study suggest that the intake of *Euglena*-containing food may regulate the autonomic nervous system balance under stress and may have favorable effects on work efficiency, psychological factors, and sleep quality. However, this study had some limitations. First, although this study was conducted with a small homogeneous cohort of individuals from one population, future studies are required to further elucidate these aspects with various ethnicities and people with different backgrounds and stress levels. Furthermore, although some of the effects were dependent on intake dose, other effects were observed for which an optimal dose may exist independent of intake, thus requiring further investigation. In the future, we would like to further investigate the mechanism underlying these beneficial effects as well.

Supplementary Materials: Supplementary materials can be found at http://www.mdpi.com/2072-6643/12/11/3243/s1, Table S1: Summary of blood test results. Table S2: Summary of blood pressure and pulse rate test results.

Author Contributions: Conceptualization, A.N., K.Y., and N.M.; methodology, A.N. and K.Y.; formal analysis, A.N., K.Y., and A.M.; resources, A.N., K.Y., A.M., and K.S.; original draft preparation and writing, A.N.; review and editing, A.N., K.Y., A.M., K.S., and N.M.; supervision, K.S. and N.M. All authors have read and agreed to the published version of the manuscript.

Funding: This research received no external funding.

Acknowledgments: The authors greatly appreciate the useful suggestions and technical support from Oneness support Co. Ltd.

Conflicts of Interest: A.N., K.Y., A.M., and K.S. are salaried employees of euglena Co. Ltd., which produced some of the *Euglena* used in this study. All research funding for this study was provided by euglena Co. Ltd. N.M. is the medical advisor of this study, and received honorarium from euglena Co. Ltd. There are no other conflicts of interest.

References

1. Selye, H. The general adaptation syndrome and the diseases of adaptation. *J. Clin. Endocrinol. Metab.* **1946**, *6*, 117–230. [CrossRef]
2. Kataoka, N.; Shima, Y.; Nakajima, K.; Nakamura, K. A central master driver of psychosocial stress responses in the rat. *Science* **2020**, *6*, 1105–1112. [CrossRef]
3. Besedovsky, H.O.; del Rey, A. Introduction: Immune-neuroendocrine network. *Front. Horm. Res.* **2002**, *29*, 1–14. [CrossRef]
4. Olini, N.; Rothfuchs, I.; Azzinnari, D.; Pryce, C.R.; Kurth, S.; Huber, R. Chronic social stress leads to altered sleep homeostasis in mice. *Behav. Brain Res.* **2017**, *327*, 167–173. [CrossRef]
5. Cacioppo, J.T. Social neuroscience: Autonomic, neuroendocrine, and immune responses to stress. *Psychophysiology* **1994**, *31*, 113–128. [CrossRef] [PubMed]
6. Suzuki, K. Large-scale cultivation of *Euglena*. *Adv. Exp. Med. Biol.* **2017**, *979*, 285–293. [CrossRef]
7. Shimada, R.; Fujita, M.; Yuasa, M.; Sawamura, H.; Watanabe, T.; Nakashima, A.; Suzuki, K. Oral administration of green algae, *Euglena gracilis*, inhibits hyperglycemia in OLETF rats, a model of spontaneous type 2 diabetes. *Food Funct.* **2016**, *7*, 4655–4659. [CrossRef]
8. Sakanoi, Y.; Shuang, E.; Yamamoto, K.; Ota, T.; Seki, K.; Imai, M.; Ota, R.; Asayama, Y.; Nakashima, A.; Suzuki, K.; et al. Simultaneous intake of *Euglena gracilis* and vegetables synergistically exerts an anti-inflammatory effect and attenuates visceral fat accumulation by affecting gut microbiota in mice. *Nutrients* **2018**, *10*, 1417. [CrossRef]
9. Okouchi, R.; Shuang, E.; Yamamoto, K.; Ota, T.; Seki, K.; Imai, M.; Ota, R.; Asayama, Y.; Nakashima, A.; Suzuki, K.; et al. Simultaneous intake of *Euglena gracilis* and vegetables exerts synergistic anti-obesity and anti-inflammatory effects by modulating the gut microbiota in diet-induced obese mice. *Nutrients* **2019**, *11*, 204. [CrossRef]
10. Aoe, S.; Yamanaka, C.; Nishioka, M.; Onaka, N.; Nishida, N.; Takahashi, M. Effects of paramylon extracted from *Euglena gracilis* EOD-1 on parameters related to metabolic syndrome in diet-induced obese mice. *Nutrients* **2019**, *11*, 1674. [CrossRef]
11. Nakashima, A.; Suzuki, K.; Asayama, Y.; Konno, M.; Saito, K.; Yamazaki, N.; Takimoto, H. Oral administration of *Euglena gracilis Z* and its carbohydrate storage substance provides survival protection against influenza virus infection in mice. *Biochem. Biophys. Res. Commun.* **2017**, *494*, 379–383. [CrossRef]
12. Ishibashi, K.; Nishioka, M.; Onaka, N.; Takahashi, M.; Yamanaka, D.; Adachi, Y.; Ohno, N. Effects of *Euglena gracilis* EOD-1 ingestion on salivary IgA reactivity and health-related quality of life in humans. *Nutrients* **2019**, *11*, 1144. [CrossRef]
13. Watanabe, T.; Shimada, R.; Matsuyama, A.; Yuasa, M.; Sawamura, H.; Yoshida, E.; Suzuki, K. Antitumor activity of the β-glucan paramylon from *Euglena* against preneoplastic colonic aberrant crypt foci in mice. *Food Funct.* **2013**, *4*, 1685–1690. [CrossRef]
14. Niijima, A. Nervous regulation of metabolism. *Prog. Neurobiol.* **1989**, *33*, 135–147. [CrossRef]
15. Bartness, T.J.; Bamshad, M. Innervation of mammalian white adipose tissue: Implications for the regulation of total body fat. *Am. J. Physiol.* **1998**, *275*, R1399–R1411. [CrossRef]
16. Shen, J.; Horii, Y.; Fujisaki, Y.; Nagai, K. Effects of L-arginine and L-lysine mixtures on splenic sympathetic nerve activity and tumor proliferation. *Auton. Neurosci.* **2009**, *147*, 86–90. [CrossRef]
17. Katafuchi, T.; Ichijo, T.; Take, S.; Hori, T. Hypothalamic modulation of splenic natural killer cell activity in rats. *J. Physiol.* **1993**, *471*, 209–221. [CrossRef]
18. McMenamin, C.A.; Travagli, R.A.; Browning, K.N. Inhibitory neurotransmission regulates vagal efferent activity and gastric motility. *Exp. Biol. Med.* **2016**, *241*, 1343–1350. [CrossRef]
19. Kuraishi, S.; Kato, M.; Tsujioka, B. Development of the "Uchida-Kuraepelin Psychodiagnostic Test" in Japan. *Psychologia* **2000**, *1*, 104–109.

20. Januszewicz, W.; Sznajderman, M.; Wocial, B.; Feltynowski, T.; Klonowicz, T. The effect of mental stress on catecholamines, their metabolites and plasma renin activity in patients with essential hypertension and in healthy subjects. *Clin. Sci.* **1979**, *57*, 229s–231s. [CrossRef]
21. Sumiyoshi, T.; Yotsutsuji, T.; Kurachi, M.; Itoh, H.; Kurokawa, K.; Saitoh, O. Effect of mental stress on plasma homovanillic acid in healthy human subjects. *Neuropsychopharmacology* **1998**, *19*, 70–73. [CrossRef]
22. Takanari, J.; Nakahigashi, J.; Sato, A.; Waki, H.; Miyazaki, S.; Uebaba, K.; Hisajima, T. Effect of enzyme-treated asparagus extract (ETAS) on psychological stress in healthy individuals. *J. Nutr. Sci. Vitaminol.* **2016**, *62*, 198–205. [CrossRef]
23. Fukuhara, S.; Ware Jr, J.E.; Kosinski, M.; Wada, S.; Gandek, B. Psychometric and clinical tests of validity of the Japanese SF-36 Health Survey. *J. Clin. Epidemiol.* **1998**, *51*, 1045–1053. [CrossRef]
24. Fukuhara, S.; Bito, S.; Green, J.; Hsiao, A.; Kurokawa, K. Translation, adaptation, and validation of the SF-36 Health Survey for use in Japan. *J. Clin. Epidemiol.* **1998**, *51*, 1037–1044. [CrossRef]
25. Fukuhara, S.; Suzukamo, Y. *Manual of SF-36v2 Japanese Version*; Institute for Health Outcomes and Process Evaluation Research: Kyoto, Japan, 2004; pp. 7–145.
26. McNair, D.M.; Lorr, M. An analysis of mood in neurotics. *J. Abnorm. Soc. Psychol.* **1964**, *69*, 62–627. [CrossRef]
27. Shacham, S. A shortened version of the profile of mood states. *J. Pers. Assess.* **1983**, *47*, 305–306. [CrossRef]
28. Oguri, M.; Shirakawa, S.; Azumi, K. Construction of standard rating scale to estimate sleep profile. *Seishin Igaku* **1985**, *27*, 791–799. [CrossRef]
29. Masafumi, N.; Hirotaka, Y.; Miyuki, K.; Naoki, M.; Koji, W.; Takuro, E. Effects of lactic acid bacteria-containing foods on the quality of sleep: A placebo-controlled, double-blinded, randomized crossover study. *Funct. Food Health Dis.* **2018**, *8*. [CrossRef]
30. Buysse, D.J.; Reynolds, C.F.; Monk, T.H.; Berman, S.R.; Kupfer, D.J. The Pittsburgh Sleep Quality Index: A new instrument for psychiatric practice and research. *Psychiatry Res.* **1989**, *28*, 193–213. [CrossRef]
31. Saito, Y.; Natsuko, M.; Teruyuki, N.; Hiroyuki, I.; Mitsunori, K.; Kimihide, N.; Tadasu, U. Relationship of a Special Acidified Milk Protein Drink with Cognitive Performance: A Randomized, Double-Blind, Placebo-Controlled, Crossover Study in Healthy Young Adults. *Nutrients* **2018**, *10*, 574. [CrossRef]
32. Kazuo, O.; Mami, K.; Takashi, N.; Toshiyuki, N.; Richard, E.; Williams, S.H. Individual differences in physiological responses and type A behavior pattern. *Appl. Human Sci.* **1999**, *3*, 101–108. [CrossRef]
33. Yoshiyasu, F.; Hitoshi, O. Hearing laughter improves the recovery process of the autonomic nervous system after a stress-loading task: A randomized controlled trial. *Biopsychosoc. Med.* **2018**, *12*. [CrossRef]
34. Kuratsune, D.; Seiki, T.; Junichi, K.; Kouzi, Y.; Tetsuya, S.; Kei, M.; Masaaki, T.; Naoko, O.; Hideki, M.; Hirokazu, T.; et al. Changes in reaction time, coefficient of variance of reaction time, and autonomic nerve function in the mental fatigue state caused by long-term computerized Kraepelin test workload in healthy volunteers. *World J. Neurosci.* **2012**, *2*, 113–118. [CrossRef]
35. Werner, G.G.; Ford, B.Q.; Mauss, I.B.; Schabus, M.; Blechert, J.; Wilhelm, F.H. High cardiac vagal control is related to better subjective and objective sleep quality. *Biol. Psychol.* **2015**, *106*, 79–85. [CrossRef]
36. Tanaka, M.; Mizuno, K.; Yamaguti, K.; Kuratsune, H.; Fujii, A.; Baba, H.; Matsuda, K.; Nishimae, A.; Takesaka, T.; Watanabe, Y. Autonomic nervous alterations associated with daily level of fatigue. *Behav. Brain Funct.* **2011**, *7*, 46. [CrossRef]
37. Fukuda, S.; Koyama, H.; Kondo, K.; Fujii, H.; Hirayama, Y.; Tabata, T.; Okamura, M.; Yamakawa, T.; Okada, S.; Hirata, S.; et al. Effects of nutritional supplementation on fatigue, and autonomic and immune dysfunction in patients with end-stage renal disease: A randomized, double-blind, placebo-controlled, multicenter trial. *PLoS ONE* **2015**, *10*, e0119578. [CrossRef]
38. Piccirillo, G.; Raffaele, Q.; Fimognari, F.; Moise, A.; Mario, M.; Lionetti, M.; Naso, C.; Di Carlo, S.; Nocco, M.; Magrì, D. Influence of L-arginine and vitamin C on the autonomic nervous system in chronic heart failure secondary to ischemic cardiomyopathy. *Am. J. Cardiol.* **2004**, *93*, 650–654. [CrossRef]
39. Kato, K.; Fukuma, N.; Kimura-Kato, Y.; Aisu, N.; Tuchida, T.; Mabuchi, K.; Takano, T. Improvement of sympathetic response to exercise by oral administration of ascorbic acid in patients after myocardial infarction. *Int. J. Cardiol.* **2006**, *111*, 240–246. [CrossRef]
40. Zheng, A.; Moritani, T. Influence of CoQ10 on autonomic nervous activity and energy metabolism during exercise in healthy subjects. *J. Nutr. Sci. Vitaminol.* **2008**, *54*, 286–290. [CrossRef]
41. Koreaki, S.; Aya, K.; Noriaki, S. The effectiveness of the Uchida-Kraepelin test for psychological stress: An analysis of plasma and salivary stress substances. *Biopsychosoc. Med.* **2006**. [CrossRef]

42. Yasuda, K.; Nakashima, A.; Murata, A.; Suzuki, K.; Adachi, T. *Euglena Gracilis* and β-Glucan Paramylon Induce Ca2+ Signaling in Intestinal Tract Epithelial, Immune, and Neural Cells. *Nutrients* **2020**, *12*, 2293. [CrossRef] [PubMed]
43. Suzuki, K.; Nakashima, A.; Igarashi, M.; Saito, K.; Konno, M.; Yamazaki, N.; Takimoto, H. *Euglena gracilis Z* and its carbohydrate storage substance relieve arthritis symptoms by modulating Th17 immunity. *PLoS ONE* **2018**, *13*, e0191462. [CrossRef] [PubMed]
44. Sugiyama, A.; Hata, S.; Suzuki, K.; Yoshida, E.; Nakano, R.; Mitra, S.; Arashida, R.; Asayama, Y.; Yabuta, Y.; Takeuchi, T. Oral administration of paramylon, a β-1,3-D-glucan isolated from *Euglena gracilis Z* inhibits development of atopic dermatitis-like skin lesions in NC/Nga mice. *J. Vet. Med. Sci.* **2010**, *72*, 755–763. [CrossRef]
45. Dimitrov, S.; Benedict, C.; Heutling, D.; Westermann, J.; Born, J.; Lange, T. Cortisol and epinephrine control opposing circadian rhythms in T cell subsets. *Blood* **2009**, *113*, 5134–5143. [CrossRef]
46. Nakashima, A.; Yamada, K.; Iwata, O.; Sugimoto, R.; Atsuji, K.; Ogawa, T.; Ishibashi-Ohgo, N.; Suzuki, K. β-glucan in foods and its physiological functions. *J. Nutr. Sci. Vitaminol.* **2018**, *64*, 8–17. [CrossRef] [PubMed]
47. Ujita, M.; Nagayama, H.; Kanie, S.; Koike, S.; Ikeyama, Y.; Ozaki, T.; Okumura, H. Carbohydrate binding specificity of recombinant human macrophage β-glucan receptor dectin-1. *Biosci. Biotechnol. Biochem.* **2009**, *73*, 237–240. [CrossRef]

nutrients

MDPI

Article

AphaMax®, an Aphanizomenon Flos-Aquae Aqueous Extract, Exerts Intestinal Protective Effects in Experimental Colitis in Rats

Maria Grazia Zizzo [1,2,*], Gaetano Caldara [1], Annalisa Bellanca [2], Domenico Nuzzo [3], Marta Di Carlo [3], Stefano Scoglio [4] and Rosa Serio [1]

1 Department of Biological, Chemical and Pharmaceutical Sciences and Technologies (STEBICEF), University of Palermo, Viale delle Scienze, 90128 Palermo, Italy; gaetanofelice.caldara@unipa.it (G.C.); rosa.serio@unipa.it (R.S.)
2 ATeN (Advanced Technologies Network) Center, Viale delle Scienze, University of Palermo, 90128 Palermo, Italy; annalisa.bellanca@unipa.it
3 Institute of Biomedicine and Molecular Immunology "Alberto Monroy" (IBIM), Consiglio Nazionale delle Ricerche (CNR), 90146 Palermo, Italy; domenico.nuzzo@irib.cnr.it (D.N.); marta.di.carlo@irib.cnr.it (M.D.C.)
4 Nutritherapy Research Center, 61029 Urbino, Italy; stefanoscoglio@me.com
* Correspondence: mariagrazia.zizzo@unipa.it; Tel.: +39-091-23897518

Received: 21 October 2020; Accepted: 24 November 2020; Published: 26 November 2020

Abstract: Background: Aphanizomenon flos-aquae (AFA) is a unicellular cyanobacterium considered to be a "superfood" for its complete nutritional profile and beneficial properties. We investigated possible beneficial effects of an AFA extract, commercialized as AphaMax®, containing concentrated amount of phycocyanins and phytochrome, in 2,4 dinitrobenzensulfonic acid(DNBS)-induced colitis in rats. Methods: Effects of preventive oral treatment of AphaMax® (20, 50 or 100 mg/kg/day) in colitic rats were assessed and then macroscopic and microscopic analyses were performed to evaluate the inflammation degree. Myeloperoxidase (MPO) activity and NF-κB, pro-inflammatory citockines, cycloxygenase-2 (COX-2), and inducible NOS (iNOS) levels of expression were determined, as Reactive Oxygen Species (ROS) and nitrite levels. Results: AphaMax® treatment attenuated the severity of colitis ameliorating clinical signs. AphaMax® reduced the histological colonic damage and decreased MPO activity, NF-κB activation, as well as iNOS and COX-2 expression. AphaMax® treatment improved the altered immune response associated with colonic inflammation reducing IL-1β, IL-6 expression. Lastly, AphaMax® reduced oxidative stress, decreasing ROS and nitrite levels. Conclusions: Preventive treatment with AphaMax® attenuates the severity of the inflammation in DNBS colitis rats involving decrease of the NF-kB activation, reduction of iNOS and COX-2 expression, and inhibition of oxidative stress. Due its anti-inflammatory and antioxidant proprieties AphaMax® could be a good candidate as a complementary drug in inflammatory bowel disease (IBD) treatment.

Keywords: inflammation; Aphanizomenon flos-aquae; blue-green algae; inflammatory bowel disease

1. Introduction

Inflammatory bowel diseases (IBDs), are chronic and non-resolving intestinal inflammatory diseases, which include two clinical entities: Crohn's disease (CD) and ulcerative colitis (UC). Their etiology is far from being fully understood, although it is widely recognized that genetic, environmental, microbial, and immune factors are functionally integrated in the IBD pathogenesis [1]. A pivotal role seems to play a worsening and inappropriate mucosal immune response. In particular,

NF-κB activation, production of pro-inflammatory cytokines such as IL-6 and IL-1β [2–4], myeloperoxidase (MPO), expression of cycloxygenase-2 (COX-2), and inducible nitric oxide synthase (iNOS) collectively enhance inflammatory response in the colon and lead to a concomitant reduction in antioxidant levels. These molecular events could culminate in a serious damage of epithelial cell and disruption of the mucosal barrier [5,6].

Conventional IBD therapies have various side effects, and many patients do not respond to these treatments [7]. Thus, to identify effective new therapeutic options for improving colitis signs with fewer or no side effects, has becoming essential.

In recent years, attention was increased on complementary medicine approaches based on natural ingredients.

Various bioactive compounds, such as phycocyanins, carotenoids, γ-linolenic acid, fibers, and plant sterols, have been reported to clinically improve IBD-related symptoms [8]. However, the efficacy and mechanism of action of these products require further studies in vitro and in vivo models. The cyanobacteria blue-green algae (BGA), are photosynthetic prokaryotes present in aquatic ecosystems, rich in these bioactive molecules [9,10]. Recent studies highlight that their consumption is associated with ameliorative effects on the components of human metabolic syndrome [11], to the decrease of inflammation markers in hypertensive patients [10], as well as in obese patients. Moreover, immunomodulatory effects of BGA supplementation has been shown in healthy volunteers [12]. Aphanizomenon flos-aquae (AFA) also known as "Klamath algae", is as cyanobacterial dominant species growing in the Upper Klamath Lake (Oregon, USA) [13].

AFA, consumed as a "superfood", is a rich source of mycosporine-like amino acids (MAAs) and phycocyanins (PCs), likely responsible for its various health benefits. AphaMax® is a unique patented extract from AFA, manufactured by Nutrigea, containing a concentrated amount of AFA-phycocyanins and AFA-phytochrome, the Klamath microalgae components showing the greatest antioxidant, anti-inflammatory, anticancer, and cardiovascular properties. These molecules, which are unique to Klamath microalgae, increased the beneficial proprieties compared to other blue-green algae. Indeed, AFA-PCs due to their peculiar structure [14] have up to 200 times higher antioxidant power compared to other PCs [15]. AFA-PCs have neuroprotective effects and ameliorate psychological stress and on menopausal well-being [16] In addition, AphaMax® is rich in some smaller molecules, endowed with similar antioxidant properties, as many as the AFA-mycosporines (MAAs, or mycosporine-like amino acids) [17], 15 carotenoids, including zeaxanthin, lutein, canthaxanthin, a wide spectrum of polyphenols as caffeic acid, and an high content of chlorophyll.

Given the well-demonstrated anti-inflammatory and antioxidant effects of the majority of the molecules present in Klamath algae, as phycocyanins that selectively inhibit key enzymes in inflammatory disease as COX-2, and iNOS, and the high concentration of these components in AphaMax®, the aim of our study was to investigate the possible protective potential of this extract in an animal model of IBD, the 2,4-dinitrofluorobenzenesulfonic acid (DNBS) -induced rat colitis. DNBS rat colitis shares many of the pathological features of human Crohn's disease and is considered useful for studying the etiopathogenesis of IBD as well as for providing an inexpensive model useful for investigating new potential treatments [18].

2. Materials and Methods

2.1. Animals

Thirty healthy adult Wistar male rats (8–9 weeks old, 250–300 g) were purchased from ENVIGO Srl (San Pietro al Natisone UD, Italy) and housed randomly in temperature-controlled rooms on a 12 h light cycle at 22–24 °C and 50–60% humidity. Animals had free access to standard pellet chow and water ad libitum throughout the experimental protocol. They were acclimatized for one week prior to experimentation. The experimental protocol, followed throughout the study, was conducted in the conformity of the Italian D.Lgs 26/2014", following the criteria outlined by the European Community

Council Directive 2010/63/UE, recognized and adopted by the Italian Government, and approved by the Ethical Committee for Animal Experimentation of the University of Palermo and by the Italian Ministry of Health (Authorization n 921/2018–released Rome, Italy).

2.2. Induction of DNBS Colitis and Treatment Protocol

AphaMax® extract (Nutrigea Research s.r.l., Borgo Maggiore, Republic of San Marino) was dissolved in 0.5 mL water, and then sonicated (twice for 60 s) immediately prior to the gavage. 20, 50 or 100 mg/kg AphaMax® solution was administered by oral gavage once a day for 14 days starting 7 days before the induction of colitis (day-7). Rats were randomly assigned into six groups ($n = 5$ animals/each): (1) control group (sham group); (2) group with colitis; (3) sham group + AphaMax® (100 mg/Kg/day); (4) group with colitis + AphaMax® (20 mg/Kg/day); (5) group with colitis + AphaMax® (50 mg/Kg/day); (6) group with colitis + AphaMax® (100 mg/Kg/day). To induce the colitis intracolonic (i.c.) instillation of 2, 4-dinitrobenzensulfonic acid (DNBS; Sigma-Aldrich Inc., St. Louis, MO, USA), was performed as already described [19–21]. Briefly, rats were fasted overnight and then, under light anesthesia with 1% isoflurane (Merial Italia Spa, Assago, MI, Italy), DNBS (15 mg) dissolved in a solution of 50% ethanol, was deposited in the colon through an 8 cm plastic catheter (PE90).

To avoid reflux actions, the rats were kept for 5 min in a Trendelenburg position, and then allowed to recover with food and water supplied. The control group (sham group) received i.c instillation of vehicle alone (50% ethanol).

None of the rats died during the study or was excluded from the study for any reason. Assessment of colitis-induced damage was performed minimizing the suffering of animals, in a blinded fashion as previously described [18–21].

2.3. Evaluation of Colitis

2.3.1. Monitoring of Change in Body Weight, Consistency of Stools and Rectal Bleeding

The animals were monitored daily during the experimental period and scored for body weight loss percentage (initial body = 100%), stool consistency and the presence of rectal blending. All these parameters were ranged and calculated as described by Cooper et al. [22].

2.3.2. Macroscopic Scores

Seven days after colitis induction, rats were sacrificed, and laparotomy was performed and the appearance of colon was then examined. Distal colon was rapidly removed, opened longitudinally and gently washed with saline. For each specimen wet weight (mg)weight/length (cm) ratio was calculated as indicator of colonic edema. For macroscopic damage, each animal was scored by Appleyard and Wallace classification system score [23], The sum of scores for the ulceration (0 = No mucosal damage, 1 = Localized hyperemia but no ulcers, 2 = Ulcers without hyperemia/bowel wall thickening, 3 = Ulcers with hyperemia/bowel wall thickening at 1 site, 4 =Two or more sites of ulceration or inflammation, 5 Area of damage (necrosis) extended >1 cm along length of colon, 6–10 Area of damage extended >2 cm along length of colon; increasing the score was by 1 for each additional cm involved) for the adhesions (0 = No adhesions, 1 = Minor adhesions (no difficult separation of the colon from the other tissue), 2 = Major adhesions) and for the evaluation of thickness (Wall thickness(x) of bowel calculated in mm) were calculated.

2.3.3. Analysis of Microscopic Inflammatory Damage

The colon tissues of rats were fixed in 4% formaldehyde (Sigma–Aldrich, Inc., St. Louis, MO, USA), embedded in paraffin for histological studies and sectioned (5 μm thick). The slices were stained with haematoxylin and eosin (Bio-Optica Milano SpA, Milano, Italy). Histological sections were examined in a blinded fashion, and photos of sections were taken by light microscope (Olympus BX50,

Olympus Optical Co., Tokio, Japan). The microscopic damage following score was calculated the Hunter et al. [24] classification method: by adding the histological finding (0 = normal, 1 = minimal, 2 = mild, 3 = severe), the degree of inflammatory infiltration (0 = normal, 1 = minimal, 2 = mild, 3 = severe), the layers infiltrated (0 = normal, 1 = minimal, 2 = mild, 3 = severe), the mucosal damage (0 = normal,1 = minimal, 2 = mild, 3 = severe), and the edema in the mucosa (0 = absent, 1 = present).

2.4. Assay of MPO Activity

The activity of myeloperoxidase (MPO) is an important marker for inflammatory damage. The MPO activity was estimated spectrophotometrically using hydrogen peroxide and o-dianisidine as substrate for MPO enzyme, as previously described [20,21] by following a method of Moreels et al. [25] The absorbance value was read at 460 nm (Beckman-Coulter Inc, Brea, CA, USA). MPO activity was expressed as units per gram tissue (U gram tissue^{-1}), taken that 1 unit of enzyme reduces 1 µmole hydrogen peroxide (H_2O_2) per minute.

2.5. ELISA Assay for Pro-Inflammatory Cytokines

Commercial ELISA Kits (Cloud-Clone Corp, Wuhan, Hubei, China) was used for the evaluation of colonic amounts of interleukin-1β and interleukin-6 according to the manufacturer's instructions, as previously described [20,21].

2.6. NFκ-B, iNOS and COX-2 mRNA Analysis by Real-Time PCR

Total RNA was extracted from colon tissues (10 mg) using RNAeasy Mini kit (Qiagen, Valencia, CA, USA). 2 ng of total RNA were reverse transcribed into cDNA using RT FirstStrand kit (Qiagen, Valencia, CA, USA). The amplification of synthesized cDNAs was performed using SYBR Premix Ex Taq II (TaKaRa, Bio Inc., Foster City, CA, USA) and StepOne Real-Time instrument (Applied Biosystems, Foster City, CA, USA). Gene expression of inducible NOS (iNOS), COX-2 and beta actin, a housekeeping gene that is not subject to regulation, was performed in triplicate, using specific primers and amplification conditions. The oligonucleotide primer sequences were reported in Table 1.

Table 1. Primers used for qRT-PCR.

Gene Description	Forward Primer	Reverse Primer	Ref.
β-actin	5′-CTAAGGCCAACCGTGAAAAG-3′	5′-GCCTGGATGGCTACGTACA-3′	[26]
COX 2	5′-TGCGATGCTCTTCCGAGCTGTGCT-3′	5′-TCAGGAAGTTCCTTATTTCCTTTC-3′	[27]
iNOS	5′-AGAAGGGGACGAACTCAGC-3′	5′-TCGAACATCGAACGTCTCAC-3′	[28]
NFκ–B p65	5′-GTCATCAGGAAGAGGTTTGGCT-3′	5′-TGATAAGCTTAGCCCTTGCAGC-3′	[29]

The program for PCR was 1 cycle of 95 °C (10 min), followed by 45 cycles of amplification. Consisting of denaturation at 95 °C (15 s), annealing at 60 °C (30 s), and extension for at 72 °C (30 s). For terminal elongation period, the samples were incubated at 72 °C (additional 10 min) at the end of the final cycle. The expression level was calculated from the PCR cycle number (CT) where the increased fluorescence curve passes across a threshold value. The relative gene expression of the target genes was calculated using $2^{-\Delta\Delta Ct}$ approximation method algorithm.

2.7. Analysis of Reactive Oxygen Species (ROS) Generation

The conversion of non-fluorescent DCFH-DA to 2′, 7′ dichlorofluorescein (DCF) was evaluated, as previously described [20,21], to monitor the amount of hydrogen peroxidase in the sample. Sample was analyzed by fluorimeter (Microplate reader WallacVictor 2-1420 Multilabel Counter; PerkinElmer, Inc., Waltham, MA, USA) at an excitation wavelength of 485 nm and an emission wavelength of 530 nm.

2.8. Nitric Oxide (NO) Concentration Assay

Nitric oxide concentration was measured using the Griess reaction [30]. Briefly after homogenization of rat colon tissue (10 mg) with 1 mL PBS, pH 7.2 and centrifugation (14,000 rpm, for 30 min, 4 °C) the supernatant was incubated with equal volume of Griess reagent (room temperature, 15 min in the dark) and the absorbance was measured at 550 nm with a spectrophotometric Microplate reader (WallacVictor 2 Multilabel Counter, Perkin Elmer, Apeldoorn, The Netherlands) NO concentration was evaluated by using a standard curve.

2.9. Data Analysis and Statistical Tests

Results are shown as the mean ± SEM: '*n*' indicates the number of animals. Statistical analysis was performed using GraphPad Prism 6.0 software, and sets were assessed by one-way ANOVA followed by Tukey's multiple comparison test. p value < 0.05 means statistically significant. To compute the sample size G*Power version 3.1.2 [31], was used, given power $(1 - \beta) = 0.8$.

3. Results

DNBS challenge in rats induced a substantial decrease in the body weight starting by day 2 following the enema, in contrast to the sham group which showed body weight gain. During the same period DNBS group experienced diarrhea and also rectal bleeding was observed. AphaMax® (20–100 mg/Kg for 14 day) dose-dependently prevented the drop in the animal weight and improved the diarrheal status and bleeding in DNBS treated rats (Table 2).

Table 2. Effects of Aphamax® on body weight change, stool consistency, rectal bleeding 2- and 7- days after colitis-induction.

Groups	% Body Weight Change 2 Days after Colitis Induction	% Body Weight Change 7 Days After Colitis Induction	Stool Consistency 2 Days after Colitis Induction	Stool Consistency 7 Days after Colitis Induction	Rectal Bleeding 2 Days after Colitis Induction	Rectal Bleeding 7 Days after Colitis Induction
Sham	102.1 ± 0.5	109 ± 0.8	0	0	0	0
Sham + 100 mg/kg Aphamax®	103.2 ± 1.2	107 ±1.2	0	0	0	0
DNBS	94.2 ± 0.8 *	95.1 ± 1.5 *	4.0 ± 0.0 *	3.8 ± 0.2 *	4.0 ± 0.0 *	3.8 ± 0.2 *
DNBS+ 20 mg/kg Aphamax®	95.4 ± 1.3 *	99.2 ± 0.9 *	3.2 ± 0.2 *	2.8 ± 0.2 *	3.2 ± 0.4 *	2.0 ± 0.3 * §
DNBS+ 50 mg/kg Aphamax®	96.7 ± 0.9 *	102.2 ± 0.5 §	2.8 ± 0.2 *	2.0 ± 0.4 * §	2.2 ± 0.4 * §	1.2 ± 0.2 §
DNBS+ 100 mg/kg Aphamax®-	97.8 ± 1.1 *	104.4 ± 1.5 §	2.7 ± 0.4 * §	1.0 ± 0.3 §	2.0 ± 0.6 * §	0.4 ± 0.2 §

Percent of body weight change compared to the original body weight (taken as 100%), consistency of the stool rectal bleeding scored following Cooper et al. [22] method in sham rats or in 20, 50, and 100 mg/kg Aphamax®- treated rats with or without colitis induction. Data are means ± S.E.M. $n = 5$ animals/ each group. * $p < 0.05$ *versus* Sham group; § $p < 0.05$ *versus* 2,4 dinitrobenzensulfonic acid (DNBS) group.

3.1. AphaMax® Effects on Macroscopic Changes in Colon of Colitis Animals

Macroscopic evaluation demonstrated that in the sham group, the distal colon showed no epithelial damage differently from colitis group in which intense mucosal damage, increased wall thickness, hyperemia, ulceration, edema and necrosis were observed associated with high macroscopic score (Figure 1A). DNBS rats revealed also colon shortening accompanied by an increased in the ratio colon weight/length, a marker of tissue inflammation, (13.27 ± 0.46 cm vs. 9.75 ± 0.41 cm colon length in Sham group and in DNBS group respectively $p < 0.05$; 1330.01 ± 120.3 mg vs. 3167.10 ± 443.21 mg colon weight in Sham group and in DNBS group respectively $p < 0.05$) (Figure 1B). AphaMax® (20, 50, or 100 mg/kg) treatment decreased the DNBS-induced macroscopic changes by improving the inflammation symptoms in colon tissues such as mucosal injury, size of ulcer area and also a reduction of the colon weight/length ratio, (11.25 ± 0.45; 12.16 ± 0.65 and 12.91 ± 0.47 cm in colon length in DNBS + AphaMax® 20, 50, or 100 mg/kg respectively ; 2447.03 ± 147.51 mg, 2141.66 ± 80.01 mg, 2095.01 ± 122.03 mg colon weight in in DNBS + AphaMax® 20, 50, or 100 mg/kg respectively, Figure 1B). Diffuse adhesions of the colon with other organs, typically observed during the acute phase of colitis, were unremarkable and only sporadically observed in AphaMax®-treated animals at all doses tested.

AphaMax® *per se*, at the highest dose used, 100 mg/Kg, had no effect on sham animals (Figure 1A,B).

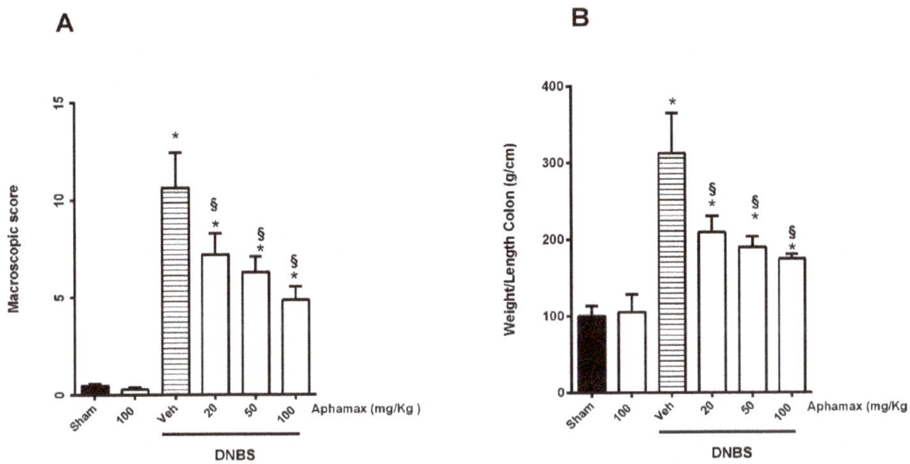

Figure 1. Effects of Aphamax® on DNBS-induced macroscopic colonic damage. (**A**) Macroscopic damage score based on Appleyard and Wallace method and (**B**) Colon weight/Length ratio in sham rats or in Aphamax®-treated rats with or without colitis induction. Data are means ± S.E.M. $n = 5$ animals /group. * $p < 0.05$ *versus* Sham animals; § $p < 0.05$ *versus* 2,4 dinitrobenzensulfonic acid (DNBS) animals.

3.2. AphaMax® Effects on Histopathological Changes in Inflamed Colon Tissue

Histological analyses showed in the sham group a colon tissue with normal mucosa, submucosa, crypts, muscularis, lamina propria, and serosa, without inflammatory cell infiltration and necrosis. A score of the histological changes in colitis-induced rats was calculated, as previously described [18–21] (Table 2). In the DNBS group, the colon showed serious histological changes with high scores of microscopic damages, characterized by large ulcers with necrosis, transmural inflammation, disrupted crypts and massive infiltration into lamina propria and submucosa of inflammatory cells, consisting of lymphocytes, macrophages, and neutrophils. A net distinction of the colon layers from one another was not allowed AphaMax® (20–100 mg/Kg) treatment induced a progressive reduction of inflammatory cell infiltration and decreased the DNBS-mediated colonic damages leading to an improvement in the microscopic damage score; however no one of the concentrations tested completely resolved the histological changes ($p < 0.05$ compared to sham group) (Figure 2A,B).

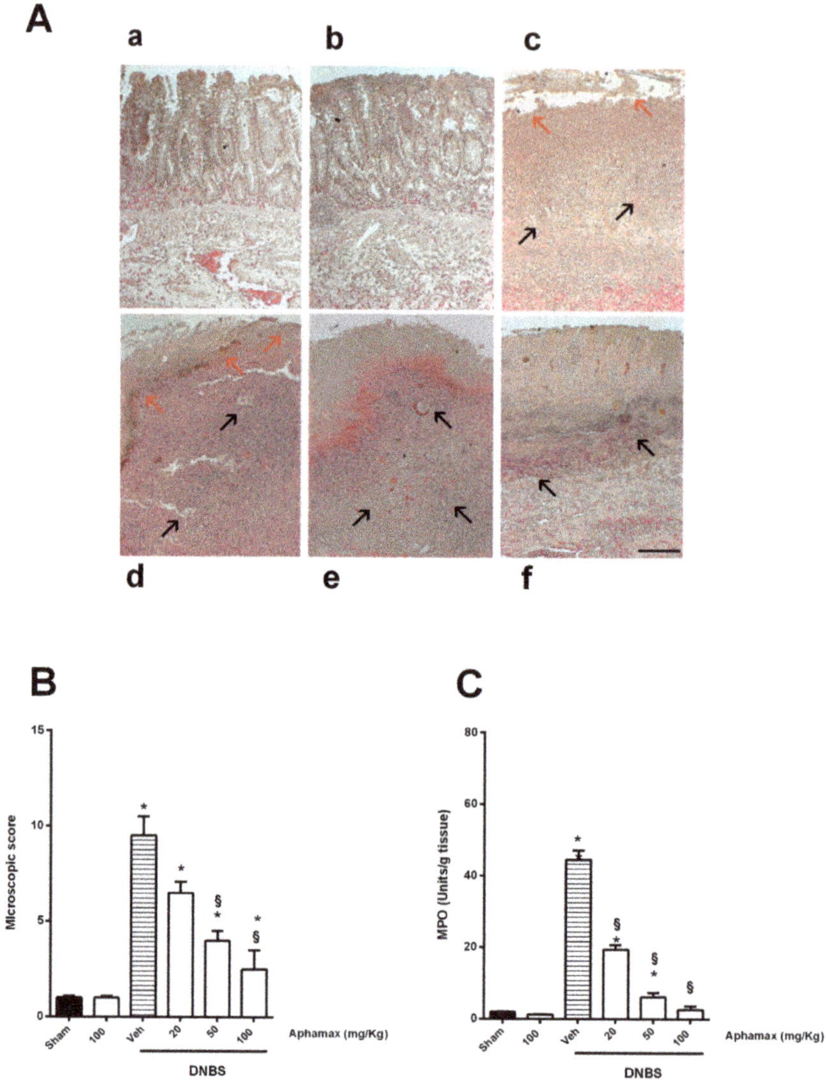

Figure 2. Effects of Aphamax® on DNBS-induced histological damage. (**A**) Photomicrographs of the colon stained with H&E from (**a**) Sham animals, (**b**) Sham rats treated with Aphamax® (100 mg/kg), (**c**) DNBS animals showing colonic damage and a diffuse infiltration of inflammatory cells in mucosa and submucosa or (**d**) DNBS rats treated with Aphamax® (20 mg/kg), (**e**) or with Aphamax® (50 mg/kg), showing progressive reduction of submucosal infiltration of inflammatory cells in submucosal layer and (**f**) DNBS rats treated with Aphamax® (100 mg/kg) showing few inflammatory cells close to the mucosal layer. (Scale bar = 100 μm, magnification 20×, red arrows = colonic damage, black arrows = inflammatory infiltrate)) (**B**) Microscopic damage scored with Hunter method and (**C**) activity of colonic myeloperoxidase in sham rats or Aphamax® treated rats with or without colitis induction. Data are means ± S.E.M. $n = 5$ animals for each group. * $p < 0.05$ *versus* Sham group; § $p < 0.05$ *versus* 2,4 dinitrobenzensulfonic acid (DNBS) group.

3.3. *AphaMax® Effects on MPO and Cytokine Levels*

To explore the possible action mechanism of AphaMax® extract, we examined, at first, the myeloperoxidase levels, index of a massive infiltration of neutrophil into the inflamed tissue. In DNBS animals, the MPO activity was significantly increased compared to the sham group. MPO activity was reduced by AphaMax® treatment in a dose-dependent manner, confirming the histological observation showing a lower infiltration of leukocytes in the colonic tissue (Figure 3C). Moreover, we evaluated the production of proinflammatory cytokines, highly involved in IBD, such as IL-1β and IL-6. In accordance with the immune cell infiltration detected in the histological examination of the DNBS colon, increased levels of IL-1β, IL-6 in comparison to sham animals were observed. Moreover, the reduction in the histological score was correlated with a significant reduction of the levels of IL-1β and IL-6 in AphaMax®-treated animals, compared with DNBS group, being however still significantly higher compared to sham group ($p < 0.05$) (Figure 3A).

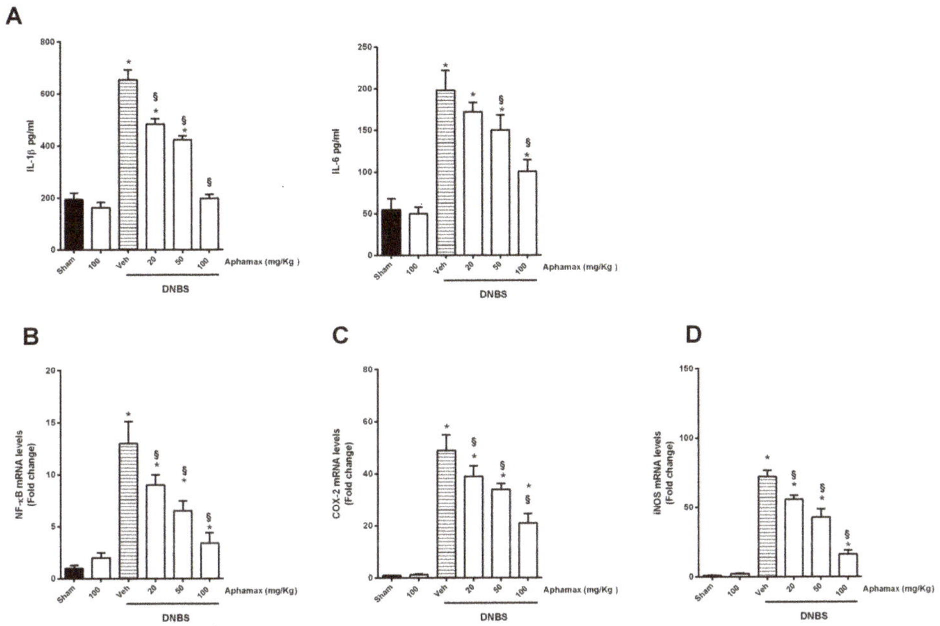

Figure 3. Effects of Aphamax® on pro-inflammatory mediator levels (**A**) Levels of IL-1β, IL-6 levels; (**B**) NF-kB, (**C**) COX-2 and (**D**) iNOS mRNA expression. Data are means ± S.E.M. $n = 5$ animals for each group. * $p < 0.05$ *versus* Sham group; § $p < 0.05$ *versus* 2,4 dinitrobenzensulfonic acid (DNBS) group.

3.4. *AphaMax® Effect on NF-κB p65, COX-2 and iNOS mRNA Expression*

We assessed the expression of the activated NF-κB p65 subunit, which plays a crucial proinflammatory role during the pathogenesis of IBD. The colon from the DNBS group showed an extensive increase in the mRNA expression of NF-κB p65. NF-κB p65 subunit mRNA expression was significant decreased in AphaMax®-treated animals suggesting that a down-regulation of these proinflammatory factors could be implicated in AphaMax® beneficial protective effects (Figure 3B).

As already mentioned, the beneficial effects of blue-green algae could be partly ascribed to the ability of their components to selectively inhibit key enzymes in inflammatory disease as COX-2 and iNOS [32]. Thus, we analyzed mRNA expression of both enzymes in the colonic tissues from the different animal groups. DNBS group showed an increased expression of both enzyme compared to control group, which was reduced after AphaMax® treatment in dose-dependent manner, ($p < 0.05$ compared to DNBS and to sham group) (Figure 3C,D).

3.5. *AphaMax® Effect on Nitrite and ROS Production*

Lastly, to investigate possible antioxidant proprieties of AphaMax® we evaluated if the treatment could affect the tissue NO and ROS generation.

DNBS induced a significant increase of the levels of nitrites and ROS in the colon of DNBS group compared with sham group ($p < 0.05$). AphaMax® at the different doses tested induced a moderate ($p < 0.05$ compared to sham groups) but significant down-regulation in the production of both nitrites and ROS in colitis rats ($p < 0.05$ compared to DNBS group) (Figure 4A,B).

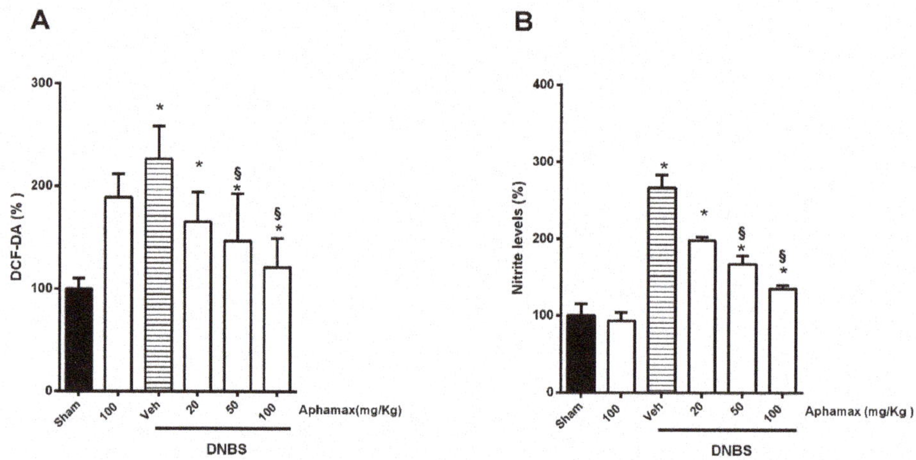

Figure 4. Effects of Aphamax® on ROS and nitrite levels. (**A**) ROS levels measured using DCFH-DA, a peroxide/redox-sensitive fluorescent method. (**B**) nitrite levels detected using Griess assay. Data are means ± S.E.M. $n = 5$ for each group. * $p < 0.05$ *versus* Sham group; § $p < 0.05$ *versus* 2,4 dinitrobenzensulfonic acid (DNBS) group.

4. Discussion

The use of microalgae recently attracted considerable attention worldwide due to their considerable application potential in the biopharmaceutical and nutraceutical industries. They are renewable, sustainable, and inexpensive sources of bioactive molecules and food with surprising pharmacological and biological qualities. In particular, blue-green algae were demonstrated to have remarkable antioxidant and anti-inflammatory properties due to the presence of many bioactive substances [9–12]. Among the blue-green algae extracts, AphaMax® contains high concentrated amount of AFA-phycocyanins and AFA-phytochrome, which are the components with the greatest antioxidant, anti-inflammatory, anticancer, and cardiovascular properties.

The aim of this study was to test the effects of AphaMax®, a commercial extract of the blue-green Klamath microalgae (AFA), on colonic inflammatory condition induced in an experimental model of colitis, the DNBS rats.

The rationale for this comes by the awareness that the current IBD therapies have significant side effects and modest results for long-term management and that recently, much evidence supports the key role of diet and nutritional factors in IBD. Our results indicate the AphaMax® supplementation is able to reduce colon injury induced by DNBS in rats, mainly due to its antiinflammatory and anti-oxidant effects.

As previously reported [18–21], intracolonic administration of DNBS causes an acute colitis in rats, characterized by marked reduction in body weight, diarrhoea with occasional blood in the stool, colonic shortening, an augmented colonic wall thickness, with a significant increase in the colonic weight/length ratio, reliable indicator of tissue oedema, and an extensive transmural, granulomatous inflammation of the distal colon.

Fourteen days of AphaMax® treatment leaded to a reduction of the aforementioned parameters as the deterioration of body weight, diarrhoea, and the increased colon weight/length ratio. Also histological analysis of the colon confirmed the macroscopic observations, suggesting that AphaMax® presents an antiulcerogenic effect, as confirmed by preservation of the colon architecture and by the attenuation of mucosal disruption, as well as of histopathological damage and oedema.

Few studies have been conducted on the possible protective effects of BGA against experimental colitis, but this is the first study about the effect of Klamath algae in such conditions. Oral administration of cyanobacterium Spirulina Platensis was reported to have protective effects in other models of experimental colitis showing both antioxidant (reduction in oxidative stress and augmented endogenous antioxidant mechanisms) and anti-inflammatory (reduction in inflammatory cytokine levels and neutrophil infiltration) effects [33]. Moreover, in ulcerative colitis patients, diet supplementation with the green algae Chlorella pyrenoidosa accelerated wound healing [34].

Furthermore, we can speculate about the mechanisms underlying the intestinal beneficial effects of the AFA extract. Among the possible mechanisms we can suppose a relation with the well-known positive properties, ascribed to the majority of the blue-green algae extracts, due to their content in compounds, as phycocyanins.

In particular, Klamath algae contains the pigment C-phycocyanin (C-PHY) bound to a structural component phycoerythrocyanin (PEC). PEC is a photosynthetic component identified only in a limited number of cyanobacterial species, likely responsible for the higher antioxidant and anti-inflammatory effects of Klamath algae compared to other cyanobacters [14,15]. C-phycocyanins can act as radical scavenger in oxidative stress-induced diseases and they have strong antioxidant and anti-inflammatory proprieties as demonstrated by various in vitro and in in vivo evidences [35,36].

Moreover, in our experimental model of colitis, AphaMax® treatment induced an attenuation of the neutrophil infiltration, as shown by both histological observations and biochemical data showing a significant reduction in colonic myeloperoxidase activity, confirming the ability of this extract to modulate the altered immune response, likely ascribable to the C-phycocyanins action. This observation is in agreement with Gonzalez et al. [37] where C-phycocyanins extracted from the Blue-green algae Arthospira maxima, have been reported to reduce significantly myeloperoxidase activity, inflammatory cell infiltration and colonic damage in acetic acid-induced colitis in rats. We are aware that the model of colitis generation in Gonzales et al. [37] is different from our model; however, it is possible to observe that in our study the highest doses of AphaMax® (25% AFA-PC concentration) was more effective in the reduction of myeloperoxidase activity of Spirulina (80% PC concentration) indicating the major anti-inflammatory and protective power of AFA-phycocyanins.

Although in our study, AphaMax®, at all doses tested, attenuated the colitis signs in the colonic tissue, we can suggest that the higher dose tested, 100 mg/Kg, containing a concentration of AFA-phycocyanins and AFA-phytochrome, carotenoids and polyphenols able to significantly improve all the analyzed scores, could be considered the best to attenuate the severity of colitis.

Moreover, the observation that AphaMax® extract, even at the highest dose tested, did not affect sham rats, might indicate that AphaMax® extract exerts its anti-inflammatory and antioxidant effects, only in the course of inflammation.

Interestingly, AphaMax® treatment is sufficient to attenuate activation of NF-κB p65. NF-κB, is a redox-sensitive transcription factor, key regulator of inflammation, innate immunity, and tissue integrity. NF-κB phosphorylation and its nuclear translocation correlate with the severity of intestinal inflammation [38,39] due to the regulation of gene expression of molecules playing a pivotal role in inflammation as molecules of adhesion, chemokines, and cytokines [40]. We can suppose that the attenuation of DNBS colitis features, observed in our study, could be the consequence of the inhibition of early steps of inflammation. A reduced activation of NF-κB would lead to the infiltrating cells to decrease the amounts of inflammatory mediators and subsequently to preserve mucosal integrity. Once more, such an effect could be also positively related to the high concentration of C-phycocyanins present in the extract, which include C-phycocyanins + PEC (phycoerythrocyanins), and C-phycocyanins by themselves have been reported to be able to suppress activation of NF-κB in RAW 264.7 macrophages stimulated with LPS [41].

The DNBS model is also commonly associated with the T-helper (Th)1 response, and with an overproduction of proinflammatory cytokines, as IL-6 and IL 1β. Many of the IBD treatments act to regulate the levels of proinflammatory cytokines [42]. These immunoregulatory cytokines are involved in the initiation of the inflammatory response in colitis, amplifying the inflammatory reaction by triggering a cascade of immune cells, impairing intestinal permeability, and causing severe colonic infiltration [43] In addition, IL-1β contributes to the induction of the epithelial cell necrosis. Our results indicated that AphaMax® treatment caused a dose-dependent reduction of both IL-6 and IL- 1β cytokine levels in colitis rats, thus ameliorating the deregulated immune response typical of experimental colitis. Once more, such an effect could be likely due to the high concentration of AFA C-phycocyanins.

Moreover, AphaMax® significantly attenuated expression of the enzymes COX-2 and iNOS which were up-regulated in colitis rats. COX-2 is a key enzyme in the pathogenesis of IBD involved in the biosynthesis of prostaglandin [44], and target of many drugs used in the treatment of human IBD, including aminosalicylates [45,46]. This inhibition is not unexpected since C-phycocyanins are reported to be selective COX-2 inhibitors [31,47]. The expression of iNOS at sites of inflammation, acting in synergy with COX-2 to produce excessive inflammatory mediators has been reported. Their products may be detrimental to the colon integrity and contribute to the intestinal hypomotility and subsequent bacterial overgrowth [48], a typical features of the inflammatory reaction [49]. Accordingly, AphaMax® treatment reduced nitrite levels in colitis rats. Considering that expression of both COX-2 and iNOS is regulated by NF-κB [50,51] the observed inhibition of NF-κB and its downstream effectors, COX-2 and iNOS, can be crucial for the protective role played by AphaMax® treatment in DNBS colitis rats.

Lastly, in colorectal biopsies of human IBD [52,53] and in different experimental models of colitis, including the DNBS model, a mucosal production of reactive oxygen mediators was described, which would contribute to tissue damage during chronic intestinal inflammation [54].

The major effects of oxidative stress reported are local intestinal cell damage and activation of several signaling pathways, initiating inflammation [54].

In our preparations, AphaMax® treatment was able to modulate the redox status by scavenging ROS and to reduce the severity of the colitis, effect that is likely ascribable to AFA-phycocyanins, which possess antioxidant activity 75 to 200 times greater than that other more common phycocyanins [14]. Phycocyanins were reported to be able to scavenge hydroxyl and alcoxyl radicals in an acetic acid-induced colitis model [36]. Thereafter, since free radical generation could contribute to the initial infiltration of neutrophils in the colonic mucosa, we can speculate that ability of our extract in preserving the colonic mucosa from oxidative insult could participate to decrease the neutrophil infiltration DNBS-induced. However, whether the antioxidative effect precedes the anti-inflammatory one or not is still uncertain and has yet to be explored.

We focused our attention on the AphaMax® effects on the local production of ROS and oxidant insult in colonic tissue; however, since a plasmatic increase of advanced oxidation protein products formed during oxidative stress in IBD patients was also reported [55], further studies investigating possible AphaMax® systemic effects are warranted.

5. Conclusions

In conclusion, taken together, data presented in this study demonstrate that AphaMax® powder has pharmacologically promising positive activity on DNBS-mediated experimental colitis.

Although several mechanisms could be proposed to explain AphaMax® beneficial effects, we highlight the reduction of inflammatory and oxidant mediators. However, further studies should be performed to resolve the exact molecular mechanisms. A limitation of this study is the lack of long-term observation following AphaMax® treatment, since DNBS rats recovered within 10 days of the experiment, making it difficult to observe and compare the long-term effects of treatment on IBD. In addition, hapten-administration shares many features with human inflammatory bowel diseases, but the "acute" inflammatory response evoked should be different from the mechanism of chronic colitis in human IBD.

Moreover, the aim of the current study did not include the analysis of the effects of AphaMax® treatment on the colon functions, we are aware that changes in colon motility and in intestinal permeability are described in DNBS-induced colitis in rats. Since it is well reported that antioxidant compounds are able to improve intestinal motility [56], additional studies could be performed to assess whether the mitigation of the inflammatory process by the AphaMax® treatment, could help the colonic mucosa and the smooth muscle to recover their functionality. We hope that results of this study can prompt further investigations on AphaMax® as a natural product in the management of intestinal inflammatory diseases.

Author Contributions: Conceptualization, M.G.Z. and R.S.; Investigation, M.G.Z., A.B., G.C. and D.N.; Data curation, M.G.Z. and D.N.; Supervision S.S. and R.S.; Writing—original draft, M.G.Z.; Writing—review & editing, M.D.C., S.S. and R.S. All authors have read and agreed to the published version of the manuscript.

Funding: This work was supported by the company Nutrigea, Borgo Maggiore, Republic of San Marino.

Conflicts of Interest: S.S. is the R&D Director of the company that owns AphaMax®'s patent.

References

1. Kaser, A.; Zeissig, S.; Blumberg, R.S. Inflammatory bowel disease. *Annu. Rev. Immunol.* **2010**, *28*, 573–621. [CrossRef]
2. Mizoguchi, A.; Mizoguchi, E.; Takedatsu, H.; Blumberg, R.S.; Bhan, A.K. Chronic intestinal inflammatory condition generates IL-10-producing regulatory B cell subset characterized by CD1d upregulation. *Immunity* **2002**, *16*, 219–230. [CrossRef]
3. Pizarro, T.T.; Cominelli, F. Cytokine therapy for Crohn's disease: Advances in translational research. *Annu. Rev. Med.* **2007**, *58*, 433–444. [CrossRef] [PubMed]
4. Mudter, J.; Amoussina, L.; Schenk, M.; Yu, J.; Brustle, A.; Weigmann, B.; Atreya, R.; Wirtz, S.; Becker, C.; Hoffman, A.; et al. The transcription factor IFN regulatory factor-4 controls experimental colitis in mice via T cell-derived IL-6. *J. Clin. Investig.* **2008**, *118*, 2415–2426. [CrossRef] [PubMed]
5. Kucharzik, T.; Walsh, S.V.; Chen, J.; Parkos, C.A.; Nusrat, A. Neutrophil transmigration in inflammatory bowel disease is associated with differential expression of epithelial intercellular junction proteins. *Am. J. Pathol.* **2001**, *159*, 2001–2009. [CrossRef]
6. Schulzke, J.D.; Ploeger, S.; Amasheh, M.; Fromm, A.; Zeissig, S.; Troeger, H.; Richter, J.; Bojarski, C.; Schumann, M.; Fromm, M. Epithelial tight junctions in intestinal inflammation. *Ann. N. Y. Acad. Sci.* **2009**, *1165*, 294–300. [CrossRef]
7. Na, S.Y.; Moon, W. Perspectives on Current and Novel Treatments for Inflammatory Bowel Disease. *Gut Liver* **2019**, *13*, 604–616. [CrossRef]
8. Gonzalez de, R.C.; Miranda-Zamora, R.; Diaz-Zagoya, J.C.; Juarez-Oropeza, M.A. Preventive effect of Spirulina maxima on the fatty liver induced by a fructose-rich diet in the rat, a preliminary report. *Life Sci.* **1993**, *53*, 57–61. [CrossRef]
9. Ciferri, O. Spirulina, the edible microorganism. *Microbiol. Rev.* **1983**, *47*, 551–578. [CrossRef]
10. Ku, C.S.; Yang, Y.; Park, Y.; Lee, J. Health benefits of blue-green algae: Prevention of cardiovascular disease and nonalcoholic fatty liver disease. *J. Med. Food* **2013**, *16*, 103–111. [CrossRef]
11. Yousefi, R.; Saidpour, A.; Mottaghi, A. The effects of Spirulina supplementation on metabolic syndrome components, its liver manifestation and related inflammatory markers: A systematic review. *Complement. Ther. Med.* **2019**, *42*, 137–144. [CrossRef] [PubMed]

12. Iyer Uma, M.; Sophia, A.; Mani Uliyar, V. Glycemic and lipemic responses of selected Spirulina supplemented rice-based recipes in normal subjects. *Int. J. Diabetes Dev. Ctries* **1999**, *19*, 17–22.
13. Xue, X.; Lv, Y.; Liu, Q.; Zhang, X.; Zhao, Y.; Zhang, L.; Xu, S. Extracellular polymeric substance from Aphanizomenon flos-aquae induces apoptosis via the mitochondrial pathway in A431 human epidermoid carcinoma cells. *Exp. Ther. Med.* **2015**, *10*, 927–932. [CrossRef] [PubMed]
14. Benedetti, S.; Rinalducci, S.; Benvenuti, F.; Francogli, S.; Pagliarani, S.; Giorgi, L.; Micheloni, M.; D'Amici, G.M.; Zolla, L.; Canestrari, F. Purification and characterization of phycocyanin from the blue-green alga Aphanizomenon flos-aquae. *J. Chromatogr. B Analyt. Technol. Biomed. Life Sci.* **2006**, *833*, 12–18.
15. Benedetti, S.; Benvenuti, F.; Pagliarani, S.; Francogli, S.; Scoglio, S.; Canestrari, F. Antioxidant properties of a novel phycocyanin extract from the blue-green alga Aphanizomenon flos-aquae. *Life Sci.* **2004**, *75*, 2353–2362. [CrossRef]
16. Scoglio, S.; Benedetti, S.; Canino, C.; Santagni, S.; Rattighieri, E.; Chierchia, E.; Canestrari, F.; Genazzani, A.D. Effect of a 2-month treatment with Klamin, a Klamath algae extract, on the general well-being, antioxidant profile and oxidative status of postmenopausal women. *Gynecol. Endocrinol.* **2009**, *25*, 235–240. [CrossRef]
17. Scoglio, S.; Benedetti, Y.; Benvenuti, F.; Battistelli, S.; Canestrari, F.; Benedetti, S. Selective monoamine oxidase B inhibition by an Aphanizomenon flos-aquae extract and by its constitutive active principles phycocyanin and mycosporine-like amino acids. *Phytomedicine* **2014**, *21*, 992–997. [CrossRef]
18. Morris, G.P.; Beck, P.L.; Herridge, M.S.; Depew, W.T.; Szewczuk, M.R.; Wallace, J.L. Hapten-induced model of chronic inflammation and ulceration in the rat colon. *Gastroenterology* **1989**, *96*, 795–803. [CrossRef]
19. Zizzo, M.G.; Auteri, M.; Amato, A.; Caldara, G.; Nuzzo, D.; Di, C.M.; Serio, R. Angiotensin II type II receptors and colonic dysmotility in 2,4-dinitrofluorobenzenesulfonic acid-induced colitis in rats. *Neurogastroenterol. Motil.* **2017**, *29*, e13019. [CrossRef]
20. Zizzo, M.G.; Caldara, G.; Bellanca, A.; Nuzzo, D.; Di, C.M.; Serio, R. Preventive effects of guanosine on intestinal inflammation in 2, 4-dinitrobenzene sulfonic acid (DNBS)-induced colitis in rats. *Inflammopharmacology* **2019**, *27*, 349–359. [CrossRef]
21. Zizzo, M.G.; Caldara, G.; Bellanca, A.; Nuzzo, D.; Di, C.M.; Serio, R. PD123319, angiotensin II type II receptor antagonist, inhibits oxidative stress and inflammation in 2, 4-dinitrobenzene sulfonic acid-induced colitis in rat and ameliorates colonic contractility. *Inflammopharmacology* **2020**, *28*, 187–199. [CrossRef] [PubMed]
22. Cooper, H.S.; Murthy, S.N.; Shah, R.S.; Sedergran, D.J. Clinicopathologic study of dextran sulfate sodium experimental murine colitis. *Lab Investig.* **1993**, *69*, 238–249. [PubMed]
23. Appleyard, C.B.; Wallace, J.L. Reactivation of hapten-induced colitis and its prevention by anti-inflammatory drugs. *Am. J. Physiol.* **1995**, *269*, G119–G125. [CrossRef] [PubMed]
24. Hunter, M.M.; Wang, A.; Hirota, C.L.; McKay, D.M. Neutralizing anti-IL-10 antibody blocks the protective effect of tapeworm infection in a murine model of chemically induced colitis. *J. Immunol.* **2005**, *174*, 7368–7375. [CrossRef]
25. Moreels, T.G.; Nieuwendijk, R.J.; De Man, J.G.; De Winter, B.Y.; Herman, A.G.; Van Marck, E.A.; Pelckmans, P.A. Concurrent infection with Schistosoma mansoni attenuates inflammation induced changes in colonic morphology, cytokine levels, and smooth muscle contractility of trinitrobenzene sulphonic acid induced colitis in rats. *Gut* **2004**, *53*, 99–107. [CrossRef]
26. Adachi, M.; Watanabe, M.; Kurata, Y.; Inoue, Y.; Notsu, T.; Yamamoto, K.; Horie, H.; Tanno, S.; Morita, M.; Miake, J.; et al. Beta-Adrenergic Blocker, Carvedilol, Abolishes Ameliorating Actions of Adipose-Derived Stem Cell Sheets on Cardiac Dysfunction and Remodeling After Myocardial Infarction. *Circ. J.* **2019**, *83*, 2282–2291. [CrossRef]
27. Bhatia, H.S.; Candelario-Jalil, E.; de Oliveira, A.C.P.; Olajide, O.A.; Martínez-Sánchez, G.; Fiebich, B.L. Mangiferin inhibits cyclooxygenase-2 expression and prostaglandin E2 production in activated rat microglial cells. *Arch. Biochem. Biophys.* **2008**, *477*, 253–258. [CrossRef]
28. Sönmez, M.F.; Dündar, M. Ameliorative effects of pentoxifylline on NOS induced by diabetes in rat kidney. *Ren. Fail.* **2016**, *38*, 605–613. [CrossRef]
29. Arab, H.H.; Al-Shorbagy, M.Y.; Abdallah, D.M.; Nassar, N.N. Telmisartan Attenuates Colon Inflammation, Oxidative Perturbations and Apoptosis in a Rat Model of Experimental Inflammatory Bowel Disease. *PLoS ONE* **2014**, *9*, e97193. [CrossRef]
30. Roediger, W.E.; Radcliffe, B.C. Role of nitrite and nitrate as a redox couple in the rat colon. Implications for diarrheal conditions. *Gastroenterology* **1988**, *94*, 915–922. [CrossRef]

31. Faul, F.; Erdfelder, E.; Lang, A.; Buchner, A. G*Power 3: A flexible statistical power analysis program for the social, behavioral, and biomedical sciences. *Behav. Res. Methods* **2007**, *39*, 175–191. [CrossRef] [PubMed]
32. Kefayat, A.; Ghahremani, F.; Safavi, A.; Hajiaghababa, A.; Moshtaghian, J. C-phycocyanin: A natural product with radiosensitizing property for enhancement of colon cancer radiation therapy efficacy through inhibition of COX-2 expression. *Sci. Rep.* **2019**, *9*, 19161. [CrossRef] [PubMed]
33. Abdel-Daim, M.M.; Farouk, S.M.; Madkour, F.F.; Azab, S.S. Anti-inflammatory and immunomodulatory effects of Spirulina platensis in comparison to Dunaliella salina in acetic acid-induced rat experimental colitis. *Immunopharmacol. Immunotoxicol.* **2015**, *37*, 126–139. [CrossRef] [PubMed]
34. Merchant, R.E.; Andre, C.A. A review of recent clinical trials of the nutritional supplement Chlorella pyrenoidosa in the treatment of fibromyalgia, hypertension, and ulcerative colitis. *Altern. Ther. Health Med.* **2001**, *7*, 79–91. [PubMed]
35. Pleonsil, P.; Soogarun, S.; Suwanwong, Y. Anti-oxidant activity of holo- and apo-c-phycocyanin and their protective effects on human erythrocytes. *Int. J. Biol. Macromol.* **2013**, *60*, 393–398. [CrossRef] [PubMed]
36. Romay, C.; Gonzalez, R.; Ledon, N.; Remirez, D.; Rimbau, V. C-phycocyanin: A biliprotein with antioxidant, anti-inflammatory and neuroprotective effects. *Curr. Protein Pept. Sci.* **2003**, *4*, 207–216. [CrossRef] [PubMed]
37. Gonzalez, R.; Rodriguez, S.; Romay, C.; Gonzalez, A.; Armesto, J.; Remirez, D.; Merino, N. Anti-Inflammatory Activity of Phycocyanin Extract in Acetic Acid-Induced Colitis in Rats. *Pharmacol. Res.* **1999**, *39*, 1055–1059. [CrossRef] [PubMed]
38. Lawrence, T. The nuclear factor NF-kappaB pathway in inflammation. *Cold Spring Harb. Perspect. Biol.* **2009**, *1*, a001651. [CrossRef]
39. Buhrmann, C.; Mobasheri, A.; Busch, F.; Aldinger, C.; Stahlmann, R.; Montaseri, A.; Shakibaei, M. Curcumin modulates nuclear factor kappaB (NF-kappaB)-mediated inflammation in human tenocytes in vitro: Role of the phosphatidylinositol 3-kinase/Akt pathway. *J. Biol. Chem.* **2011**, *286*, 28556–28566. [CrossRef]
40. Liu, Z.; Kong, F.; Vallance, J.E.; Harmel-Laws, E.; Amarachintha, S.; Steinbrecher, K.A.; Rosen, M.J.; Bhattacharyya, S. Activation of TGF-beta activated kinase 1 promotes colon mucosal pathogenesis in inflammatory bowel disease. *Physiol. Rep.* **2017**, *5*, e13181. [CrossRef]
41. Cherng, S.C.; Cheng, S.N.; Tarn, A.; Chou, T.C. Anti-inflammatory activity of c-phycocyanin in lipopolysaccharide-stimulated RAW 264.7 macrophages. *Life Sci.* **2016**, *81*, 1431–1435. [CrossRef] [PubMed]
42. Yadav, V.; Varum, F.; Bravo, R.; Furrer, E.; Bojic, D.; Basit, A.W. Inflammatory bowel disease: Exploring gut pathophysiology for novel therapeutic targets. *Transl. Res.* **2016**, *176*, 38–68. [CrossRef] [PubMed]
43. Hering, N.A.; Schulzke, J.D. Therapeutic options to modulate barrier defects in inflammatory bowel disease. *Dig. Dis.* **2009**, *27*, 450–454. [CrossRef] [PubMed]
44. Norregaard, R.; Kwon, T.H.; Frokiaer, J. Physiology and pathophysiology of cyclooxygenase-2 and prostaglandin E2 in the kidney. *Kidney Res. Clin. Pract.* **2015**, *34*, 194–200. [CrossRef]
45. Sheibanie, A.F.; Yen, J.H.; Khayrullina, T.; Emig, F.; Zhang, M.; Tuma, R.; Ganea, D. The proinflammatory effect of prostaglandin E2 in experimental inflammatory bowel disease is mediated through the IL-23–>IL-17 axis. *J. Immunol.* **2007**, *178*, 8138–8147. [CrossRef]
46. Eliakim, R.; Karmeli, F.; Chorev, M.; Okon, E.; Rachmilewitz, D. Effect of drugs on colonic eicosanoid accumulation in active ulcerative colitis. *Scand. J. Gastroenterol.* **1992**, *27*, 968–972. [CrossRef]
47. Reddy, M.C.; Subhashini, J.; Mahipal, S.V.; Bhat, V.B.; Srinivas, R.P.; Kiranmai, G.; Madyastha, K.M.; Reddanna, P. C-Phycocyanin, a selective cyclooxygenase-2 inhibitor, induces apoptosis in lipopolysaccharide-stimulated RAW 264.7 macrophages. *Biochem. Biophys. Res. Commun.* **2003**, *304*, 385–392. [CrossRef]
48. Szalai, Z.; Szasz, A.; Nagy, I.; Puskas, L.G.; Kupai, K.; Kiraly, A.; Berko, A.M.; Posa, A.; Strifler, G.; Barath, Z.; et al. Anti-inflammatory effect of recreational exercise in TNBS-induced colitis in rats: Role of NOS/HO/MPO system. *Oxid. Med. Cell. Longev.* **2014**, *2014*, 925981. [CrossRef]
49. Pallone, F.; Monteleone, G. Mechanisms of tissue damage in inflammatory bowel disease. *Curr. Opin. Gastroenterol.* **2001**, *17*, 307–312. [CrossRef]
50. Barnes, P.J.; Karin, M. Nuclear factor-kappaB: A pivotal transcription factor in chronic inflammatory diseases. *N. Engl. J. Med.* **1997**, *336*, 1066–1107. [CrossRef]
51. Ghosh, S.; Karin, M. Missing pieces in the NF-kappaB puzzle. *Cell* **2002**, *109*, S81–S96. [CrossRef]
52. Thomson, A.; Hemphill, D.; Jeejeebhoy, K.N. Oxidative stress and antioxidants in intestinal disease. *Dig. Dis.* **1998**, *16*, 152–158. [CrossRef] [PubMed]

53. Lih-Brody, L.; Powell, S.R.; Collier, K.P.; Reddy, G.M.; Cerchia, R.; Kahn, E.; Weissman, G.S.; Katz, S.; Floyd, R.A.; McKinley, M.J.; et al. Increased oxidative stress and decreased antioxidant defenses in mucosa of inflammatory bowel disease. *Dig. Dis. Sci.* **1996**, *41*, 2078–2086. [CrossRef] [PubMed]
54. Balmus, I.M.; Ciobica, A.; Trifan, A.; Stanciu, C. The Implications of Oxidative Stress and Antioxidant Therapies in Inflammatory Bowel Disease: Clinical Aspects and Animal Models Saudi. *J. Gastroenterol.* **2016**, *22*, 3–17. [CrossRef]
55. Krzystek-Korpacka, M.; Neubauer, K.; Berdowska, I.; Boehm, D.; Zielinski, B.; Petryszyn, P.; Terlecki, G.; Paradowski, L.; Gamian, A. Enhanced formation of advanced oxidation protein products in IBD. *Inflamm. Bowel. Dis.* **2008**, *14*, 794–802. [CrossRef]
56. Jabri, M.A.; Rtibi, K.; Ben-Said, A.; Aouadhi, C.; Hosni, K.; Sakly, M.; Sebai, H. Antidiarrhoeal, antimicrobial and antioxidant effects of myrtle berries (Myrtus communis L.) seeds extract. *J. Pharm. Pharmacol.* **2016**, *68*, 264–274. [CrossRef]

Publisher's Note: MDPI stays neutral with regard to jurisdictional claims in published maps and institutional affiliations.

nutrients

MDPI

Article

Oral Bovine Milk Lactoferrin Administration Suppressed Myopia Development through Matrix Metalloproteinase 2 in a Mouse Model

Shin-Ichi Ikeda [1,2], Toshihide Kurihara [1,2,*], Masataro Toda [1], Xiaoyan Jiang [1,2], Hidemasa Torii [1,2] and Kazuo Tsubota [2,3,*]

1 Laboratory of Photobiology, Keio University School of Medicine, 35 Shinanomachi, Shinjuku-ku, Tokyo 160-8582, Japan; shin-ikeda@keio.jp (S.-I.I.); vcfmasa@gmail.com (M.T.); jiaxiangya@yahoo.co.jp (X.J.); htorii@2004.jukuin.keio.ac.jp (H.T.)

2 Department of Ophthalmology, Keio University School of Medicine, 35 Shinanomachi, Shinjuku-ku, Tokyo 160-8582, Japan

3 Tsubota Laboratory, Inc., 34 Shinanomachi, Shinjuku-ku, Tokyo 160-0016, Japan

* Correspondence: kurihara@z8.keio.jp (T.K.); tsubota@z3.keio.jp (K.T.); Tel.: +81-3-5315-4132 (T.K.); +81-3-5363-3269 (K.T.); Fax: +81-3-5363-3274 (T.K.); +81-3-3358-5961 (K.T.)

Received: 30 October 2020; Accepted: 2 December 2020; Published: 5 December 2020

Abstract: Recent studies have reported an association between myopia development and local ocular inflammation. Lactoferrin (LF) is an iron-binding protein present in saliva, tears, and mother's milk. Furthermore, sequestering iron by LF can cause its antibacterial property. Moreover, LF has an anti-inflammatory effect. We aimed to determine the suppressive effect of LF against the development and progress of myopia using a murine lens-induced myopia (LIM) model. We divided male C57BL/6J mice (3 weeks old) into two groups. While the experimental group was orally administered LF (1600 mg/kg/day, from 3-weeks-old to 7-weeks-old), a similar volume of Ringer's solution was administered to the control group. We subjected the 4-week-old mice to −30 diopter lenses and no lenses on the right and left eyes, respectively. We measured the refraction and the axial length at baseline and 3 weeks after using a refractometer and a spectral domain optical coherence tomography (SD-OCT) system in both eyes. Furthermore, we determined the matrix metalloproteinase-2 (MMP-2) activity, and the amount of interleukin-6 (IL-6), MMP-2, and collagen 1A1 in the choroid or sclera. The eyes with a minus lens showed a refractive error shift and an axial length elongation in the control group, thus indicating the successful induction of myopia. However, there were no significant differences in the aforementioned parameters in the LF group. While LIM increased IL-6 expression and MMP-2 activity, it decreased collagen 1A1 content. However, orally administered LF reversed these effects. Thus, oral administration of LF suppressed lens-induced myopia development by modifying the extracellular matrix remodeling through the IL-6–MMP-2 axis in mice.

Keywords: myopia; lactoferrin; IL-6; MMP-2; collagen

1. Introduction

There has been a significant rise in the incidence of myopia in the past five decades. Furthermore, myopia is rapidly becoming the primary cause of visual impairment worldwide [1–3]. Thus, myopia prevention will avert visual impairment and retinal complications. The molecular basis of myopia has not been completely elucidated. Nonetheless, several researchers have reported on the role of inflammation [4] and remodeling of the scleral extracellular matrix (ECM) in the onset and progress of myopia [5]. Patients with inflammatory diseases, such as type 1 diabetes, uveitis, and systemic lupus erythematous, show a higher incidence of myopia. Topical application of cyclosporine A, an immunosuppressive agent, slows down the progress of the monocular form of deprivation-induced myopia. In contrast, it is accelerated by the topical application of lipopolysaccharide, an inflammation inducer [6]. The concentrations of interleukine-6 (IL-6) and matrix metalloproteinase-2 (MMP-2) in the aqueous humor are greater in highly myopic eyes than in ametropic or mildly myopic eyes. Moreover, the contents are positively associated with axial length [7]. While the content of collagen I, the most abundant structural collagen in the sclera, decreases via suppression of the *COL1A1* gene, it increases because of collagen degradation enzymes, such as MMP-2 and -9 [8–11]. The latter reportedly leads to collagen fiber thinning, followed by scleral thinning and weakening, thus causing axial elongation.

Lactoferrin (LF) is an iron-binding glycoprotein not only abundant in cow's milk but also abundant in body fluids, such as colostrum, saliva, and tears [12,13]. It possesses various biological functions, such as immunomodulation and the prevention of oxidative damage and photo damage [14–17]. LF reportedly affects collagen synthesis and degradation. Bovine LF enhances the transcription of the *COL1A1* gene and collagen I synthesis in human dermal fibroblasts [18]. In addition, LF hydrolyzate attenuates interleukin-1β-induced expression of MMP-1, -3, and -13 in human articular chondrocytes [19]. Furthermore, it has beneficial effects against eye dysfunctions, including dry eye, choroidal neovascularization, and age-related lachrymal gland dysfunction [15,20–22]. Considering the association between inflammation and collagen turnover, and between myopia onset and progress, LF can counteract these effects. Thus, we hypothesized that daily LF supplementation prevents or suppresses myopia development. We showed that oral administration of LF attenuated minus-lens-induced myopia development in C57BL6J mice. Furthermore, LF suppressed lens-induced myopia (LIM)-induced MMP-2 activation, besides increasing *COL1A1* expression.

2. Materials and Methods

Experimental animals: All animal experiments in this study were approved by the Animal Experimental Committee of the Keio University (Permit number: 16017-3). Our study adhered to the Institutional Guidelines on Animal Experimentation at the Keio University, the Association for Research in Vision and Ophthalmology (ARVO) Statement for the Use of Animals in Ophthalmic and Vision Research, and the Animal Research: Reporting of In Vivo Experiments (ARRIVE) guidelines for the use of animals in research. We purchased male C57BL/6J mice (3-week-old) from CLEA Japan (Yokohama, Japan). We maintained five mice in one cage by free intake with standard chow and water. We kept them in an environment with a 12 h/12 h light/dark cycle (the dark cycle extended from 8:00 p.m. to 8:00 a.m.) at 23 ± 3 °C. We maintained the light cycle using a 50-lux background according to a previous report on experimental myopia induction [23,24].

Lactoferrin (LF) administration: We purchased LF from bovine milk from FUJIFILM Wako Pure Chemical Corporation (Osaka, Japan). LF was dissolved in Ringer's solution at a concentration of 160 mg/mL and stored it at -30 °C until use. We orally administered 1600 mg/kg/day LF once a day to 3-week-old mice for 4 weeks. We administered a similar volume of Ringer's solution to the control group.

Lens-induced myopia (LIM): We measured the refraction, choroidal thickness, and axial length of all eyes before (4-week-old) and after inducing myopia (7-week-old) using a refractometer (Steinberis Transfer Center, Tübingen, Germany) and spectral domain optical coherent tomography (SD-OCT; Envisu R4310, Leica, Wetzlar, Germany), respectively. We anesthetized the mice with 0.75 mg/kg medetomidine (Sandoz K.K., Tokyo, Japan), 4 mg/kg midazolam (Domitor; Orion Corporation, Espoo, Finland), and 5 mg/kg butorphanol tartrate (Meiji Seika Pharma Co., Ltd., Tokyo, Japan) dissolved in normal saline. Several eyes could not be measured because of corneal abnormalities, which were observed in a certain percentage of cases and not specifically in this study. At baseline for myopia induction, we fixed −30 diopter (D) lenses and frames without lenses onto the right and left eyes (as a control), respectively. Furthermore, we maintained this setup for 3 weeks, until the mice were 7 weeks old. Figure 1 outlines the protocol for LF administration and LIM.

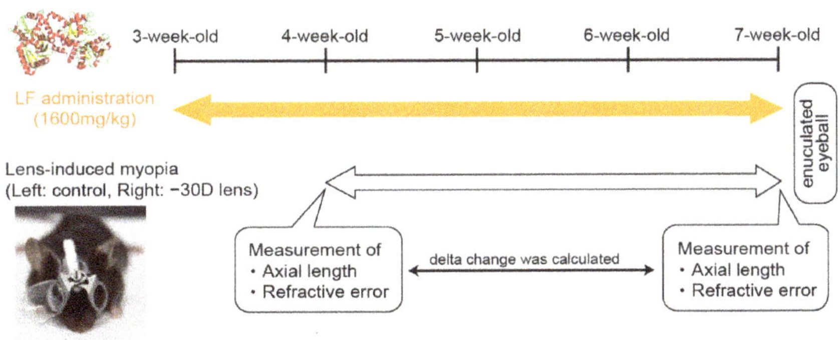

Figure 1. Procedure for oral lactoferrin (LF) administration and minus-lens-induced myopia.

Tissue preparation: After myopia induction, we harvested the complete choroid and sclera of each eye for gel zymography and cytokine analysis. We pooled the issues from the LIM and control eyes of three mice in 1 tube and subsequently homogenized them to obtain a protein lysate (three samples each of LIM and control eyes). We homogenized the tissues in an RIPA buffer (50 mM HEPES (pH 7.5), 150 mM NaCl, 1% NP-40, 0.1% sodium deoxycholate, 1 mM EDTA, 5 mM benzamidine, 10 mM β-glycerophosphate, 1 mM Na_3VO_4, 50 mM NaF, and 1 mM PMSF) containing a Halt protease inhibitor cocktail (Thermo Fisher Scientific, MA, USA) using Precellys (M&S Instruments Inc., Osaka, Japan). After centrifugation, we collected the supernatant and stored it at −80 °C until use.

Gelatin zymography: We performed zymography as previously described with some modifications. We added a 6× non-reducing sample buffer (Nacalai Tesque, Kyoto, Japan) to the supernatant and added a 1× non-reducing sample buffer, thus adjusting the concentration to 0.5 g/L. The samples were loaded onto 10% SDS-PAGE gels containing 1 mg/mL gelatin (Sigma-Aldrich Japan, Tokyo, Japan). After electrophoresis, we rinsed the gels twice for 30 min in a washing buffer (2.5% Triton X-100; 50 mM Tris-HCl pH 7.5; 5 mM $CaCl_2$; 1 μM $ZnCl_2$, all chemicals were purchased from Wako chemical, Tokyo, Japan), incubated them for 24 h in an incubation buffer (1% Triton X-100; 50 mM Tris-HCl pH 7.5; 5 mM $CaCl_2$; 1 μM $ZnCl_2$) at 37 °C, used a staining solution (0.5% Coomasie Blue; 40% methanol; 10% acetic acid, all chemicals were purchased from Wako chemical, Tokyo, Japan) for 1 h, and destained the samples to expose gelatinolytic bands. We obtained the gel images and quantified them using GELSCAN-2 (iMeasure, Nagano, Japan) and ImageJ software (version 1.63), respectively.

Western blotting: We added a reducing sample buffer (Nacalai Tesque, Kyoto, Japan) to the remaining supernatant and boiled it at 95 °C for 5 min. The samples were loaded onto a 4–15% gradient SDS-PAGE gel (Thermofisher Scientific, Waltham, MA, USA), transferred to a PVDF membrane (Merck Millipore, Burlington, MA, USA), and blocked by Blocking One (Nacalai Tesque). Furthermore, we incubated them overnight with anti-MMP2, anti-IL-6, anti-Collagen1A1, and anti-GAPDH (Cell Signaling Technologies Japan, Tokyo, Japan) at 4 °C. After washing with TBS-T, we incubated the membranes with a horseradish peroxidase-conjugated secondary antibody. We visualized the bands using EzWestLumi plus (ATTA, Tokyo, Japan) and LAS-4000 (GE Healthcare, Chicago, IL, USA). After visualization, densitometric analysis was performed using ImageJ software (version 1.63).

Statistics: All data are expressed as means ± standard deviations (SDs). We analyzed the differences between groups by one-way analysis of variance (ANOVA) or Student's *t* test. A p-value < 0.05 was considered statistically significant.

3. Results

3.1. Oral LF Supplementation Suppressed Minus-Lens-Induced Myopia Development in C57BL/6J Mice

There was no impact of LF administration on body weight. A −30 D lens induced refractive shift and axial elongation, thus indicating successful myopia development in the control group (Figure 2a,b). In contrast, LIM failed to induce axial elongation and refractive shift in the LF group (Figure 2a,b). Therefore, LF supplementation suppressed lens-induced myopia in C57BL6J mice.

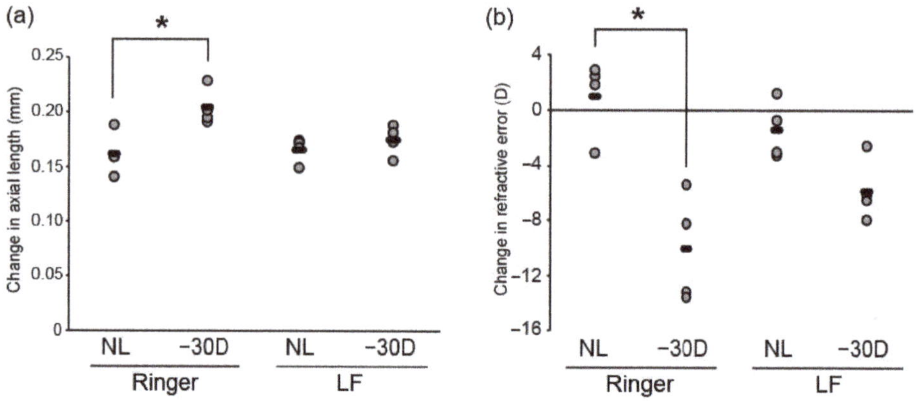

Figure 2. Scleral ER stress suppression or induction was sufficient to modulate axial elongation. (**a**) lens-induced myopia (LIM)-induced axial elongation was inhibited by oral administration of LF (lactoferrin) for 3 weeks ($n = 4$ per group); * $p < 0.05$. (**b**) LIM-induced myopic shift in refraction was inhibited by oral LF administration for 3 weeks ($n = 4$ per group); * $p < 0.05$.

3.2. While LIM Increased Active MMP-2 Activity and IL-6 Expression in the Choroid and Sclera, LF Administration Reversed This Effect

Axial length elongation was accompanied by scleral remodeling, such as a suppression of collagen production and an increase in the activity of collagen degradative enzymes. Thus, we assessed pro-MMP-9, pro-MMP-2, and active MMP activity by gelatin zymography from the choroid and sclera (Figure 3a). The gelatin digesting activity of pro-MMP-9 and active-MMP-2 was enhanced in LIM in the right eyes of the control group. In contrast, the MMP activity was comparable between LIM and control eyes in the LF-treated group (Figure 3b). Consistent with the results of zymography, the expression level of MMP-2 protein was also higher in the myopia-induced group than in the control group, and there was no difference in the expression level of MMP-2 protein between the control and myopia-induced groups in mice treated with LF (Figure 3c,d). Thus, LF administration inhibited LIM-induced proteolytic activity of MMP-2 and MMP-9, followed by the suppression of myopia development.

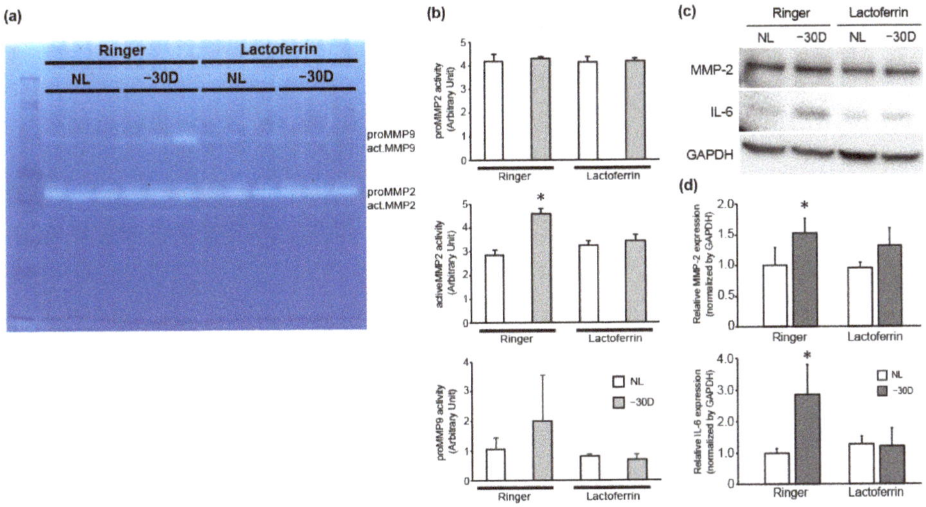

Figure 3. LIM activated MMP-2 activity and oral LF administration reversed this effect. (**a**) Gel image of gelatin zymography. (**b**) Quantified data of pro-MMP-2, active MMP-2, and pro-MMP-9; * $p < 0.05$. (**c**) Immunoblots showing that LIM increased IL-6 and MMP-2 expression and that LF suppressed them; representative blots from 3 independent experiments are shown. (**d**) The densitometry quantitation of MMP-2 (upper) and IL-6 (lower) blots; the Ringer's–no lens (NL) group was assigned a value of 1.0; all other values are expressed relative to this value; * $p < 0.05$.

LF has anti-inflammatory properties and its administration decreased the expression level of inflammatory cytokines, such as tumor necrosis factor-α and IL-6 [13]. As IL-6 is a positive regulator of MMP2 expression and is increased in myopic conditions [25,26], we hypothesized that IL-6 is an upstream factor in the increased MMP-2 activity in myopic eyes and examined the effect of LIM and LF on IL-6 expression level. As shown in Figure 3c,d, IL-6 expression level was higher in −30 D lens-wearing eyes in the Ringer's group, however, the expression level of IL-6 was comparable between the no-lens (NL) and −30 D eyes in the LF-treated group. Taken together, it is suggested that myopic stimuli induced IL-6 expression, followed by enhanced MMP-2 expression/activity, and that LF can reverse it.

3.3. LIM Decreased the Content of Collagen 1A1 Protein in Choroid and Sclera, and LF Administration Reversed This Effect

Collagen 1A1 is the most abundant structural collagen in the sclera. Furthermore, its content is decreased by form-deprived and minus-lens-induced myopia, which reduces *COL1A1* expression and increases its degradation by MMP. Therefore, we assessed the effects of LIM and orally administered LF on the protein expression of collagen 1A1. LIM decreased the level of its expression in the choroid and sclera (but not statistically significantly; $p = 0.062$). In contrast, there was an increase in the expression level in the LIM sclera of the LF-treated group (Figure 4a,b), suggesting the protective effect of LF against LIM-induced collagen 1A1 degradation.

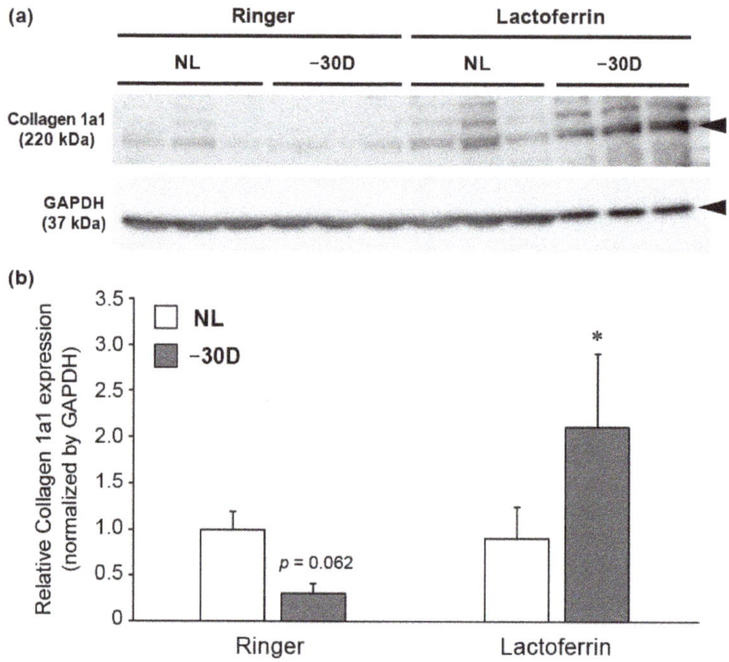

Figure 4. LIM decreased collagen 1A1 expression and oral administration of LF reversed its effect. (**a**) Immunoblotting showed a decrease in collagen 1A1 expression in the right eye in the control group; this decrease was absent in the LF group. (**b**) The densitometry quantitation of blots: the Ringer's–NL group was assigned a value of 1.0; all other values are expressed relative to this value; * $p < 0.05$.

4. Discussion

In the present study, we demonstrated that LF suppresses myopia in a mouse LIM model. This was concomitant with the inhibition of an LIM-induced IL-6 increase, MMP-2 activation, and a decrease in collagen 1A1 in the sclera and choroid.

Scleral ECM remodeling reportedly plays an essential role in myopia development. Furthermore, MMP-2 is a key player in scleral remodeling and myopia development. In myopic schlera, the expression of MMP-2 was increased when compared with control eyes in mice, chicks, tree shrews, and guinea pigs [27–29]. An increase in MMP-2 levels in the human aqueous humor was positively correlated with axial length [27]. Forced expression of MMP-2 in the sclera was sufficient to induce a myopic shift in refraction. Moreover, an injection of Adeno-associated virus 8 packaging with shRNA to MMP-2 suppressed the form of deprivation-induced increases in the latter's expression and in myopia development [29]. We demonstrated that while LIM enhanced MMP-2 activity, oral administration of LF reversed this effect. Therefore, LF suppressed LIM-induced MMP-2 activation. Inoculation of *Escherichia coli* enhanced MMP-2, -3, and -9 activities in cervical tissue. In contrast, pre-treatment with LF before *E. coli* inoculation suppressed the increase [30]. Apolactoferrin inhibited the catalytic domain of MMP-2 in the metal-free forms of LF in vitro [31]. Our results suggest that orally administered LF inhibited MMP-2 activation during LIM, followed by a suppression of myopia development.

Inflammation is a critical factor in myopia onset and development [6]. Consistent with previous studies, IL-6 expression was increased by myopia-inducing stimuli. We also found that the increase did not occur in the LF group. The changes in MMP2 activity and expression levels with myopia induction and LF administration were consistent with this IL-6 expression pattern, and IL-6 is a positive regulator of MMP-2 expression [25,26], suggesting that LF suppresses the activation of MMP-2 by myopia induction through its anti-inflammatory effects, including IL-6 expression.

MMP-2 is an ECM degradation enzyme. Furthermore, its activation induces collagen degradation, followed by a weakening of scleral stiffness. Myopia development is concomitant with an increase in MMP-2 expression and a decrease in collagen 1A1 content [5,8,32,33]. We reproduced a decrease in collagen 1A1 by LIM. Moreover, oral administration of LF could prevent this decrease. LF reportedly has a suppressive effect on MMPs [30,31] that is consistent with the present study (Figure 3). Furthermore, LF stimulates *COL1A1* transcription and collagen I synthesis in human dermal fibroblasts [18]. Fibroblasts are the dominant cell population in the sclera. Thus, LF can counteract myopia development by suppressing collagen I degradation by MMP-2 inhibition, and by enhancing collagen I synthesis in scleral fibroblasts.

In conclusion, oral administration of LF can prevent negative-lens-induced myopia in mice by suppressing the IL-6–MMP-2 axis and collagen 1a1 degradation. Hence, LF is considered beneficial to prevent myopia development in humans.

5. Patents

Patent registered in Japan for the myopia induction model (#WO2018/164113 by K.T., T.K., S.-i.I., and X.J.) and patent pending internationally. Patent registered in Japan for inhibition of myopia by altering the gut microbiota, including lactoferrin (#WO/2019/093262 by K.T., T.K., and S.I.) and patent pending internationally. K.T. reports his position as CEO of Tsubota Laboratory, Inc., a company aimed at developing products for the treatment for myopia. H.T., T.K, and K.T. own unlisted stocks of Tsubota Laboratory.

Author Contributions: K.T. and T.K. conceived the concept of the study and supervised the study. S.-I.I. carried out all experiments and wrote the manuscript. X.J., M.T., and H.T. developed and prepared the LIM mice and measured their axial length and refraction. All authors have read and agreed to the published version of the manuscript.

Funding: This work was supported by grants from KAKENHI (Grant-in-Aid for Scientific Research (B): 16H03258, Grant-in-Aid for Scientific Research (C): 18K09424, and Grant-in-Aid for Scientific Research (C): 20K09834) from the Ministry of Education, Culture, Sports, Science and Technology of Japan.

Acknowledgments: We are grateful to the Collaborative Research Resources, School of Medicine, Keio University, for their technical support and reagents.

Conflicts of Interest: Hidemasa Torii and Toshihide Kurihara report receiving personal fees from Tsubota Laboratory Inc. outside the submitted work. Outside the submitted work, Kazuo Tsubota reports his position as CEO of Tsubota Laboratory, Inc., a company aimed at developing products for the treatment for myopia. Torii, Kurihara and Tsubota own the unlisted stocks of Tsubota Laboratory. The funders had no role in the design of the study; in the collection, analyses, or interpretation of data; in the writing of the manuscript; or in the decision to publish the results.

References

1. Dolgin, E. The myopia boom. *Nature* **2015**, *519*, 276–278. [CrossRef] [PubMed]
2. Holden, B.A.; Fricke, T.R.; Wilson, D.A.; Jong, M.; Naidoo, K.S.; Sankaridurg, P.; Wong, T.Y.; Naduvilath, T.J.; Resnikoff, S. Global prevalence of myopia and high myopia and temporal trends from 2000 through 2050. *Ophthalmology* **2016**, *123*, 1036–1042. [CrossRef] [PubMed]
3. Tideman, J.W.; Snabel, M.C.; Tedja, M.S.; van Rijn, G.A.; Wong, K.T.; Kuijpers, R.W.; Vingerling, J.R.; Hofman, A.; Buitendijk, G.H.S.; Keunen, J.E.E.; et al. Association of axial length with risk of uncorrectable visual impairment for Europeans with myopia. *JAMA Ophthalmol.* **2016**, *134*, 1355–1363. [CrossRef] [PubMed]
4. Herbort, C.P.; Papadia, M.; Neri, P. Myopia and inflammation. *J. Ophthalmic Vis. Res.* **2011**, *6*, 270–283. [PubMed]
5. Harper, A.R.; Summers, J.A. The dynamic sclera: Extracellular matrix remodeling in normal ocular growth and myopia development. *Exp. Eye Res.* **2015**, *133*, 100–111. [CrossRef] [PubMed]

6. Lin, H.J.; Wei, C.C.; Chang, C.Y.; Chen, T.H.; Hsu, Y.A.; Hsieh, Y.C.; Chen, H.-J.; Wan, L. Role of Chronic Inflammation in Myopia Progression: Clinical Evidence and Experimental Validation. *EBioMedicine* **2016**, *10*, 269–281. [CrossRef]
7. Yuan, J.; Wu, S.; Wang, Y.; Pan, S.; Wang, P.; Cheng, L. Inflammatory cytokines in highly myopic eyes. *Sci. Rep.* **2019**, *9*, 3517. [CrossRef]
8. Gentle, A.; Liu, Y.; Martin, J.E.; Conti, G.L.; McBrien, N.A. Collagen gene expression and the altered accumulation of scleral collagen during the development of high myopia. *J. Biol. Chem.* **2003**, *278*, 16587–16594. [CrossRef]
9. Schippert, R.; Brand, C.; Schaeffel, F.; Feldkaemper, M.P. Changes in scleral MMP-2, TIMP-2 and TGFbeta-2 mRNA expression after imposed myopic and hyperopic defocus in chickens. *Exp. Eye Res.* **2006**, *82*, 710–719. [CrossRef]
10. Siegwart, J.T., Jr.; Norton, T.T. The time course of changes in mRNA levels in tree shrew sclera during induced myopia and recovery. *Investig. Ophthalmol. Vis. Sci.* **2002**, *43*, 2067–2075.
11. Siegwart, J.T., Jr.; Norton, T.T. Selective regulation of MMP and TIMP mRNA levels in tree shrew sclera during minus lens compensation and recovery. *Investig. Ophthalmol. Vis. Sci.* **2005**, *46*, 3484–3492. [CrossRef] [PubMed]
12. Baker, E.N. Lactoferrin: A multi-tasking protein par excellence. *Cell. Mol. Life Sci.* **2005**, *62*, 2529–2530. [CrossRef] [PubMed]
13. Legrand, D. Lactoferrin, a key molecule in immune and inflammatory processes. *Biochem. Cell Biol.* **2012**, *90*, 252–268. [CrossRef] [PubMed]
14. Actor, J.K.; Hwang, S.A.; Kruzel, M.L. Lactoferrin as a natural immune modulator. *Curr. Pharm. Des.* **2009**, *15*, 1956–1973. [CrossRef] [PubMed]
15. Kawashima, M.; Kawakita, T.; Inaba, T.; Okada, N.; Ito, M.; Shimmura, S.; Watanabe, M.; Shinmura, K.; Tsubota, K. Dietary lactoferrin alleviates age-related lacrimal gland dysfunction in mice. *PLoS ONE* **2012**, *7*, e33148. [CrossRef] [PubMed]
16. Kuhara, T.; Tanaka, A.; Yamauchi, K.; Iwatsuki, K. Bovine lactoferrin ingestion protects against inflammation via IL-11 induction in the small intestine of mice with hepatitis. *Br. J. Nutr.* **2014**, *111*, 1801–1810. [CrossRef]
17. Yamauchi, K.; Wakabayashi, H.; Hashimoto, S.; Teraguchi, S.; Hayasawa, H.; Tomita, M. Effects of orally administered bovine lactoferrin on the immune system of healthy volunteers. *Adv. Exp. Med. Biol.* **1998**, *443*, 261–265.
18. Saito, S.; Takayama, Y.; Mizumachi, K.; Suzuki, C. Lactoferrin promotes hyaluronan synthesis in human dermal fibroblasts. *Biotechnol. Lett.* **2011**, *33*, 33–39. [CrossRef]
19. Yan, D.; Chen, D.; Shen, J.; Xiao, G.; van Wijnen, A.J.; Im, H.J. Bovine lactoferricin is anti-inflammatory and anti-catabolic in human articular cartilage and synovium. *J. Cell. Physiol.* **2013**, *228*, 447–456. [CrossRef]
20. Higuchi, A.; Inoue, H.; Kaneko, Y.; Oonishi, E.; Tsubota, K. Selenium-binding lactoferrin is taken into corneal epithelial cells by a receptor and prevents corneal damage in dry eye model animals. *Sci. Rep.* **2016**, *6*, 36903. [CrossRef]
21. Ibuki, M.; Shoda, C.; Miwa, Y.; Ishida, A.; Tsubota, K.; Kurihara, T. Lactoferrin has a therapeutic effect via HIF inhibition in a murine model of choroidal neovascularization. *Front. Pharmacol.* **2020**, *11*, 174. [CrossRef] [PubMed]
22. Kawashima, M.; Nakamura, S.; Izuta, Y.; Inoue, S.; Tsubota, K. Dietary supplementation with a combination of lactoferrin, fish oil, and *Enterococcus faecium* WB2000 for treating dry eye: A Rat Model and Human Clinical Study. *Ocul. Surf.* **2016**, *14*, 255–263. [CrossRef] [PubMed]
23. Jiang, X.; Kurihara, T.; Ikeda, S.-I.; Kunimi, H.; Mori, K.; Torii, H.; Tsubota, K. Inducement and evaluation of a murine model of experimental myopia. *J. Vis. Exp.* **2019**, *2019*, e58822. [CrossRef] [PubMed]
24. Jiang, X.; Kurihara, T.; Kunimi, H.; Miyauchi, M.; Ikeda, S.-I.; Mori, K.; Tsubota, K.; Torii, H.; Tsubota, K. A highly efficient murine model of experimental myopia. *Sci. Rep.* **2018**, *8*, 2026. [CrossRef] [PubMed]
25. Sun, W.; Liu, D.-B.; Li, W.-W.; Zhang, L.-L.; Long, G.-X.; Wang, J.; Mei, Q.; Hu, G. Interleukin-6 promotes the migration and invasion of nasopharyngeal carcinoma cell lines and upregulates the expression of MMP-2 and MMP-9. *Int. J. Oncol.* **2014**, *44*, 1551–1560. [CrossRef] [PubMed]

26. Kossakowska, A.E.; Edwards, D.R.; Prusinkiewicz, C.; Zhang, M.C.; Guo, D.; Urbanski, S.J.; Grogran, T.; Marquez, L.A.; Janowska-Wieczorek, A. Interleukin-6 regulation of matrix metalloproteinase (MMP-2 and MMP-9) and tissue inhibitor of metalloproteinase (TIMP-1) expression in malignant non-Hodgkin's lymphomas. *Blood* **1999**, *94*, 2080–2089. [CrossRef] [PubMed]
27. Jia, Y.; Shun-Fa, Y.; Zhu, D.; Zhang, L.; Gu, P.; Fan, X.; Zhou, J. MMP-2, MMP-3, TIMP-1, TIMP-2, and TIMP-3 protein levels in human aqueous humor: Relationship with axial length. *Investig. Ophthalmol. Vis. Sci.* **2014**, *55*, 3922–3928. [CrossRef] [PubMed]
28. Rada, J.A.; Perry, C.A.; Slover, M.L.; Achen, V.R. Gelatinase A and TIMP-2 expression in the fibrous sclera of myopic and recovering chick eyes. *Investig. Ophthalmol. Vis. Sci.* **1999**, *40*, 3091–3099.
29. Zhao, F.; Zhou, Q.; Reinach, P.S.; Yang, J.; Ma, L.; Wang, X.; Wen, Y.; Srinivasalu, N.; Qu, J.; Qu, J. Cause and effect relationship between changes in scleral matrix metallopeptidase-2 expression and myopia development in mice. *Am. J. Pathol.* **2018**, *188*, 1754–1767. [CrossRef]
30. Nakayama, K.; Otsuki, K.; Yakuwa, K.; Hasegawa, A.; Sawada, M.; Mitsukawa, K.; Chiba, H.; Nagatsuka, M.; Okai, T. Recombinant human lactoferrin inhibits matrix metalloproteinase (MMP-2, MMP-3, and MMP-9) activity in a rabbit preterm delivery model. *J. Obstet. Gynaecol. Res.* **2008**, *34*, 931–934. [CrossRef]
31. Newsome, A.L.; Johnson, J.P.; Seipelt, R.L.; Thompson, M.W. Apolactoferrin inhibits the catalytic domain of matrix metalloproteinase-2 by zinc chelation. *Biochem. Cell Biol.* **2007**, *85*, 563–567. [CrossRef] [PubMed]
32. McBrien, N.A.; Cornell, L.M.; Gentle, A. Structural and ultrastructural changes to the sclera in a mammalian model of high myopia. *Investig. Ophthalmol. Vis. Sci.* **2001**, *42*, 2179–2187.
33. Norton, T.T.; Rada, J.A. Reduced extracellular matrix in mammalian sclera with induced myopia. *Vis. Res.* **1995**, *35*, 1271–1281. [CrossRef]

nutrients

MDPI

Article

Dietary Cameroonian Plants Exhibit Anti-Inflammatory Activity in Human Gastric Epithelial Cells

Achille Parfait Atchan Nwakiban [1,2], Marco Fumagalli [2], Stefano Piazza [2], Andrea Magnavacca [2], Giulia Martinelli [2], Giangiacomo Beretta [3], Paolo Magni [2,4], Armelle Deutou Tchamgoue [5], Gabriel Agbor Agbor [5], Jules-Roger Kuiaté [1], Mario Dell'Agli [2,*] and Enrico Sangiovanni [2]

[1] Department of Biochemistry, Faculty of science, University of Dschang, P.O. Box 96 Dschang, Cameroon; achilestyle@yahoo.fr or achille.parfait@guest.unimi.it (A.P.A.N.); jrkuiate@yahoo.com (J.-R.K.)

[2] Department of Pharmacological and Biomolecular Sciences, Università degli Studi di Milano, 20133 Milan, Italy; marco.fumagalli3@unimi.it (M.F.); stefano.piazza@unimi.it (S.P.); andrea.magnavacca@unimi.it (A.M.); giulia.martinelli@unimi.it (G.M.); paolo.magni@unimi.it (P.M.); enrico.sangiovanni@unimi.it (E.S.)

[3] Department of Environmental Science and Policy, Università degli Studi di Milano, 20133 Milan, Italy; giangiacomo.beretta@unimi.it

[4] IRCCS MultiMedica, Sesto San Giovanni, Via Milanese, 300, Sesto San Giovanni, 20099 Milan, Italy

[5] Centre for Research on Medicinal Plants and Traditional Medicine, Institute of Medical Research and Medicinal Plants Studies, 4124 Yaoundé, Cameroon; armelle_d2002@yahoo.fr (A.D.T.); agogae@yahoo.fr (G.A.A.)

* Correspondence: mario.dellagli@unimi.it; Tel.: +39-0250-318-398

Received: 17 November 2020; Accepted: 8 December 2020; Published: 10 December 2020

Abstract: In Cameroon, local plants are traditionally used as remedies for a variety of ailments. In this regard, several papers report health benefits of Cameroonian spices, which include antioxidant and anti-microbial properties, whereas gastric anti-inflammatory activities have never been previously considered. The present study investigates the antioxidant and anti-inflammatory activities of hydro-alcoholic extracts of eleven Cameroonian spices in gastric epithelial cells (AGS and GES-1 cells). The extracts showed antioxidant properties in a cell-free system and reduced H_2O_2-induced ROS generation in gastric epithelial cells. After preliminary screening on TNFα-induced NF-κB driven transcription, six extracts from *Xylopia parviflora*, *Xylopia aethiopica*, *Tetrapleura tetraptera*, *Dichrostachys glomerata*, *Aframomum melegueta*, and *Aframomum citratum* were selected for further studies focusing on the anti-inflammatory activity. The extracts reduced the expression of some NF-κB-dependent pro-inflammatory mediators strictly involved in the gastric inflammatory process, such as IL-8, IL-6, and enzymes such as PTGS2 (COX-2), without affecting PTGS1 (COX-1). In conclusion, the selected extracts decreased pro-inflammatory markers by inhibiting the NF-κB signaling in gastric cells, justifying, in part, the traditional use of these spices. Other molecular mechanisms cannot be excluded, and further studies are needed to better clarify their biological activities at the gastric level.

Keywords: gastric inflammation; ethnopharmacology; Cameroonian plants; interleukin 8; interleukin 6; NF-κB; COX-2; antioxidant

1. Introduction

Gastritis is an inflammatory-based pathology which can be classified as acute or chronic. Acute gastritis is provoked by several risk factors which include stress, alcohol abuse, the use of non-steroidal anti-inflammatory drugs (NSAIDs), and bile reflux; chronic conditions are mainly due to *Helicobacter pylori* (*H. pylori*) infection. Infected people can also develop more severe pathologies, such as peptic ulcer or gastric cancer [1,2]. In 1994, the WHO assessed *H. pylori* as a class I carcinogen for gastric cancer. Several transcription factors are involved in gastric inflammatory conditions, including NF-κB [3]; according to the literature, in vitro studies demonstrate that *H. pylori* and some pro-inflammatory cytokines (i.e., TNFα) are able to induce the activity of this transcription factor in gastric epithelial cells [4,5]. NF-κB plays a pivotal role in the expression and release of some pro-inflammatory mediators, such as IL-8, IL-6 and enzymes such as prostaglandin endoperoxide synthase 2 (PTGS2) (COX-2), which, in turn, lead to the amplification of the gastric phlogistic process [6,7]. IL-8 is considered a key element involved in the development of gastritis. At this regard, Crabtree et al. demonstrated the increased expression of this chemokine in the epithelium of the infected gastric mucosa [8,9]. In this context, an over-production of oxygen reactive species (ROS) leads to increased oxidative stress contributing to exacerbate the inflammatory process [10]. Indeed, it has been demonstrated that several ROS can enhance IL-8 expression in gastric epithelial cells through the NF-κB activation [11].

NSAIDs act on prostaglandin synthesis, through the inhibition of cyclooxygenase enzymes, PTGS1 (COX-1) and PTGS2 (COX-2). The latter is encoded by an NF-κB-dependent gene, strictly involved in the inflammatory process, whose expression can be rapidly up-regulated by cytokines and growth factors; otherwise, *PTGS1* gene is constitutively expressed in human epithelial cells, contributing to preserve the protective mucus layer by cytoprotective prostaglandin production; therefore, the blockage of PTGS1 (COX-1) activity is the main factor responsible for the gastric side effects (e.g., peptic ulcers) of NSAID chronic drug treatment [12–14]. These findings underline the importance of selective inhibitor agents for COX-2, able to preserve COX-1 activity.

In Cameroon, the traditional use of natural products for the treatment of several diseases is largely diffused and constitutes the first health approach among the population [15]. The efficacy of the traditional use of plants for human health is not generally fully supported by scientific evidence. Many plants are traditionally employed in Cameroon for the treatment of different ailments, such as diabetes, hypertension, malaria, and gastric disorders [16–18]. Nkui and Nahpoh are two traditional soups of the eastern region which contain many spices, among which there are plants used in the present study. These plants are widely distributed in eastern and central Africa, and their fruits and seeds, normally boiled with the help of a thread or a stick of bamboo, are pounded and used in cooked foods or as spices for sauces or beverages. As a traditional remedy, they are mainly employed in association with other botanicals for the treatment of various ailments, including stomach disorders [18–20]. According to the literature, several papers report the potential health benefits of these Cameroonian spices, including anti-microbial, anti-inflammatory, and hypoglycemic properties [19,21–23]. However, no study has investigated the anti-inflammatory activity of Cameroonian spices at the gastric level.

The aim of this study is to investigate the potential anti-inflammatory and antioxidant effects of the hydroethanolic extracts from eleven selected Cameroonian spices in human gastric epithelial cells. Human GES-1 and AGS cells are used as reliable in vitro models of human gastric epithelium. The extracts are assessed on the TNFα-induced expression of different NF-κB-dependent mediators, such as IL-8, IL-6 and PTGS2 (COX-2), and the effect on the TNFα-induced NF-κB-driven transcription is assessed as well. Among the extracts tested, six showed promising activity as anti-inflammatory agents, confirming their use in Cameroonian traditional medicine for the treatment of gastric disorders.

2. Materials and Methods

2.1. Preparation of the Hydroethanolic Extracts

The material from eleven different species of plants was harvested in different areas of the West Cameroon in September 2017, as previously described [24,25]. Plants collected were: *Xylopia aethiopica* (Dunal) A.Rich. (**XA**), *Xylopia parviflora* Spruce (**XP**), *Scorodophloeus zenkeri* Harms (**SZ**), *Monodora myristica* (Gaertn.) Dunal (**MM**), *Tetrapleura tetraptera* (Schum. and Thonn.) Taub. (**TT**), *Echinops giganteus* A.

Rich. (**EG**), *Dichrostachys glomerata* (Forssk.) Chiov. syn. *Dichrostachys cinerea* (L.) Wight and Arn. (**DG**), *Afrostyrax lepidophyllus* Mildbr. (**AL**), *Aframomum melegueta* K.Schum. (**AM**), *Aframomum citratum* (C. Pereira) K. Schum. (**AC**), and *Zanthoxylum leprieurii* Guill. and Perr. (**ZL**).

Selected samples consisted of different fruits, seeds, or roots, and were identified in the National Herbarium of Cameroon (http://irad.cm/national-herbarium-of-cameroun/) in Yaoundè (Cameroon), by comparison with preserved specimens [24,25]. The plant material (100 g) of each species was powdered and extracted, under stirring, with 100 mL of an hydroalcoholic (ethanol:water, 70:30) mixture for 4 h at room temperature, in dark conditions. Then, the extract was filtered, and the plant material was recovered and subjected to a second overnight extraction with fresh solvent. The solvent was removed through rotary evaporator (Laborota 4000 efficient, Heidolph Instruments, Schwabach, Germany), and subjected to lyophilization. All the extracts were dissolved in pure DMSO, aliquoted, and stored at −80 °C. Parts used, aspect, color, and extraction yield were previously reported in [25]. The extracts were characterized by gas chromatography coupled with mass spectrometry, as previously reported [24].

2.2. ORAC Assay

The oxygen radical absorbance capacity (ORAC) assay was carried out according to Ou B. et al. and Dávalos A. et al. [26,27], with minor modifications. Briefly, 20 μL of stock solution of each extract (1 μg/mL) was distributed into a black 96-well plate. Then, 120 μL of fluorescein solution (70 nM final concentration), previously prepared with a phosphate buffer (pH 7.4, 75 mM), was added to each well. Peroxyl radicals were generated by adding 60 μL of AAPH 40 mM (Sigma-Aldrich, St. Louis, MO, USA). The final concentration of each extract in the assay well was 0.1 μg/mL. The plate was put in a spectrophotometer (Victor X3, PerkinElmer, USA) and the fluorescence detector was set at excitation and emission wavelengths of 484 and 528 nm, respectively. The fluorescence was read, after shaking, every 2 min for 60 min at 37 °C. Trolox (0–50 μM) was used as reference inhibitor. The area under the curve (AUC), of each extract, was calculated and the results were expressed as μM Trolox equivalent.

2.3. Cell Culture

Human adenocarcinoma gastric epithelial cells (AGS, CRL-1739, LGC Standard S.r.l., Milano, Italy) were grown in DMEM/F12 medium (Gibco, Thermo Fisher Scientific, Waltham, MA, USA) supplemented with penicillin 100 units/mL (Gibco, Thermo Fisher Scientific, Waltham, MA, USA), streptomycin 100 mg/mL (Gibco, Thermo Fisher Scientific, Waltham, MA, USA), L-glutamine 2 mM (Gibco, Thermo Fisher Scientific, Waltham, MA, USA) and 10% heat-inactivated fetal bovine serum (Euroclone S.p.A, Pero, Italy), at 37 °C in humidified atmosphere containing 5% CO_2. Human normal gastric epithelial cells (GES-1, kindly provided by Dr. Dawit Kidane-Mulat (University of Texas, Austin) were grown in RPMI 1640 medium, supplemented as previously described, and in the same atmospheric conditions. When cells reached 80–90% of confluence, usually every 4 days, they were detached from the flask (Euroclone S.p.A., Milan, Italy) using trypsin-EDTA 0.25% (Gibco, Thermo Fisher Scientific, Waltham, MA, USA), counted, and replaced in a new flask (1×10^6 cell density per flask) to promote cell growth.

2.4. Cytotoxicity Assay

Cell morphology was checked by light microscope inspection. Cell viability was measured, after 6 h co-treatment with the stimulus (TNFα, 10 ng/mL) and the extracts, by the 3,4,5-dimethylthiazol-2-yl-2,5-diphenyltetrazolium bromide (MTT) method (Sigma-Aldrich, St. Louis, MA, USA) [28]. This method evaluates the succinate dehydrogenase activity, which is an index of cell viability. Six hours later, the medium was removed from each well and 0.1 mg/mL of MTT solution (200 μL) was added for 45 min at 37 °C, in dark conditions. Violet formazan salt was extracted from the cells with 200 μL of a mixture isopropanol:DMSO (90:10), and the absorbance was read at 570 nm (Envision, PerkinElmer, Waltham, MA, USA). The extracts were assessed in the range 0.1–20 μg/mL.

2.5. Cell Treatment

To investigate the release and gene expression of the pro-inflammatory mediators, the NF-κB driven transcription, and IL-8 promoter activity, cells were seeded in 24-well plates (DB Falcon™) at the density of 30,000 cells/well. After 72 h, cells were co-treated with the pro-inflammatory stimulus (TNFα, 10 ng/mL) and the extracts for 6 h, using serum-free medium: DMEM/F12 or RPMI 1640 medium (Gibco, Thermo Fisher Scientific, Waltham, USA), supplemented with L-glutamine 2 mM (Gibco, Thermo Fisher Scientific, Waltham, USA), penicillin 100 units/mL (Gibco, Thermo Fisher Scientific, Waltham, MA, USA), and streptomycin 100 mg/mL (Gibco, Thermo Fisher Scientific, Waltham, USA). Then, the medium or the cell lysate was collected for the biological assays. To assess the effect of the extracts on ROS generation, AGS and GES-1 cells were seeded in black 96-well plates (PerkinElmer, USA) at the density of 10,000 cells/well; after 72 h, cells were pre-treated with the extracts for 24 h in serum-free medium, and subsequently challenged with H_2O_2 for 2 h.

2.6. ROS Production

The intracellular ROS level was measured, in GES-1 and AGS cells, with the oxidant-sensitive fluorescence probe CM-H2DCFDA (Invitrogen, Thermo Fisher Scientific, Waltham, MA, USA). After the treatment, cells were incubated for 30 min with the fluorescent probe (10 μM), washed with PBS and stimulated with H_2O_2 for 2 h. The fluorescence intensity was read at an excitation wavelength of 485 nm and an emission wavelength of 535 nm using a plate reader (Envision, Perkin Elmer, Waltham, MA, USA). Trolox (500 μM) was used as reference compound. Data were expressed as mean ± SD of at least three experiments.

2.7. Transient Transfection Assays

GES-1 and AGS cells were transiently transfected in 24-well plates with two different reporter plasmids, the NF-κB-Luc (50 ng/well) or the IL-8-Luc (100 ng/well) [29]. The NF-κB-Luc contains the luciferase gene under control of the E-selectin promoter characterized by three κB responsive elements, while IL-8-Luc contains the luciferase gene under control of a fragment of the native promoter of the human IL-8. The plasmid NF-κB-Luc was a kind gift of Dr. N. Marx (Department of Internal Medicine-Cardiology, University of Ulm, Ulm, Germany)m whereas the native IL-8-Luc promoter was kindly provided by Dr. T. Shimohata (Department of Preventive Environment and Nutrition, University of Tokushima Graduate School, Tokushima, Japan). GES-1 cells were transfected using Lipofectamine® (Invitrogen, Thermo Fisher Scientific, Waltham, MA, USA), whereas AGS cells by the calcium phosphate method. Sixteen hours later, the cells were treated for 6 h with extracts in the presence of the pro-inflammatory mediators (TNFα 10 ng/mL). Six hours later, cells were harvested and the luciferase assay was carried out using the Britelite™ Plus reagent (PerkinElmer Inc., Waltham, MA, USA), according to the manufacturer's instructions. The results (mean ± SD of at least three experiments) were expressed as percentage, relative to stimulated control, which was arbitrarily assigned the value 100%.

2.8. IL-8 and IL-6 Release

IL-8 and IL-6 were quantified in cell media after the treatment with TNFα and plant extracts, using two different ELISA kits, a Human Interleukin-8 ELISA Development Kit and a Human Interleukin-6 ELISA Development Kit (Peprotech Inc., London, UK). Briefly, Corning 96-well EIA/RIA plates (Sigma-Aldrich, St. Louis, USA) were coated with the corresponding antibody provided by the ELISA Kit (Peprotech Inc., London, UK) and incubated overnight at room temperature. The non-specific binding sites were blocked with albumin 1%. A total of 200 μL of samples was transferred in duplicate into wells at room temperature for 2 h. The amount of IL-8 and IL-6 in the samples was detected at 450 nm by measuring the colorimetric reaction between horseradish peroxidase enzyme and 3,3′,5,5′-tetramethylbenzidine substrate (Sigma-Aldrich, St. Louis, MO, USA). The absorbance was read at 450 nm using a spectrophotometer (Victor X3, PerkinElmer, USA). IL-8 and IL-6 were quantified by optimization of the standard curve provided with the ELISA kit (8.0–1000 pg/mL for IL-8 and 32–2000 pg/mL for IL-6). Epigallocatechin-3-O-gallate (EGCG, 20 μM) was used as a reference molecule able to inhibit TNFα-induced IL-8 and IL-6 secretion. The results (mean ± SD of at least three experiments) were expressed as percentage, relative to stimulated control, which was arbitrarily assigned the value 100% (values around 1000 and 1500 pg/mL, respectively for IL-8 and IL-6 under stimulated conditions).

2.9. Gene Expression

GES-1 and AGS cells were lysed with QIAzol Lysis Reagent (Qiagen, Hilden, Germany), RNA was purified using the miRNeasy Mini Kit (Qiagen, Hilden, Germany) and quantified by spectrophotometric analysis at 260 nm (NanoDrop ND-100, Thermo Fisher Scientific, Waltham, MA, USA). Primers for *IL-8*, *IL-6*, *PTGS1* (*COX-1*), *PTGS2* (*COX-2*) and *GAPDH* (housekeeping gene) genes Are reported in Table S1. *IL-8*, *IL-6*, *PTGS1* (*COX-1*), and *PTGS2* (*COX-2*) mRNA levels were investigated through SYBR Green method (iTaq Universal SYBR Green One-Step Kit, Bio-Rad Laboratories Srl, Segrate, Italy). The total RNA (10 ng/μL) from each sample was mixed with SYBR Green, corresponding primers and the reverse transcription enzyme, the real-time PCR was assessed using the CFX384™ Real-Time PCR Detection System (coupled to C1000™ Thermal Cycler; Bio-Rad Laboratories Srl, Segrate, Italy). The threshold values were set manually, the relative expression of each gene was calculated by normalizing the data on the basis of the housekeeping gene (GAPDH). The experiments were repeated, at least, three times.

2.10. Statistical Analysis

All results are expressed as mean (±SD) of at least three experiments. Data were analyzed by unpaired one-way analysis of variance (ANOVA), followed by Bonferroni post hoc test. Gene expression results were calculated using the ΔΔCt method. ΔΔCt values were analyzed by unpaired t-test or unpaired one-way analysis of variance (ANOVA), followed by Bonferroni post hoc test, and graphed as mean relative expression (2^ΔΔCt) values (± SEM). Statistical analyses were calculated using GraphPad Prism 8.0 software (GraphPad Software Inc., San Diego, CA, USA) as well as IC_{50}s. $p < 0.05$ was considered statistically significant.

3. Results

3.1. Cytotoxicity of the Extracts in Gastric Epithelial Cells

The cytotoxicity of the extracts was assessed in the concentration range 0.1–20 μg/mL in both human gastric epithelial cells (GES-1 and AGS cells) by means of the MTT assay. After 6 h treatment, XA extract showed cytotoxic effects only at 20 μg/mL in both the cell cultures (data not shown), and this concentration was not used for assessing the biological activity of the extract.

3.2. Hydroalcoholic Extracts Inhibit ROS Production in AGS and GES-1 Cells

The extracts were investigated for their ability to block peroxyl radicals in a cell-free system (ORAC) assay, and to reduce ROS generation in AGS and GES-1 cells. XP and AM extracts were the most active in ORAC assay (around 9.5 and 8.5 μmol Trolox equivalents, respectively), followed by EG and AC extracts (around 6 μmol Trolox equivalents) (Figure 1A).

ROS generation in AGS and GES-1 cells stimulated with hydrogen peroxide (200 μM in AGS and 100 μM in GES-1 cells) was close to 2.5-fold and 3.5-fold compared to the unstimulated controls, respectively. Most of the extracts inhibited ROS production induced by H_2O_2.

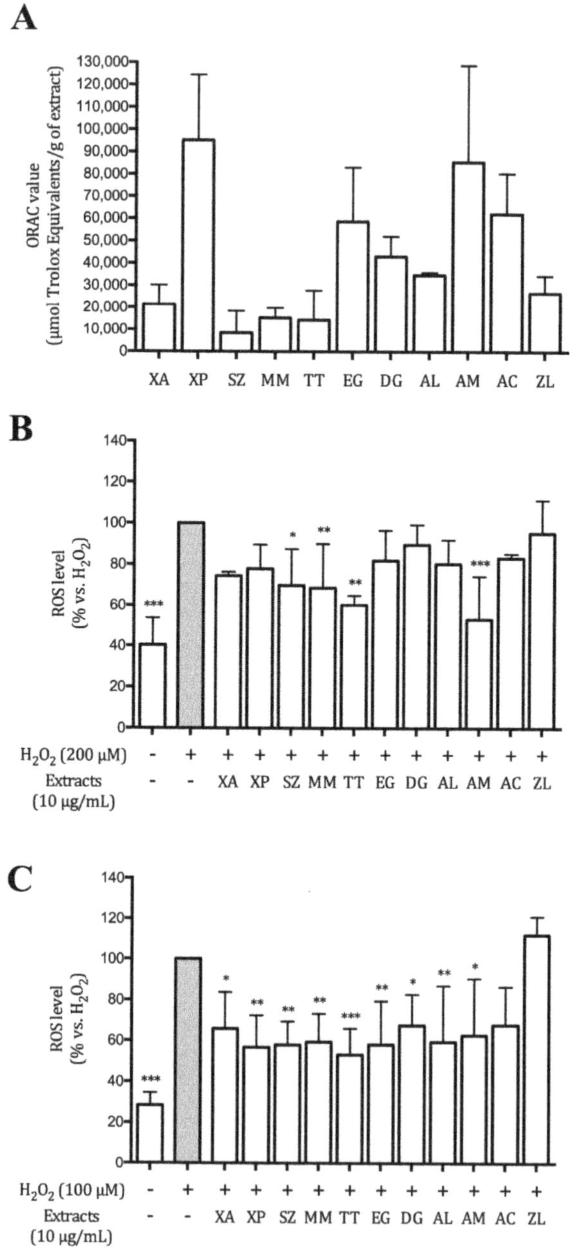

Figure 1. Antioxidant activity of the extracts. Oxygen radical absorbance capacity (ORAC) assay (**A**). Data are expressed as ORAC value (µmol Trolox Equivalent/g of sample). Effect of the extracts on intracellular ROS production induced by hydrogen peroxide in human gastric adenocarcinoma (AGS) (**B**) and gastric epithelial (GES-1) (**C**) cells. Antioxidant activity is expressed as µmol Trolox equivalent. Data reported in panels B and C are expressed as percentage versus the stimulated control (grey bar), to which is arbitrarily assigned the value 100%. * $p < 0.05$; ** $p < 0.01$; *** $p < 0.001$.

SZ, MM, TT, and AM exhibited a statistically significant inhibition in AGS cells ($p < 0.05$, Figure 1B), while in GES-1 cells the effects were more pronounced since all the extracts, except for ZL, showed a significant effect approximately halving the levels of ROS (Figure 1C).

3.3. Effect of the Extracts on the TNFα-Induced NF-κB-Driven Transcription in AGS and GES-1 Cells

The eleven hydroalcoholic extracts were preliminary screened at 10 μg/mL for their ability to impair the TNFα-induced NF-κB driven transcription, in AGS and GES-1 cells. TNFα increased the NF-κB driven transcription around 10-folds compared to unstimulated cells, in both cell lines (Figure 2); XP, TT, DG, AM, and AC extracts significantly impaired the activation in both cell lines, while XA only in AGS cells. The effect was more pronounced in AGS than in GES-1 cells.

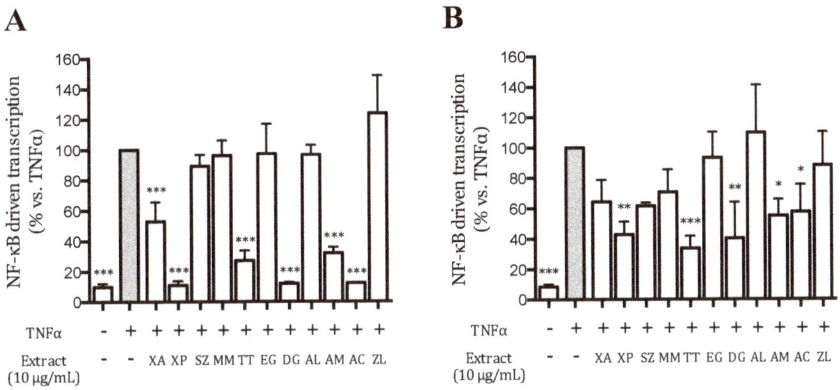

Figure 2. Effect of the extracts (10 μg/mL) on the NF-κB-driven transcription in human gastric adenocarcinoma (AGS) (**A**) and gastric epithelial (GES-1) (**B**) cells. Data are expressed as percentage versus the stimulated control, which is arbitrarily assigned the value 100% (grey bar). * $p < 0.05$; ** $p < 0.01$; *** $p < 0.001$.

XA, XP, TT, DG, AM, and AC extracts were further investigated for their ability to inhibit the TNFα-induced NF-κB driven transcription in a concentration-dependent manner (Figures S1 and S2). The extracts were tested in the range 0–10 μg/mL, and the corresponding IC_{50} values were calculated. In AGS cells, all the extracts, except for XA, showed an IC_{50} below 10 μg/mL (the highest concentration tested) and DG was the most active (IC_{50} 2.1 μg/mL). In GES-1 cells, the IC_{50}s were higher compared to those obtained in AGS cells, and only TT and DG extracts showed IC_{50} values below 10 μg/mL (Table 1).

Table 1. Half-maximal Inhibitory Concentration (IC_{50}) (μg/mL) of the extracts on the TNFα-induced NF-κB driven transcription in human gastric adenocarcinoma (AGS) and gastric epithelial (GES-1) cells.

	AGS Cells	**GES-1 Cells**
	IC_{50} (μg/mL)	
Xylopia parviflora Spruce (XP)	4.1	>10
Xylopia aethiopica (Dunal) A. Rich. (XA)	>10	>10
Tetrapleura tetraptera (Schum. and Thonn.) Taub (TT)	9.7	5.9
Dichrostachys glomerata (Forssk.) Chiov. (DG)	2.1	8.8
Aframomum melegueta K.Schum (AM)	9.9	>10
Aframomum citratum (C.Pereira) K.Schum (AC)	6.8	>10

3.4. *Plant Extracts Inhibit TNFα-Induced IL-8 Release and Expression in AGS and GES-1 Cells*

IL-8 is a well-known NF-κB-dependent chemokine strictly involved in the inflammatory process at the gastric level [5,7]; for this reason, the selected extracts (XP, XA, TT, DG, AM, and AC) were assessed for their ability to reduce its release as well as its promoter activity induced by TNFα in AGS and GES-1 cells. All the extracts, at concentrations ranging between 0.1 and 10 μg/mL, were able to reduce IL-8 secretion with different IC_{50}s (Figures S3 and S4), whose values were lower in AGS cells compared to GES-1 cells, as previously observed on the NF-κB driven transcription. The extracts also inhibited the IL-8 promoter activity, in AGS and GES-1 cells (Figures S5 and S6), suggesting that the effects showed on IL-8 release could be due, at least in part, to the impairment of the promoter activity (Table 2).

Table 2. Half-maximal Inhibitory Concentration (IC_{50}) (μg/mL) of the extracts on the TNFα-induced IL-8 release and expression in human gastric adenocarcinoma (AGS) and gastric epithelial (GES-1) cells.

	AGS Cells		**GES-1 Cells**	
	IC_{50} (μg/mL)			
	IL-8 release	IL-8 promoter activity	IL-8 release	IL-8 promoter activity
Xylopia parviflora Spruce (XP)	0.3	1.0	8.4	>10
Xylopia aethiopica (Dunal) A. Rich. (XA)	8.1	>10	>10	>10
Tetrapleura tetraptera (Schum. and Thonn.) Taub (TT)	1.4	1.4	5.7	4.6
Dichrostachys glomerata (Forssk.) Chiov. (DG)	0.2	0.3	4.2	>10
Aframomum melegueta K.Schum (AM)	1.2	1.9	>10	>10
Aframomum citratum (C.Pereira) K.Schum (AC)	0.4	2.0	2.3	>10

Since IC_{50}s on IL-8 release were not always comparable with those on promoter activity, we investigated the inhibitory mechanism at mRNA level; concentrations corresponding to the IC_{50} values obtained on IL-8 release by each extract were tested for this purpose.

In GES-1 cells, the extracts with an IC_{50} >10 μg/mL (XA and AM) were tested at 10 μg/mL. As shown in Figure 3, in AGS cells, the *IL-8* mRNA levels were around 40–50% lower with respect to stimulated control; however, this effect was not statistically significant. In GES-1 cells, XP, DG, and TT extracts were able to reduce the *IL-8* mRNA levels of around 60%, while the activity of AC extract was around 40% lower. These results confirmed that the activity on IL-8 release could be attributed to an effect at transcriptional level (Figure 3).

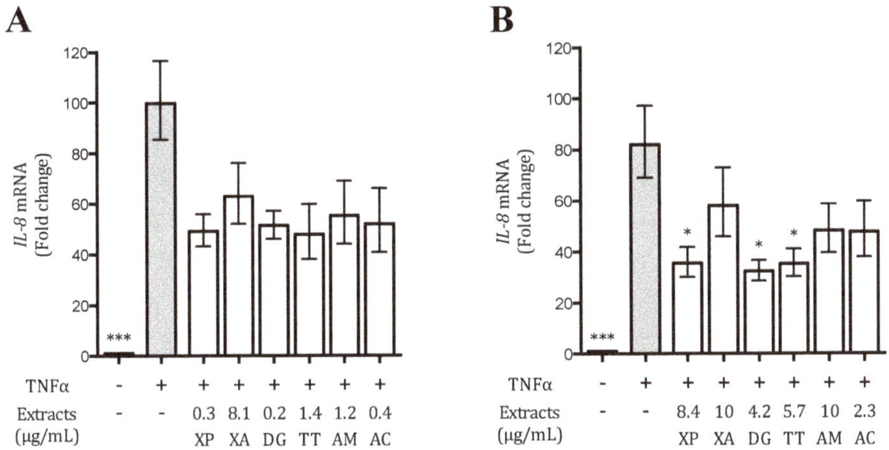

Figure 3. Effect of the extracts on *IL-8* mRNA levels in human gastric adenocarcinoma (AGS) (**A**) and gastric epithelial (GES-1) (**B**) cells. Data are expressed as fold changes versus stimulated control (grey bar). * $p < 0.05$; *** $p < 0.001$.

3.5. Effect of the Extracts on the TNFα-Induced IL-6 Release and Expression in AGS and GES-1 Cells

IL-6 is a cytokine dependent on the NF-κB activation involved in *H. pylori*-induced gastric inflammation [30]. Thus, the possible ability of the extracts to counteract gastric inflammation through the decrease of IL-6 release and expression was investigated as well. In these studies, just GES-1 cells were considered since TNFα did not induce any detectable IL-6 secretion in AGS cells (data not shown). The extracts were tested on IL-6 release in the range 0–10 μg/mL (Figure S7), while the activity on *IL-6* gene expression was screened at 10 μg/mL. XP and DG extracts inhibited IL-6 release both with an IC_{50} of 3.5 μg/mL, whereas the effect of TT and AC was lower (4.9 and 5.1 μg/mL, respectively); XA and AM exhibited an IC_{50} higher than 10 μg/mL (Table 3).

Table 3. Half-maximal Inhibitory Concentration (IC_{50}) (μg/mL) of the extracts on the IL-6 release in gastric epithelial (GES-1) cells.

	GES-1 Cells IC_{50} (μg/mL)
Xylopia parviflora Spruce (XP)	3.5
Xylopia aethiopica (Dunal) A. Rich. (XA)	>10
Tetrapleura tetraptera (Schum. and Thonn.) Taub (TT)	4.9
Dichrostachys glomerata (Forssk.) Chiov. (DG)	3.5
Aframomum melegueta K.Schum (AM)	>10
Aframomum citratum (C.Pereira) K.Schum (AC)	5.1

The effect of DG extract (10 μg/mL) on *IL-6* gene expression was statistically significant, as displayed in Figure 4; XP, DG, and TT also reduced *IL-6* mRNA levels; however, this effect was not statistically significant. XA and AM extracts did not show any activity (Figure 4).

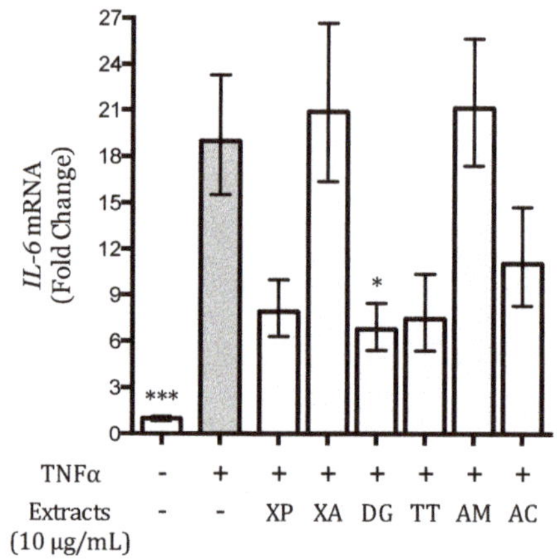

Figure 4. Effect of the extracts on IL-6 mRNA levels in gastric epithelial (GES-1) cells. Results are expressed as fold changes versus stimulated control (grey bar). * $p < 0.05$; *** $p < 0.001$.

3.6. *Tetrapleura Tetraptera Extract Inhibits PTGS2 (COX-2) Gene Expression in AGS and GES-1 Cells*

The gene expression, induced by TNFα, of the two cyclooxygenases, *PTGS1* (*COX-1*) and *PTGS2* (*COX-2*), was investigated at different times (1, 3, and 6 h) in AGS and GES-1 cells. *PTGS1* (*COX-1*) expression, as expected, was not induced by the stimulus in both cell lines; *PTGS2* (*COX-2*) mRNA was considerably increased in GES-1 cells at 6 h, while a mild increased after 1 h stimulation was observed in AGS cells (data not shown). The activity of the extracts on the gene expression of the two isoforms of cyclooxygenase was evaluated in GES-1 cells after 6 h treatment, in basal and inflammatory conditions. The *PTGS2* (*COX-2*) mRNA level of the cells treated with TT extract and TNFα was the only condition non-statistically different with respect to unstimulated control (Figure 5), suggesting the ability of this extract to reduce TNFα-induced *PTGS2* (*COX-2*) gene expression; the effect was exerted without affecting the gene expression of the *PTGS1* (*COX-1*) isoform. TT extract was also inactive on *PTGS2* (*COX-2*) and *PTGS1* (*COX-1*) mRNA levels in the basal conditions (Figure 5). Interestingly, TT extract did not affect the enzymatic activity of PTGS1 (COX-1) and PTGS2 (COX-2) isoforms in a cell-free system (data not shown).

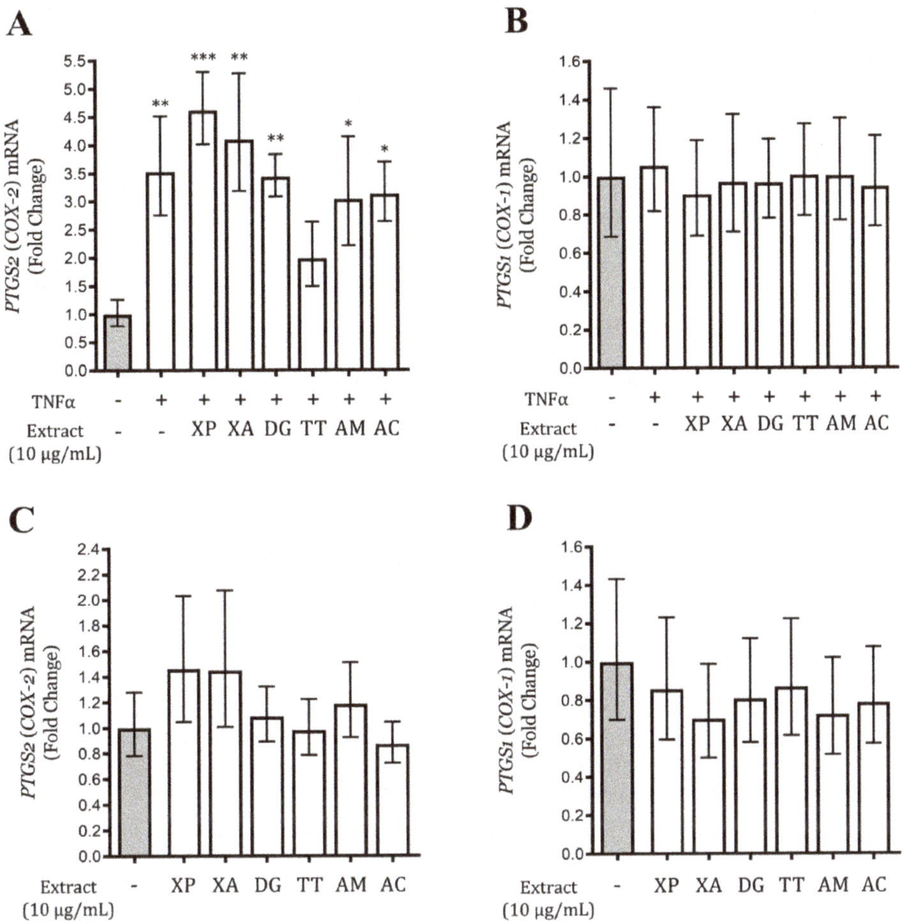

Figure 5. Effect of the extracts on prostaglandin endoperoxide synthase 2 *PTGS2* (*COX-2*) and prostaglandin endoperoxide synthase 1 *PTGS1* (*COX-1*) mRNA levels in the presence (**A**,**B**) or in the absence (**C**,**D**) of TNFα, in gastric epithelial (GES-1) cells. Data are expressed as fold changes versus control (grey bar). * $p < 0.05$; ** $p < 0.01$; *** $p < 0.001$.

4. Discussion

Gastritis is an inflammatory-based condition promoted by a variety of risk factors, which include stress, alcohol or drug abuse, and others; the most severe chronic form of gastritis is mainly caused by *Helicobacter pylori* infection. Unfortunately, the infection is diffused in several countries with unsustainable economic situation, and the conventional therapy against gastric inflammatory processes or for *H. pylori* eradication is not easily available.

This work investigates the anti-inflammatory and radical scavenger activities at the gastric level of eleven hydroalcoholic extracts obtained from plants widely used in Cameroon as spices and in the traditional medicine against a variety of diseases, including gastric disorders.

The study was carried out using two gastric cell models (AGS and GES-1 cells) stimulated with TNFα, which contributes to the inflammatory process in infected gastric epithelium [31]. AGS cells, a tumor gastric epithelial cell line, is a well-established in vitro model, while GES-1 cells are considered an in vitro model closer to the gastric epithelium of healthy subjects [32].

A prolonged oxidative stress may cause lipid peroxidation and DNA damages, leading to an increased risk to develop gastric cancer [33,34]; several extracts included in this study prevented H_2O_2-induced ROS generation in AGS and GES-1 cells, this effect could be explained with the presence of compounds able to counteract the excessive amount of ROS, derived from the unbalance between ROS production and endogenous antioxidant systems, acting as scavengers. The extracts displaying the highest antioxidant effect were those with the highest phenol content, as previously reported [25].

The investigation on the anti-inflammatory activity identified six extracts (XA, XP, DG, TT, AM, and AC) able to reduce the release and the gene expression of two NF-κB-dependent pro-inflammatory mediators, IL-8 and IL-6, which contribute to the amplification of the gastric inflammatory process [6,8,31]; the effect on the NF-κB pathway suggests that this transcription factor could be involved in the molecular mechanism responsible for the observed activity. In addition, the IC_{50}s on IL-8 release in our experiments reflected in halved IL-8 mRNA level, suggesting the important contribution of the transcription, although not always statistically confirmed. The different activity on the IL-8 promoter, especially evident in GES-1 cells, may be partly justified by the intervention of post-transcriptional regulatory mechanisms. The chemical analysis of the extracts has been previously reported by our group [24]; on the basis of this characterization, the hydroalcoholic extracts contain previously quantified secondary metabolites able to act as NF-κB inhibitors, such as pimaric acid in XP (8.73%), gingerol and shogaol in AM (0.7% and 2.05% respectively), chlorogenic acid and catechins, identified in some extracts [35–40]. Although further investigation is needed, these individual compounds could contribute, at least in part, to the anti-inflammatory activity displayed by the extracts. According to the literature, the anti-inflammatory effects could be linked to the antioxidant properties of the extracts, since it has been demonstrated that ROS are able to over-express IL-8 by activating oxidant-sensitive nuclear factors, such as NF-κB, in gastric epithelial cells [11].

Although the six extracts inhibited IL-8 release in both gastric cell models, only XP, TT, DG, and AC reduced IL-6 levels. Again, the mRNA analysis showed that all the extracts influenced the IL-6 at the transcriptional level, in particular DG, which obtained a statistically significant difference with respect to the stimulated control. Our results allow to speculate that XA and AM do not inhibit other important transcription factors for IL-6, such as CREB, but further studies are needed to confirm this hypothesis.

NSAIDs are not used in the therapy of gastric inflammatory diseases, since they act on prostaglandin-endoperoxide synthase (cyclooxygenase) enzymes blocking also the PTGS1 (COX-1) isoform, which is involved in the protection of the gastric mucosa [14]. In this context, TT extract could be considered the most promising extract, for the ability to inhibit TNFα-induced *PTGS2 (COX-2)* gene expression without affecting the *PTGS1 (COX-1)* mRNA basal levels. Indeed, *Tetrapleura tetraptera* (TT) extract did not reduce the enzymatic activity of PTGS1 (COX-1) and PTGS2 (COX-2) isoforms in a cell-free system; however, these findings need further investigation.

In the literature, different studies have demonstrated the beneficial effects of the fruits of this plant including antimalarial, anti-inflammatory, hypotensive, anti-insulin resistance, and antilipidemic properties [22,41,42].

This study demonstrates the antioxidant and anti-inflammatory activity, at the gastric level, of different hydroalcoholic extracts from Cameroonian plants, selecting those with the most promising effects. XA, XP, TT, DG, AM, and AC are largely used and diffused in some African countries as spices, thus making these plants interesting as functional foods, especially XP, TT, DG, and AC. Further studies are clearly needed to collect more evidence, especially for TT extract, although our findings provide a sound scientific support to the traditional use of these plants against gastric disorders. In particular, studies on the stability of the extracts in the gastric environment should reinforce the translational value of this work. Considering the order of the inhibitory concentrations (0.1–10 μg/mL) and the yields of extraction [23], it is plausible that the dietary consumption of 1–10 mg of Cameroonian spices may achieve the bioactivity at gastric level. Thus, these products could be useful to alleviate gastric inflammation in countries where conventional therapy is not easily available.

Supplementary Materials: The following are available online at http://www.mdpi.com/2072-6643/12/12/3787/s1, Figure S1: Effect of the extracts on the NF-κB driven transcription in AGS cells. Data are expressed as percentage versus the stimulated control, which is arbitrarily assigned the value 100%., Figure S2: Effect of the extracts on the NF-κB driven transcription in GES-1 cells. Data are expressed as percentage versus the stimulated control, which is arbitrarily assigned the value 100%, Figure S3: Effect of the extracts on the IL-8 release in AGS cells. Data are expressed as percentage versus the stimulated control, which is arbitrarily assigned the value 100%, Figure S4: Effect of the extracts on the IL-8 release in GES-1 cells. Data are expressed as percentage versus the stimulated control, which is arbitrarily assigned the value 100%, Figure S5: Effect of the extracts on the IL-8 promoter activity in AGS cells. Data are expressed as percentage versus the stimulated control, which is arbitrarily assigned the value 100%, Figure S6: Effect of the extracts on the IL-8 promoter activity in GES-1 cells. Data are expressed as percentage versus the stimulated control, which is arbitrarily assigned the value 100%, Figure S7: Effect of the extracts on the IL-6 release in GES-1 cells. Data are expressed as percentage versus the stimulated control, which is arbitrarily assigned the value 100%, Table S1: List of the primers used in the study.

Author Contributions: Conceptualization, A.P.A.N., M.D., J.-R.K., G.A.A. and E.S.; methodology, A.P.A.N., M.F., S.P., A.M., G.M. and G.B.; software, E.S.; validation, P.M., G.B. and M.D.; formal analysis, A.P.A.N.; investigation, A.P.A.N., A.D.T., M.F., A.M., G.M., S.P.; resources, A.P.A.N., M.F., A.M., G.M., S.P.; data curation, M.D., E.S.; writing—original draft preparation, A.P.A.N., M.F. and M.D.; writing—review and editing, M.D. and P.M.; supervision, M.D., J.-R.K. and G.A.A.; project administration, E.S.; funding acquisition, M.D. All authors have read and agreed to the published version of the manuscript.

Funding: This research was partially funded by MIUR Progetto di Eccellenza.

Acknowledgments: The authors thank Dawit Kidane-Mulat, University of Texas, Austin, for providing GES-1 cells.

Conflicts of Interest: The authors declare no conflict of interest. The funder had no role in the design of the study; in the collection, analyses, or interpretation of data; in the writing of the manuscript, or in the decision to publish the results.

References

1. Peek, R.M., Jr.; Crabtree, J.E. Helicobacter infection and gastric neoplasia. *J. Pathol.* **2006**, *208*, 233–248. [CrossRef] [PubMed]
2. Noto, J.M.; Peek, R.M., Jr. Helicobacter pylori: An overview. *Methods Mol. Biol.* **2012**, *921*, 7–10. [CrossRef] [PubMed]
3. Isomoto, H.; Mizuta, Y.; Miyazaki, M.; Takeshima, F.; Omagari, K.; Murase, K.; Nishiyama, T.; Inoue, K.; Murata, I.; Kohno, S. Implication of nf-kappab in helicobacter pylori-associated gastritis. *Am. J. Gastroenterol.* **2000**, *95*, 2768–2776. [CrossRef] [PubMed]
4. Aihara, M.; Tsuchimoto, D.; Takizawa, H.; Azuma, A.; Wakebe, H.; Ohmoto, Y.; Imagawa, K.; Kikuchi, M.; Mukaida, N.; Matsushima, K. Mechanisms involved in helicobacter pylori-induced interleukin-8 production by a gastric cancer cell line, mkn45. *Infect. Immun.* **1997**, *65*, 3218–3224. [CrossRef] [PubMed]
5. Yasumoto, K.; Okamoto, S.; Mukaida, N.; Murakami, S.; Mai, M.; Matsushima, K. Tumor necrosis factor alpha and interferon gamma synergistically induce interleukin 8 production in a human gastric cancer cell line through acting concurrently on ap-1 and nf-kb-like binding sites of the interleukin 8 gene. *J. Biol. Chem.* **1992**, *267*, 22506–22511.
6. Aggarwal, B.B. Nuclear factor-kappab: The enemy within. *Cancer Cell* **2004**, *6*, 203–208. [CrossRef]
7. Sharma, S.A.; Tummuru, M.K.; Blaser, M.J.; Kerr, L.D. Activation of il-8 gene expression by helicobacter pylori is regulated by transcription factor nuclear factor-kappa b in gastric epithelial cells. *J. Immunol.* **1998**, *160*, 2401–2407.
8. Crabtree, J.E.; Wyatt, J.I.; Trejdosiewicz, L.K.; Peichl, P.; Nichols, P.H.; Ramsay, N.; Primrose, J.N.; Lindley, I.J. Interleukin-8 expression in helicobacter pylori infected, normal, and neoplastic gastroduodenal mucosa. *J. Clin. Pathol.* **1994**, *47*, 61–66. [CrossRef]
9. Crabtree, J.E.; Peichl, P.; Wyatt, J.I.; Stachl, U.; Lindley, I.J. Gastric interleukin-8 and iga il-8 autoantibodies in helicobacter pylori infection. *Scand. J. Immunol.* **1993**, *37*, 65–70. [CrossRef]
10. Naito, Y.; Yoshikawa, T. Molecular and cellular mechanisms involved in helicobacter pylori-induced inflammation and oxidative stress. *Free Radic. Biol. Med.* **2002**, *33*, 323–336. [CrossRef]

11. Kim, Y.; Seo, J.H.; Kim, H. Beta-carotene and lutein inhibit hydrogen peroxide-induced activation of nf-kappab and il-8 expression in gastric epithelial ags cells. *J. Nutr. Sci. Vitaminol.* **2011**, *57*, 216–223. [CrossRef]
12. Hawkey, C.J. Cox-1 and cox-2 inhibitors. *Best Pract. Res. Clin. Gastroenterol.* **2001**, *15*, 801–820. [CrossRef] [PubMed]
13. Yamamoto, K.; Arakawa, T.; Ueda, N.; Yamamoto, S. Transcriptional roles of nuclear factor kappa b and nuclear factor-interleukin-6 in the tumor necrosis factor alpha-dependent induction of cyclooxygenase-2 in mc3t3-e1 cells. *J. Biol. Chem.* **1995**, *270*, 31315–31320. [CrossRef]
14. Harirforoosh, S.; Asghar, W.; Jamali, F. Adverse effects of nonsteroidal antiinflammatory drugs: An update of gastrointestinal, cardiovascular and renal complications. *J. Pharm. Pharm. Sci.* **2013**, *16*, 821–847. [CrossRef] [PubMed]
15. Abdou Bouba, A.; Njintang, Y.N.; Scher, J.; Mbofung, C.M.F. Phenolic compounds and radical scavenging potential of twenty cameroonian spices. *Agric. Biol. J. N. Am.* **2010**, *1*, 213–224. [CrossRef]
16. Fokunang, C.N.; Ndikum, V.; Tabi, O.Y.; Jiofack, R.B.; Ngameni, B.; Guedje, N.M.; Tembe-Fokunang, E.A.; Tomkins, P.; Barkwan, S.; Kechia, F.; et al. Traditional medicine: Past, present and future research and development prospects and integration in the national health system of cameroon. *Afr. J. Tradit. Complement. Altern. Med.* **2011**, *8*, 284–295. [CrossRef] [PubMed]
17. Tsabang, N.; Yedjou, C.G.; Tsambang, L.; Tchinda, A.T.; Donfagsiteli, N.; Agbor, G.A.; Tchounwou, P.; Nkongmeneck, B.A. Treatment of diabetes and/or hypertension using medicinal plants in cameroon. *Med. Aromat. Plants* **2015**. [CrossRef]
18. Noumi, E.; Dibakto, T.W. Medicinal plants used for peptic ulcer in the bangangte region, western cameroon. *Fitoterapia* **2000**, *71*, 406–412. [CrossRef]
19. Dzoyem, J.P.; Tchuenguem, R.T.; Kuiate, J.R.; Teke, G.N.; Kechia, F.A.; Kuete, V. In vitro and in vivo antifungal activities of selected cameroonian dietary spices. *BMC Complement. Altern. Med.* **2014**, *14*, 58. [CrossRef]
20. Kuete, V.; Krusche, B.; Youns, M.; Voukeng, I.; Fankam, A.G.; Tankeo, S.; Lacmata, S.; Efferth, T. Cytotoxicity of some cameroonian spices and selected medicinal plant extracts. *J. Ethnopharmacol.* **2011**, *134*, 803–812. [CrossRef]
21. Fankam, A.G.; Kuete, V.; Voukeng, I.K.; Kuiate, J.R.; Pages, J.M. Antibacterial activities of selected cameroonian spices and their synergistic effects with antibiotics against multidrug-resistant phenotypes. *BMC Complement. Altern. Med.* **2011**, *11*, 104. [CrossRef]
22. Ojewole, J.A.; Adewunmi, C.O. Anti-inflammatory and hypoglycaemic effects of tetrapleura tetraptera (taub) [fabaceae] fruit aqueous extract in rats. *J. Ethnopharmacol.* **2004**, *95*, 177–182. [CrossRef]
23. Woguem, V.; Fogang, H.P.; Maggi, F.; Tapondjou, L.A.; Womeni, H.M.; Quassinti, L.; Bramucci, M.; Vitali, L.A.; Petrelli, D.; Lupidi, G.; et al. Volatile oil from striped african pepper (xylopia parviflora, annonaceae) possesses notable chemopreventive, anti-inflammatory and antimicrobial potential. *Food Chem.* **2014**, *149*, 183–189. [CrossRef]
24. Atchan Nwakiban, A.P.; Sokeng, A.J.; Dell'Agli, M.; Bossi, L.; Beretta, G.; Gelmini, F.; Deutou Tchamgoue, A.; Agbor Agbor, G.; Kuiate, J.R.; Daglia, M.; et al. Hydroethanolic plant extracts from cameroon positively modulate enzymes relevant to carbohydrate/lipid digestion and cardio-metabolic diseases. *Food Funct.* **2019**, *10*, 6533–6542. [CrossRef] [PubMed]
25. Nwakiban, A.P.A.; Cicolari, S.; Piazza, S.; Gelmini, F.; Sangiovanni, E.; Martinelli, G.; Bossi, L.; Carpentier-Maguire, E.; Tchamgoue, A.D.; Agbor, G.; et al. Oxidative stress modulation by cameroonian spice extracts in hepg2 cells: Involvement of nrf2 and improvement of glucose uptake. *Metabolites* **2020**, *10*, 182. [CrossRef]
26. Ou, B.; Hampsch-Woodill, M.; Prior, R.L. Development and validation of an improved oxygen radical absorbance capacity assay using fluorescein as the fluorescent probe. *J. Agric. Food Chem.* **2001**, *49*, 4619–4626. [CrossRef] [PubMed]
27. Davalos, A.; Gomez-Cordoves, C.; Bartolome, B. Extending applicability of the oxygen radical absorbance capacity (orac-fluorescein) assay. *J. Agric. Food Chem.* **2004**, *52*, 48–54. [CrossRef]
28. Denizot, F.; Lang, R. Rapid colorimetric assay for cell growth and survival. Modifications to the tetrazolium dye procedure giving improved sensitivity and reliability. *J. Immunol. Methods* **1986**, *89*, 271–277. [CrossRef]
29. Dell'Agli, M.; Galli, G.V.; Bosisio, E.; D'Ambrosio, M. Inhibition of nf-kb and metalloproteinase-9 expression and secretion by parthenolide derivatives. *Bioorg. Med. Chem. Lett.* **2009**, *19*, 1858–1860. [CrossRef]

30. Bagheri, N.; Azadegan-Dehkordi, F.; Shirzad, H.; Rafieian-Kopaei, M.; Rahimian, G.; Razavi, A. The biological functions of il-17 in different clinical expressions of helicobacter pylori-infection. *Microb. Pathog.* **2015**, *81*, 33–38. [CrossRef]
31. Tanahashi, T.; Kita, M.; Kodama, T.; Yamaoka, Y.; Sawai, N.; Ohno, T.; Mitsufuji, S.; Wei, Y.P.; Kashima, K.; Imanishi, J. Cytokine expression and production by purified helicobacter pylori urease in human gastric epithelial cells. *Infect. Immun.* **2000**, *68*, 664–671. [CrossRef] [PubMed]
32. Ke, Y.; Ning, T.; Wang, B. [establishment and characterization of a sv40 transformed human fetal gastric epithelial cell line-ges-1]. *Zhonghua Zhong Liu Za Zhi* **1994**, *16*, 7–10. [PubMed]
33. Suzuki, M.; Miura, S.; Mori, M.; Kai, A.; Suzuki, H.; Fukumura, D.; Suematsu, M.; Tsuchiya, M. Rebamipide, a novel antiulcer agent, attenuates helicobacter pylori induced gastric mucosal cell injury associated with neutrophil derived oxidants. *Gut* **1994**, *35*, 1375–1378. [CrossRef] [PubMed]
34. Butcher, L.D.; den Hartog, G.; Ernst, P.B.; Crowe, S.E. Oxidative stress resulting from helicobacter pylori infection contributes to gastric carcinogenesis. *Cell Mol. Gastroenterol. Hepatol.* **2017**, *3*, 316–322. [CrossRef]
35. Suh, S.J.; Kwak, C.H.; Chung, T.W.; Park, S.J.; Cheeeei, M.; Park, S.S.; Seo, C.S.; Son, J.K.; Chang, Y.C.; Park, Y.G.; et al. Pimaric acid from aralia cordata has an inhibitory effect on tnf-alpha-induced mmp-9 production and hasmc migration via down-regulated nf-kappab and ap-1. *Chem. Biol. Interact.* **2012**, *199*, 112–119. [CrossRef]
36. Saha, A.; Blando, J.; Silver, E.; Beltran, L.; Sessler, J.; DiGiovanni, J. 6-shogaol from dried ginger inhibits growth of prostate cancer cells both in vitro and in vivo through inhibition of stat3 and nf-kappab signaling. *Cancer Prev. Res.* **2014**, *7*, 627–638. [CrossRef]
37. Liang, N.; Sang, Y.; Liu, W.; Yu, W.; Wang, X. Anti-inflammatory effects of gingerol on lipopolysaccharide-stimulated raw 264.7 cells by inhibiting nf-kappab signaling pathway. *Inflammation* **2018**, *41*, 835–845. [CrossRef]
38. Ruifeng, G.; Yunhe, F.; Zhengkai, W.; Ershun, Z.; Yimeng, L.; Minjun, Y.; Xiaojing, S.; Zhengtao, Y.; Naisheng, Z. Chlorogenic acid attenuates lipopolysaccharide-induced mice mastitis by suppressing tlr4-mediated nf-kappab signaling pathway. *Eur. J. Pharmacol.* **2014**, *729*, 54–58. [CrossRef]
39. Luo, K.W.; Wei, C.; Lung, W.Y.; Wei, X.Y.; Cheng, B.H.; Cai, Z.M.; Huang, W.R. Egcg inhibited bladder cancer sw780 cell proliferation and migration both in vitro and in vivo via down-regulation of nf-kappab and mmp-9. *J. Nutr. Biochem.* **2017**, *41*, 56–64. [CrossRef]
40. Contreras, T.C.; Ricciardi, E.; Cremonini, E.; Oteiza, P.I. (-)-epicatechin in the prevention of tumor necrosis alpha-induced loss of caco-2 cell barrier integrity. *Arch. Biochem. Biophys.* **2015**, *573*, 84–91. [CrossRef]
41. Kuate, D.; Kengne, A.P.; Biapa, C.P.; Azantsa, B.G.; Abdul Manan Bin Wan Muda, W. Tetrapleura tetraptera spice attenuates high-carbohydrate, high-fat diet-induced obese and type 2 diabetic rats with metabolic syndrome features. *Lipids Health Dis.* **2015**, *14*, 50. [CrossRef] [PubMed]
42. Okokon, J.E.; Udokpoh, A.E.; Antia, B.S. Antimalaria activity of ethanolic extract of tetrapleura tetraptera fruit. *J. Ethnopharmacol.* **2007**, *111*, 537–540. [CrossRef] [PubMed]

nutrients MDPI

Article

The Effect of Gum Arabic (*Acacia senegal*) on Cardiovascular Risk Factors and Gastrointestinal Symptoms in Adults at Risk of Metabolic Syndrome: A Randomized Clinical Trial

Amjad H. Jarrar [1], Lily Stojanovska [1,2], Vasso Apostolopoulos [2], Jack Feehan [2,3], Mo'ath F. Bataineh [4], Leila Cheikh Ismail [5,6] and Ayesha S. Al Dhaheri [1,*]

1 Nutrition and Health Department, College of Medicine and Health Sciences, United Arab Emirates University, Al Ain 15551, UAE; amjadj@uaeu.ac.ae (A.H.J.); lily.stojanovska@uaeu.ac.ae (L.S.)
2 Institute for Health and Sport, Victoria University, Melbourne 3011, Australia; vasso.apostolopoulos@vu.edu.au (V.A.); jfeehan@student.unimelb.edu.au (J.F.)
3 Department of Medicine–Western Health, The University of Melbourne, Melbourne 3010, Australia
4 Department of Sport and Rehabilitation, The Hashemite University, 13115 Zarqa, Jordan; mfmbataineh@hu.edu.jo
5 Clinical Nutrition and Dietetics Department, College of Health Sciences, University of Sharjah, Sharjah 27272, UAE; lcheikhismail@sharjah.ac.ae
6 Nuffield Department of Women's & Reproductive Health, University of Oxford, Oxford OX1 2JD, UK
* Correspondence: ayesha_aldhaheri@uaeu.ac.ae; Tel.: +971-50-6167671; Fax: +971-37-134593

Abstract: Gum Arabic (GA) is a widely-used additive in food processing, but is also historically used in a number of traditional therapies. It has been shown to have a broad range of health benefits, particularly in improving important cardiovascular risk indicators. Metabolic syndrome and its associated cardiac outcomes are a significant burden on modern healthcare systems, and complementary interventions to aid in its management are required. We aimed to examine the effect of GA on those with, or at risk of, metabolic syndrome to identify an effect on improving important disease parameters related to cardiovascular outcomes. A single-blind, randomized, placebo-controlled trial was conducted to identify the effects of daily GA supplementation on metabolic and cardiovascular risk factors. A total of 80 participants were randomized to receive 20 g of GA daily (n = 40) or placebo (1 g pectin, n = 40) for 12 weeks. Key endpoints included body-anthropometric indices, diet and physical activity assessment, and blood chemistry (HbA1c, fasting glucose, and blood lipids). Of the 80 enrolled, 61 completed the study (intervention: 31, control: 30) with 19 dropping out due to poor treatment compliance. After 12 weeks, the participants receiving the GA showed significant decreases in systolic and diastolic blood pressure, fat-free body mass, energy and carbohydrate consumption, and fasting plasma glucose, as well as increased intake of dietary fiber. They also reported improvements in self-perceived bloating and quality of bowel movements, as well as a decreased appetite score following GA consumption. These results suggest that GA could be a safe and beneficial adjunct to other treatments for those with, or at risk of, metabolic syndrome.

Keywords: metabolic syndrome; Gum Arabic; *Acacia senegal*; *Acacia seyal*

Citation: Jarrar, A.H.; Stojanovska, L.; Apostolopoulos, V.; Feehan, J.; Bataineh, M.F.; Ismail, L.C.; Al Dhaheri, A.S. The Effect of Gum Arabic (*Acacia senegal*) on Cardiovascular Risk Factors and Gastrointestinal Symptoms in Adults at Risk of Metabolic Syndrome: A Randomized Clinical Trial. *Nutrients* **2021**, *13*, 194. https://doi.org/10.3390/nu13010194

Received: 13 December 2020
Accepted: 6 January 2021
Published: 9 January 2021

Publisher's Note: MDPI stays neutral with regard to jurisdictional claims in published maps and institutional affiliations.

1. Introduction

Gum Arabic (GA) or *Acacia* gum is a soluble dietary fiber obtained from the stems and branches of the *Acacia senegal* and *Acacia seyal* plants, which grow mainly in the African region of Sahe in Sudan [1]. It is often associated with health benefits relevant to reducing the risk of metabolic syndrome (MetS). GA contains three different fractions of highly-branched carbohydrate structures that vary in molecular mass and protein content, which are believed to underlie its physiological effects. These are commonly known as the arabinogalactan-protein, arabinogalactan, and glycoprotein fractions [2].

However, the composition of GA may change depending on the source, climate, and soil [3]. Because of the physical properties of GA, it has been widely used in various industries including cosmetics, textiles, ceramics, pharmaceuticals and foods [4]. GA is commonly used in industrial food production as an emulsifier, a stabilizer, and a thickener due to its nondigestibility, low-solution viscosity, and safety [5]. Used as a traditional remedy for many years, several studies have described the antioxidant properties of GA and its capacity to neutralize reactive oxygen substances [6–8]. Research also suggests it may have an effect on lipid metabolism [9], as well as renal function [10–12] and satiety [13], lending support to its use as an adjunct in the prevention and treatment of metabolic syndrome. Broadly, GA appears to have a hypocholesterolemic effect, decreasing low-density lipoproteins (LDL) and very low density lipoproteins (VLDL) without affecting high-density lipoproteins (HDL) and triglycerides in animal models [6]. GA has shown potential to relieve the effects of chronic renal failure by improving creatinine clearance as well as excretion of magnesium and calcium [14]. It has also been reported to decrease blood pressure in mice, and has been shown to lower caloric intake significantly, potentially due to increased dietary-fiber intake increasing satiety [5]. This reduction in energy intake makes GA a strong candidate for adjunct weight-control therapies.

Metabolic syndrome describes a cluster of conditions including increased blood pressure, high blood glucose, excess body fat, and dyslipidemia occurring simultaneously [15]. Metabolic syndrome is one of the most significant risk factors for a wide range of noncommunicable diseases (NCDs) such as cardiovascular diseases and diabetes [16]. According to the World Health Organization (WHO), NCDs are responsible for 71 percent of deaths globally, with cardiovascular disease being the leading cause of death, followed by cancer, respiratory diseases, and diabetes [17]. Most NCDs share common behavioral risk factors such as poor diet, physical inactivity, and smoking, as well as the key MetS risk factors such as overweight, obesity, high blood sugar, and hypercholesterolemia [18]. This makes reducing the burden of MetS a key element in the prevention of NCDs [16].

Given the significant burden of MetS and its associated risk of NCDs, there is a need for novel interventions to help prevent its onset. This study aimed to investigate the effect of consuming 20 g of Gum Arabic-Acacia Senegal (GA-AS) per day based on key metabolic parameters in adults with, or at risk of, metabolic syndrome. It is the hope that the results of this study will provide insight into the dietary effects of consistent consumption of Gum Arabic-Acacia Senegal. It is hypothesized that GA-AS will provide benefits to the metabolic health of the participants with regular use and will be well tolerated by individuals taking it.

2. Materials and Methods

2.1. Study Design

The present study was a controlled, randomized, single-blind, parallel-design study comparing an intervention group receiving 20 g of GA-AS daily for 12 weeks with a control group receiving a daily placebo containing 1 g of pectin for the same period. The primary endpoints of the study were blood glucose, lipid profile, blood pressure, body composition, gastrointestinal motion, and satiety. The study was conducted at the department of Nutrition and Health, College of Food and Agriculture at United Arab Emirates University (UAEU) during the period from January to May 2018. This study was conducted according to the principles of the Helsinki declaration on human research ethics and was approved by the UAEU scientific research ethics committee (ref. no. ERH_2016_4372).

2.2. Study Participants

Participants were recruited from the UAEU (students and staff) through face-to-face interviews, email, social media, and printed advertisements on the campus and in the dormitories of the university. Participants were given both verbal and written information about the aim of the study, data to be collected, and the duration of intervention, and they were required to sign a written informed-consent form to participate in the study prior to screening.

Participants were screened for eligibility at the nutrition clinic at UAEU. Participant eligibility was based on the presence of metabolic-syndrome risk factors. Inclusion and exclusion criteria are summarized in Table 1. Risk factors assessed included waist circumference (females > 80 cm and males > 94 cm), systolic blood pressure (≥130 mm Hg), diastolic blood pressure (≥80 mm Hg), blood-fasting glucose (≥100 mg/dL), HDL cholesterol level (for female < 50 mg/dL and males < 40 mg/dL), and triglycerides level (≥150 mg/dL). Participants with more than three risk factors, or with two risk factors and one borderline, were included in the study. Participants were excluded from the study if they were smokers, pregnant women, lactating women, or were taking permanent medication.

Table 1. Nutrient composition of GA per 20 g dose.

Nutrient	Per 20 g Dose
Energy (kcal)	1.8
Protein (g)	0.4
Carbohydrate (g)	17.1
Fat (g)	0.02
Total dietary fiber (g)	17.1
Sodium (mg)	2.8
Calcium (mg)	214.8
Magnesium (mg)	78.0
Potassium (mg)	182.8
Iron (mg)	0.2

Participants were asked to complete a health-screening questionnaire that contained questions about medical conditions and medications that might influence glucose control, appetite, and energy expenditure. All participants signed an informed-consent form before taking part in the study. This study was conducted according to the guidelines in the Declaration of Helsinki. All procedures involving human subjects were approved by the United Arab Emirates University (UAEU) Scientific Research Ethics Committee.

Participants were randomly assigned to control and intervention groups via computer software, with the experimental group receiving 20 g of GA-AS powder per day and the control group receiving 1 g of placebo (pectin) powder per day for a period of 12 weeks. This dose was selected based on previous research showing metabolic effects with doses of 10–30 g per day for 4–12 weeks, with lipid effects being most significant after 5 weeks [3,13,19,20]. GA powder and placebo were provided in premeasured sachets, and participants were asked to consume the GA powder or the placebo two times per day by adding it to hot water, tea, milk, or on any meal. The nutrient composition of the study dose of Gum Arabic is presented in Table 1. Body weight (kg), height (cm), waist circumference (cm), body composition, blood glucose, blood-lipid profile, glycated hemoglobin A1c (HbA1c), and blood pressure were measured at baseline (week 0) and 12 weeks. In addition, participants were asked to complete a bowel-movement questionnaire and satiety scale at the baseline and endpoint of the study.

2.3. Research Parameters

2.3.1. Anthropometric Measurements

Body weight was recorded to the nearest 0.01 kg while the subject was wearing minimal clothes (as per local cultural requirements) and no shoes. Body composition was assessed via a bioimpedance device (InBody720, InBody, CA, USA), providing measurement of percentage body fat (%BF), fat mass (kg), and fat free mass (kg). Waist circumference was measured using measuring tapes, according to standard methods at the mid-point between inferior costal margin and superior border of the iliac crest. In obese individuals, the measurement was taken at the level of the umbilicus [21]. Body mass index (BMI) was calculated as BMI = kg/m^2. All measurements were taken at baseline and after 12 weeks of intervention.

2.3.2. Diet and Physical Activity Assessment

During the study period, the participants were asked to maintain their normal lifestyle. Participants were asked to record their dietary intake at baseline (week 0) and at week 12 of the study period. Food records were taken over three days including two weekdays and one weekend day. Photographs of food with different portion sizes were used to help participants estimate the correct portion size consumed. The Food Processor® Nutrition and Fitness Software, ESHA food-analysis program (version 10.4), and the Kuwaiti Food Composition database were used to analyze the energy and nutrient contents of the consumed foods [22].

Physical activity level was assessed using the International Physical Activity Questionnaire (IPAQ) (Arabic and English versions) at baseline and week 12 [23].

2.3.3. Bowel Movement and Satiety Questionnaires

Participants were asked to answer a bowel-movement questionnaire including frequency and intensity of constipation, bloating, diarrhea, and heartburn. A satiety questionnaire with scoring points was used to assess satiety after 60 min of ingestion of either GA-AS or the placebo intervention. The questionnaires were administered at both baseline and at the end of the study (adapted from [24,25]). In the satiety questionnaire, participants answered the following questions: How did the meal (with the study treatment or placebo) you just ate make you feel? Did it satisfy your hunger, or did you feel like you needed to snack later? Then they rated their feelings of satiety for 60 min using a score of 100 [24,25].

2.3.4. Biochemical Parameters

A fasting, venous blood sample was collected (5 mL) by a certified phlebotomist at baseline and at end of the study period. Fasting blood glucose (FBG), HbA1c, serum triglyceride (TG), high-density lipoprotein cholesterol (HDLC), low-density lipoprotein cholesterol (LDLC), and total cholesterol (TC) were analyzed using Cobas C111 automated biochemical analyzer (Roche Diagnostics, Indianapolis, IN, USA). All data were collected at the laboratory facilities of the Nutrition and Health Department.

2.4. *Statistical Analysis*

G*Power 3.1.9.2 software was used for sample-size calculation of repeated measures ANOVA with parallel design. Power calculation identified a sample size of at least 54 participants to detect a medium-effect size (0.25) with 95% power with significance level set at 0.05. The statistical analysis was performed using SPSS version 24.0 and results presented as (Mean ± Standard Deviation). Repeated measures ANOVA was used to detect the main effects of time and group on study measures. Paired t-tests and independent t-tests were employed to compare the effect of time and groups (Control vs. Intervention), respectively. Binary data was assessed for statistical significance with the N-1 chi square test. Results were considered statistically significant at p-value < 0.05.

3. Results

The study sample consisted of 80 participants from the United Arab Emirates University, aged 18–50 years with a mean age of 25.51 ± 9.5 years, mean BMI of 33.9 ± 5.4, and with 62.3% being female. Nineteen participants dropped out of the study due to poor treatment compliance, with 61 ultimately completing the 12-week intervention. Dropouts were largely due to failure to take prescribed GA or placebo, or failure to present for follow up testing. The control group consisted of 30 participants, and the experimental group 31 (Figure 1). There were no significant differences in baseline characteristics between groups (Table 2).

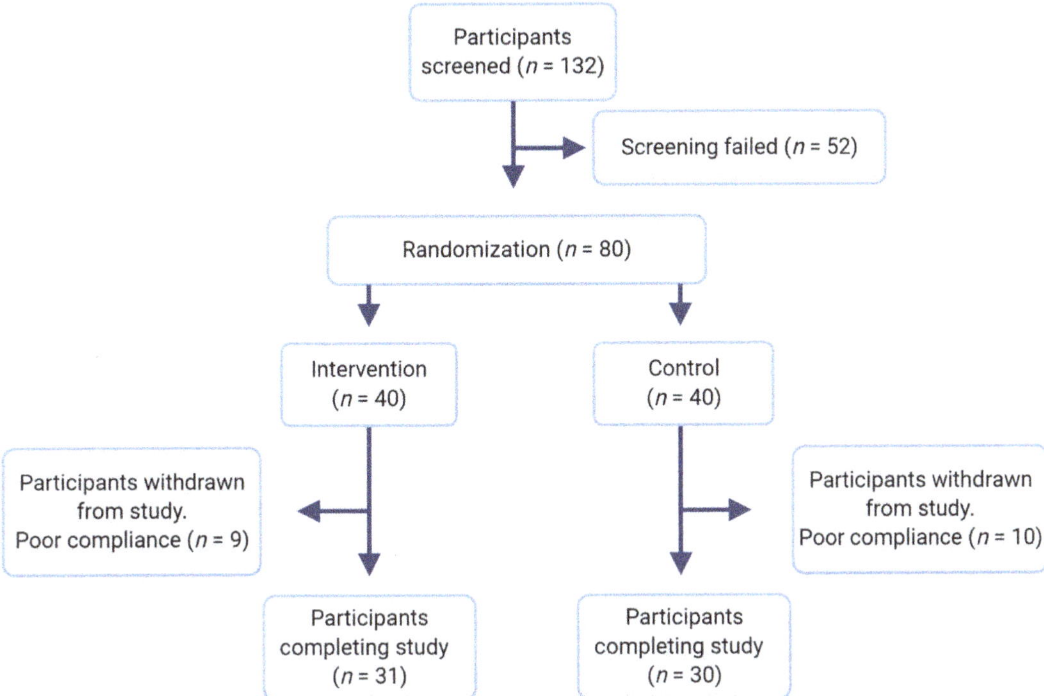

Figure 1. Summary of study recruitment.

Table 2. Baseline characteristics of the subjects (N = 61).

Variable	Control (n = 30) Mean ± SD	Intervention (n = 31) Mean ± SD	p-Value
Age (years)	25.6 ± 9.9	28.3 ± 11.8	0.452
Weight (kg)	91.7 ± 20.8	92.1 ± 17.4	0.948
Height (cm)	168.3 ± 11.6	164.3 ± 7.5	0.163
BMI (kg/m^2)	31.9 ± 4.7	34.1 ± 5.9	0.174
Waist circumference (cm)	100.5 ± 16.1	101.2 ± 12.7	0.867
Body Fat (%)	39.7 ± 8.4	43.7 ± 7.6	0.088
Fat free mass (kg)	53.8 ± 14.5	50.9 ± 9.4	0.382
Systolic Blood Pressure (mm Hg)	114.8 ± 16.4	118.3 ± 17.0	0.482
Diastolic Blood Pressure (mm Hg)	75.8 ± 9.9	81.1 ± 9.2	0.066
Physical Activity Levels			
Vigorous (min/week)	11.1 ± 3.0	14.4 ± 3.2	0.683
Moderate (min/week)	59.4 ± 5.7	62.1 ± 8.9	0.117
Light (min/week)	190.2 ± 13.1	193.5 ± 3.2	0.939
Sedentary Activity (h/day)	9.20 ± 0.4	10.5 ± 0.35	0.155
Nutritional Intake			
Energy (kcal)	2142 ± 551.8	2036.9 ± 601.5	0.534
Carbohydrate (g)	256 ± 57.7	239.4 ± 84.3	0.446
Fat (g)	82.2 ± 34.1	80.5 ± 30.7	0.852
Protein (g)	81.7 ± 28.9	86.9 ± 38.4	0.609
Dietary fiber (g)	15.0 ± 9.8	17.1 ± 15.2	0.648

Table 2. *Cont.*

Variable	Control (n = 30) Mean ± SD	Intervention (n = 31) Mean ± SD	p-Value
Biochemical Parameters			
HbA1c (%)	6.1 ± 0.9	6.0 ± 1.7	0.763
Glucose (mg/dL)	101.5 ± 14.0	105.6 ± 36.0	0.635
Triglycerides (mg/dL)	94.7 ± 41.6	100.9 ± 53.9	0.661
Total Cholesterol (mg/dL)	150.7 ± 34.3	157.9 ± 28.2	0.413
LDLC (mg/dL)	2.5 ± 0.9	2.50 ± 0.7	0.926
HDLC (mg/dL)	46.3 ± 12.1	45.1 ± 12.1	0.747

After 12 weeks of GA treatment, the experimental group showed significant decreases in both systolic (p = 0.008) and diastolic blood pressure (0.009), as well as fat free mass (p = 0.03), with no intragroup difference in the control group. No significant inter- or intragroup differences were observed in BMI, waist circumference, or body fat between baseline and week 12 (Table 3).

Table 3. Changes in physical characteristics for study population after 12 weeks.

	Control			Intervention			Intergroup Difference (Week 12)
Variable	Baseline (Mean ± SD)	Week 12 (Mean ± SD)	p-Value	Baseline (Mean ± SD)	Week 12 (Mean ± SD)	p-Value	p-Value
Weight (kg)	91.74 ± 20.8	93.0 ± 22.3	0.37	92.09 ± 17.4	91.43 ± 17.0	0.116	0.778
BMI (kg/m^2)	31.92 ± 4.7	32.6 ± 5.6	0.288	34.07 ± 5.9	33.90 ± 6.0	0.465	0.43
Waist circ. (cm)	100.50 ± 16.1	100.6 ± 18.9	0.937	101.20 ± 12.7	99.07 ± 13.0	0.155	0.728
Body Fat (%)	39.72 ± 8.4	40.90 ± 8.90	0.206	43.70 ± 7.6	44.10 ± 7.8	0.094	0.198
Fat Free Mass (kg)	53.79 ± 14.5	58.80 ± 9.70	0.185	50.85 ± 9.4	55.38 ± 8.9	0.030 *	0.206
Systolic (mm Hg)	114.80 ± 16.4	117.0 ± 15.0	0.242	118.30 ± 17.0	111.30 ± 19.8 *	0.008 *	0.273
Diastolic (mm Hg)	75.8 ± 9.90	79.50 ± 9.7	0.102	81.1 ± 9.2	76.70 ± 13.2 *	0.009 *	0.419

* p = < 0.05.

3.1. Diet and Physical-Activity Assessment

After 12 weeks of GA treatment, the intervention group showed a decrease in carbohydrate (p = 0.008) and calorie (p = 0.014) intake and an increase in dietary fiber consumption ($p \leq 0.001$), with no intergroup change in the controls (Table 4). There were also intergroup differences in carbohydrate (p = 0.004) and dietary-fiber ($p \leq 0.001$) consumption at 12 weeks. The intervention group also showed a trend toward lower energy intake compared to controls after 12 weeks; however, this did not reach significance (p = 0.069). There were no inter- or intra-group changes to physical activity parameters either in fat or protein consumption (Table 4).

Table 4. Dietary and physical-activity characteristics of the study population.

	Control			Intervention			Intergroup Difference (Week 12)
Variable	Baseline (Mean ± SE [1])	Week 12 (Mean ± SE [1])	p-Value	Baseline (Mean ± SE [1])	Week 12 (Mean ± SE [1])	p-Value	p-Value
Vigorous (min/week)	11.1 ± 3.0	10.7 ± 2.8	0.101	14.4 ± 3.2	18.4 ± 6.0	0.91	0.55
Moderate (min/week)	59.4 ± 5.7	71.5 ± 16.6	0.525	62.1 ± 8.9	59.0 ± 10.2	0.665	0.709
Light (min/week)	190.2 ± 13.1	206.5 ± 14.2	0.408	193.5 ± 3.2	198.4 ± 20.6	0.923	0.863
Sedentary Activity (h/day)	9.20 ± 0.4	10.9 ± 0.4	0.089	10.5 ± 0.35	10.7 ± 0.43	0.677	0.847
Energy (kcal)	2142 ± 62.4	2092 ± 55.0	0.644	2036.9 ± 68.0	1810 ± 62.5 *	0.014	0.069
Carbohydrate (g)	256 ± 6.5	256 ± 6.7	0.976	239.4 ± 9.5	194.1 ± 8.8 *	0.008	0.004
Fat (g)	82.2 ± 3.9	73.2 ± 3.11	0.259	80.5 ± 3.5	71.3 ± 3.2	0.139	0.815
Protein (g)	81.7 ± 3.3	87.0 ± 3.6	0.453	86.9 ± 4.3	84.9 ± 5.2	0.836	0.854
Dietary fiber (g)	15.0 ± 1.1	16.1 ± 1.2	0.829	17.1 ± 1.7	31.9 ± 1.64 *	0.001	<0.001

[1] Standard Error, * p = <0.05.

3.2. Biochemical Assessment

After 12 weeks of GA treatment, the intervention group had significantly decreased in fasting plasma glucose (p = 0.046), with the control group showing no change; however, there was no difference between the groups at the endpoint (p = 0.101). There were no changes to HbA1c or blood lipid profile in either group (Table 5).

Table 5. Biochemical measurements of study population.

	Control			Intervention			Intergroup Difference (Week 12)
Variable	Baseline (Mean ± SD)	Week 12 (Mean ± SD)	p-Value	Baseline (Mean ± SD)	Week 12 (Mean ± SD)	p-Value	p-Value
HbA1c (%)	6.10 ± 0.90	6.0 ± 0.4	0.662	6.0 ± 1.7	6.00 ± 0.8	0.938	0.145
Glucose (mg/dL)	101.50 ± 14.00	99.5 ± 14.2	0.489	105.6 ± 36.0	92.90 ± 13.20 *	0.046	0.55
Triglycerides (mg/dL)	94.70 ± 41.60	94.9 ± 41.0	0.955	100.9 ± 53.9	93.90 ± 44.0	0.936	0.832
Total Cholesterol (mg/dL)	150.7 ± 34.30	151.2 ± 37.8	0.941	157.9 ± 28.20	152.60 ± 30.6	0.892	0.709
LDLC (mg/dL)	2.50 ± 0.90	2.3 ± 0.8	0.433	2.50 ± 0.70	2.4 ± 0.7	0.625	0.925
HDLC (mg/dL)	46.3 ± 12.1	46.2 ± 13.2	0.951	45.1 ± 12.1	44.8 ± 13.9	0.736	<0.001

* p = <0.05.

3.3. Bowel Movement

The intervention group reported significant reductions (p = 0.005) in bloating and improvement in bowel movement (0.047) compared to the control group. Although there were no statistically significant differences in abdominal pain, better digestion, or reduction in nausea, the intervention rate responses were almost double those of the control group (Table 6).

Table 6. The effect of GA on the bowel movement at week 12 in the Control and Intervention Groups.

Response Rate for	Week 12		
	Control (Yes, %)	Intervention (Yes, %)	p-Value
Improved bowel movements	33.3	54.8	0.047 *
Reduction in bloating feelings	20.0	51.6	0.005 *
Reduction in abdominal pain	10.0	22.6	0.094
Feeling of better digestion	23.3	41.9	0.062
Reduction in nausea	10.0	25.8	0.056
Satiety Scores	Baseline (Mean ±SD)	60 min (Mean ±SE)	
Control	44.9 ± 24.7	51.6 ± 24.2	0.174
Gum Arabic	48.7 ± 22.3	62.5 ± 27.5	0.011

* p = <0.05.

3.4. Feeling of Satiety

One hour after taking the GA treatment, the participants in the intervention group showed a significant increase in appetite score (reflective of a decreased appetite) compared to controls receiving the placebo treatment (p = 0.01) (Table 6).

4. Discussion

To the best of our knowledge, the current study is the first to examine the effect of Gum Arabic on individuals with, or at risk of developing, metabolic syndrome. The daily treatment with GA caused improvement in a number of parameters important to modifying outcomes and risk of metabolic syndrome, including fasting plasma glucose, blood pressure, and energy intake. However, contrary to some other studies, our trial found no significant reduction in body weight or BMI. In one study, 30 g of GA per day for 6 weeks caused significant reductions in BMI, body-fat percentage, and weight in adult females compared to placebo-treated controls [3]. Another study on daily GA administration (30 g) for three months also found that BMI, visceral-adiposity index, and body-adiposity index were significantly lowered in the intervention group [26]. The difference in findings may reflect the relative size of the studies, as, while in our study these parameters did not reach significance, there was some evidence of a trend. With a larger group, the results may have become significant. The differences may also be due to differences in patient demographic. In the first study, the participants were younger and generally healthier, and while in the second study they were significantly older, they had a much lower BMI.

The demonstrated decreases in energy and carbohydrate consumption are significant in the context of metabolic syndrome, characterized in large part by impaired glucose metabolism and obesity. Reducing the intake of these through nonpharmacological supplementation could provide a means of management or prevention of the syndrome. While the exact mechanism underlying the behavior change is unclear, it may be due to the increased feelings of satiety following GA administration. These findings are echoed in another study, which showed that supplementation with two blends of GA (EmulGold1 (EG) and PreVitae1 (PV)) decreased the caloric intake significantly three hours after consumption and increased subjective ratings of satiety [19]. An increase in fat free mass (FFM) was also identified in this study. While interesting and potentially of benefit to the population studied, these results should be interpreted with caution. The bioimpedance device used in this study reports two-compartment FFM which measures bone, muscle, connective tissue, and water as one. While this is a reliable measure of body fat, it is unable to discriminate between beneficial changes in muscle or bone mass from increases in water retention [27]. To evaluate a potential change in bone or muscle parameters, measuring body composition by either four-compartment or dual-Xray absorptiometry (DXA) is required. This provides some support for the clinical use of GA in the management of obesity, potentially providing a means to lower the burden of invasive bariatric surgery or harsh pharmacotherapeutic avenues.

Another significant finding of the study at hand was its significant impact on the blood pressure of the participants receiving GA. As the most significant outcomes of metabolic syndrome are cardiovascular diseases, such as myocardial infarction and stroke, interventions lowering cardiac risk are particularly important. Again, the specific mechanism for this is unclear; however, it has been reported that intake of dietary fiber, including GA, was associated with a significant fall in mean systolic blood pressure [SBP] in normal individuals who neither had hypertension nor diabetes [10]. Another study of GA treatment in people living with diabetes also found a decrease in blood pressure with SBP decreasing by 5.9% and diastolic blood pressure (DBP) by 5.4% [26].

Gum Arabic is a soluble fermentable fiber that has shown hypoglycemic, antioxidant effects and improved lipid metabolism in previous studies [6,26,28]. Our data showed a significant reduction in blood glucose in the intervention group, a key parameter of metabolic syndrome. An animal model showed that GA has a glucose-lowering effect in rabbits with alloxan–induced diabetes. They showed that GA (at doses of 2, 3, and 4 mg/kg) significantly reduced the blood-glucose concentration of normal but not diabetic rabbits. They therefore concluded that GA initiated the release of insulin from pancreatic b cells in normal rabbits [29]. Another study suggested that the glycemic effects of GA may be due to its viscosity, which slows gastric emptying and alters the absorption kinetics in the intestine [30]. It has also been suggested that GA may have a prebiotic effect, which may underlie some of its metabolic effects [20], as it has recently been identified as a novel modulator of lipid profiles in vivo. Studies of GA have also demonstrated that consumption of 10 g/d for four weeks is associated with higher numbers of bifidobacteria and lactobacilli [19], both of which have been associated with beneficial effects on health in vivo.

In this study, 51% of the participants in the intervention group reported a significant improvement in reducing bloated feelings after 12 weeks of intake of GA-AS. The intervention group reported better responses in bowel movement, reduction in abdominal pain, better digestion, and reduction in nausea, but the results were insignificant compared with the control group. These findings may be due to the high dietary-fiber content in GA-AS (85%), which aids in healthy digestion and bowel movement. In addition to what has been previously reported, GA is not degraded in the stomach and small intestine, but undergoes complete fermentation within the cecum of rats [31,32] and humans [33]. Such fermentation promotes bacterial proliferation, which contribute to the prebiotic effect of GA [13,34].

5. Conclusions

Daily ingestion of 20 g of GA-AS for 12 weeks was shown to improve satiety and significantly reduce energy and carbohydrate intakes. It also improves blood pressure, blood glucose, and bowel movement, while increasing perceived satiety. This positions GA as a potential addition to the management of those with or at risk of developing metabolic syndrome. GA is a strong candidate for supplementation, as it is edible, safe, and already widely available in industry settings. It appears to have beneficial effects in a number of key areas relevant to improving the outcomes of those with metabolic syndrome, and could be a means of lowering the disease burden of NCDs globally. Future large-scale trials should evaluate the long-term use of GA in patients with metabolic syndrome to clarify its effects as well as identify optimal dose strategies and long-term efficacy.

Author Contributions: Conceptualization, A.S.A.D.; Formal analysis, A.H.J., J.F., M.F.B.; Investigation, A.H.J.; Methodology, A.S.A.D.; Resources, A.S.A.D.; Supervision, A.H.J. and A.S.A.D.; Writing—original draft, A.H.J. and A.S.A.D.; Writing—review & editing, A.H.J., L.S., V.A., J.F., M.F.B., L.C.I. and A.S.A.D. All authors have read and agreed to the published version of the manuscript.

Funding: This research received no external funding.

Institutional Review Board Statement: The study was conducted according to the guidelines of the Declaration of Helsinki, and approved by the UAEU scientific research ethics committee (ref. no. ERH_2016_4372, 8 June 2016).

Informed Consent Statement: Informed consent was obtained from all subjects involved in the study.

Data Availability Statement: The data presented in this study are available on request from the corresponding author. The data are not publicly available due to university privacy guidelines.

Acknowledgments: Authors would like to express their sincere gratitude to everyone who participated in the study, as well as to the Perfect Life Food Company who provided the Gum Arabic powder.

Conflicts of Interest: The authors declare no conflict of interest.

References

1. Islam, A.; Phillips, G.; Sljivo, A.; Snowden, M.; Williams, P. A review of recent developments on the regulatory, structural and functional aspects of gum arabic. *Food Hydrocoll.* **1997**, *11*, 493–505. [CrossRef]
2. Williams, P.A.; Phillips, G. Gum arabic. In *Handbook of Hydrocolloids*; Elsevier: Amsterdam, The Netherlands, 2009; pp. 252–273.
3. Babiker, R.; Merghani, T.H.; Elmusharaf, K.; Badi, R.M.; Lang, F.; Saeed, A.M. Effects of gum Arabic ingestion on body mass index and body fat percentage in healthy adult females: Two-arm randomized, placebo controlled, double-blind trial. *Nutr. J.* **2012**, *11*, 111. [CrossRef] [PubMed]
4. Verbeken, D.; Dierckx, S.; Dewettinck, K. Exudate gums: Occurrence, production, and applications. *Appl. Microbiol. Biotechnol.* **2003**, *63*, 10–21. [CrossRef] [PubMed]
5. Patel, S.; Goyal, A. Applications of natural polymer gum arabic: A review. *Int. J. Food Prop.* **2015**, *18*, 986–998. [CrossRef]
6. Ali, B.H.; Ziada, A.; Blunden, G. Biological effects of gum arabic: A review of some recent research. *Food Chem. Toxicol.* **2009**, *47*, 1–8. [CrossRef] [PubMed]
7. Gado, A.M.; Aldahmash, B.A. Antioxidant effect of Arabic gum against mercuric chloride-induced nephrotoxicity. *Drug Des. Dev. Ther.* **2013**, *7*, 1245. [CrossRef] [PubMed]
8. Kaddam, L.; Fadl-Elmula, I.; Eisawi, O.A.; Abdelrazig, H.A.; Salih, M.A.; Lang, F.; Saeed, A.M. Gum Arabic as novel anti-oxidant agent in sickle cell anemia, phase II trial. *BMC Hematol.* **2017**, *17*, 4. [CrossRef]
9. Tiss, A.; Carrière, F.; Verger, R. Effects of gum arabic on lipase interfacial binding and activity. *Anal. Biochem.* **2001**, *294*, 36–43. [CrossRef]

10. Glover, D.A.; Ushida, K.; Phillips, A.O.; Riley, S.G. Acacia (sen) SUPERGUM™(Gum Arabic): An evaluation of potential health benefits in human subjects. *Food Hydrocoll.* **2009**, *23*, 2410–2415. [CrossRef]
11. Ali, B.H.; Ziada, A.; Al Husseni, I.; Beegam, S.; Al-Ruqaishi, B.; Nemmar, A. Effect of Acacia gum on blood pressure in rats with adenine-induced chronic renal failure. *Phytomedicine* **2011**, *18*, 1176–1180. [CrossRef]
12. Matsumoto, N.; Riley, S.; Fraser, D.; Al-Assaf, S.; Ishimura, E.; Wolever, T.; Phillips, G.; Phillips, A.O. Butyrate modulates TGF-β1 generation and function: Potential renal benefit for Acacia (sen) SUPERGUM™(gum arabic)? *Kidney Int.* **2006**, *69*, 257–265. [CrossRef] [PubMed]
13. Calame, W.; Thomassen, F.; Hull, S.; Viebke, C.; Siemensma, A.D. Evaluation of satiety enhancement, including compensation, by blends of gum arabic. A methodological approach. *Appetite* **2011**, *57*, 358–364. [CrossRef] [PubMed]
14. Salih, N.K.M. 23-Applications of Gum Arabic in Medical and Health Benefits. In *Gum Arabic*; Mariod, A.A., Ed.; Academic Press: Cambridge, MA, USA, 2018; pp. 269–281. [CrossRef]
15. Kaur, J. A comprehensive review on metabolic syndrome. *Cardiol. Res. Pract.* **2014**, *2014*, 1–21. [CrossRef] [PubMed]
16. Esmailnasab, N.; Moradi, G.; Delaveri, A. Risk factors of non-communicable diseases and metabolic syndrome. *Iran. J. Public Health* **2012**, *41*, 77–85. [PubMed]
17. World Health Organisation. *Noncommunicable Diseases*; World Health Organization: Geneva, Switzerland, 2018.
18. Collaborators, G.R.F. Global, regional, and national comparative risk assessment of 79 behavioural, environmental and occupational, and metabolic risks or clusters of risks, 1990–2015: A systematic analysis for the Global Burden of Disease Study 2015. *Lancet* **2016**, *388*, 1659.
19. Calame, W.; Weseler, A.R.; Viebke, C.; Flynn, C.; Siemensma, A.D. Gum arabic establishes prebiotic functionality in healthy human volunteers in a dose-dependent manner. *Br. J. Nutr.* **2008**, *100*, 1269–1275. [CrossRef]
20. Kaddam, L.; Fadl-Elmula, I.; Eisawi, O.A.; Abdelrazig, H.A.; Saeed, A.M. *Acacia senegal* (Gum Arabic) Supplementation Modulate Lipid Profile and Ameliorated Dyslipidemia among Sickle Cell Anemia Patients. *J. Lipids* **2019**, *2019*, 3129461. [CrossRef]
21. WHO. *Waist Circumference and Waist-Hip Ratio: Report of a WHO Expert Consultation, Geneva, 8–11 December 2008*; World Health Organization: Geneva, Switzerland, 2011.
22. Dashti, B.H.; Al-Awadi, F.; Khalafawi, M.S.; Al-Zenki, S.; Sawaya, W. Nutrient contents of some traditional Kuwaiti dishes: Proximate composition, and phytate content. *Food Chem.* **2001**, *74*, 169–175. [CrossRef]
23. Helou, K.; El Helou, N.; Mahfouz, M.; Mahfouz, Y.; Salameh, P.; Harmouche-Karaki, M. Validity and reliability of an adapted arabic version of the long international physical activity questionnaire. *BMC Public Health* **2018**, *18*, 49. [CrossRef]
24. Holt, S.H.; Miller, J.C.; Petocz, P.; Farmakalidis, E. A satiety index of common foods. *Eur. J. Clin. Nutr.* **1995**, *49*, 675–690.
25. Jarrar, A.H.; Beasley, J.M.; Ohuma, E.O.; Ismail, L.C.; Qeshta, D.A.; Mohamad, M.N.; Al Dhaheri, A.S. Effect of High Fiber Cereal Intake on Satiety and Gastrointestinal Symptoms during Ramadan. *Nutrients* **2019**, *11*, 939. [CrossRef] [PubMed]
26. Babiker, R.; Elmusharaf, K.; Keogh, M.B.; Saeed, A.M. Effect of Gum Arabic (*Acacia senegal*) supplementation on visceral adiposity index (VAI) and blood pressure in patients with type 2 diabetes mellitus as indicators of cardiovascular disease (CVD): A randomized and placebo-controlled clinical trial. *Lipids Health Dis.* **2018**, *17*, 56. [CrossRef] [PubMed]
27. Alemán-Mateo, H.; Huerta, R.H.; Esparza-Romero, J.; Méndez, R.O.; Urquidez, R.; Valencia, M.E. Body composition by the four-compartment model: Validity of the BOD POD for assessing body fat in mexican elderly. *Eur. J. Clin. Nutr.* **2007**, *61*, 830–836. [CrossRef] [PubMed]
28. Pouteau, E.; Ferchaud-Roucher, V.; Zair, Y.; Paintin, M.; Enslen, M.; Auriou, N.; Macé, K.; Godin, J.-P.; Ballèvre, O.; Krempf, M. Acetogenic fibers reduce fasting glucose turnover but not peripheral insulin resistance in metabolic syndrome patients. *Clin. Nutr.* **2010**, *29*, 801–807. [CrossRef] [PubMed]
29. Wadood, A.; Wadood, N.; Shah, S. Effects of Acacia arabica and Caralluma edulis on blood glucose levels of normal and alloxan diabetic rabbits. *J. Pak. Med. Assoc.* **1989**, *39*, 208–212. [PubMed]
30. Leclere, C.J.; Champ, M.; Boillot, J.; Guille, G.; Lecannu, G.; Molis, C.; Bornet, F.; Krempf, M.; Delort-Laval, J.; Galmiche, J. Role of viscous guar gums in lowering the glycemic response after a solid meal. *Am. J. Clin. Nutr.* **1994**, *59*, 914–921. [CrossRef]
31. Ross, A.H.M.; Eastwood, M.A.; Brydon, W.G.; Busuttil, A.; McKay, L.F.; Anderson, D.M. A study of the effects of dietary gum arabic in the rat. *Br. J. Nutr.* **1984**, *51*, 47–56. [CrossRef]
32. Walter, D.; Eastwood, M.; Brydon, W.; Elton, R. Fermentation of wheat bran and gum arabic in rats fed on an elemental diet. *Br. J. Nutr.* **1988**, *60*, 225–232. [CrossRef]
33. Idris, O.; Williams, P.; Phillips, G. Characterisation of gum from *Acacia senegal* trees of different age and location using multidetection gel permeation chromatography. *Food Hydrocoll.* **1998**, *12*, 379–388. [CrossRef]
34. Cherbut, C.; Michel, C.; Lecannu, G. The prebiotic characteristics of fructooligosaccharides are necessary for reduction of TNBS-induced colitis in rats. *J. Nutr.* **2003**, *133*, 21–27. [CrossRef]

nutrients

MDPI

Review

Risks Associated with the Use of Garcinia as a Nutritional Complement to Lose Weight

Naroa Andueza [1], Rosa M. Giner [2,3] and Maria P. Portillo [1,3,4,5,*]

1 Nutrition and Obesity Group, Department of Pharmacy and Food Science, Lucio Lascaray Research Institute, University of the Basque Country (UPV/EHU), 01006 Vitoria, Spain; naroaandueza97@gmail.com
2 Department of Pharmacology, Faculty of Pharmacy, University of Valencia, 46100 Burjassot, Spain; Rosa.M.Giner@uv.es
3 Spanish Agency of Food Safety and Nutrition (AESAN), 28014 Madrid, Spain
4 Bioaraba Health Research Institute, 01009 Vitoria, Spain
5 CIBEROBN Physiopathology of Obesity and Nutrition, Institute of Health Carlos III, 01006 Vitoria, Spain
* Correspondence: mariapuy.portillo@ehu.eus; Tel.: +34-945013067; Fax: +34-945013014

Abstract: Nowadays, obesity is one of the great nutritional problems facing public health. The prevalence of this pathology has increased in a worrying way over recent years, currently reaching epidemic proportions. In this context, nutritional supplements are presented as a therapeutic alternative to which more and more people are turning to. Nutritional supplements to lose weight based on the Garcinia plant, specifically on *Garcinia cambogia*, are commonly used. The active principle of this plant to which these properties have been attributed, is hydroxycitric acid (HCA). The aim of the present review is to gather reported data concerning the effectiveness of nutritional supplements based on Garcinia extracts on weight loss and their possible negative effects. Contradictory results have been observed regarding the effectiveness of the supplements. While statistically significant weight loss was observed in some studies, no changes were found in others. Regarding safety, although Garcinia supplements have been revealed as safe in the vast majority of the studies carried out in animal models and humans, some cases of hepatotoxicity, serotonin toxicity and mania have been reported. In conclusion, the results suggest that Garcinia-based supplements could be effective in short-term weight loss, although the data are not conclusive. In addition, the safety of the complement should be further studied.

Keywords: *Garcinia cambogia*; nutritional supplements; weight loss; hepatotoxicity; serotonin toxicity

Citation: Andueza, N.; Giner, R.M.; Portillo, M.P. Risks Associated with the Use of Garcinia as a Nutritional Complement to Lose Weight. *Nutrients* **2021**, *13*, 450. https://doi.org/10.3390/nu13020450

Academic Editor: Antonella Amato
Received: 4 December 2020
Accepted: 24 January 2021
Published: 29 January 2021

1. Introduction

Obesity is one of the most common nutritional problems worldwide, currently reaching epidemic proportions. This pathology was previously considered typical in developed countries, but nowadays it also shows a high prevalence in underdeveloped countries. According to figures from the World Health Organization (WHO), in 2019 more than 1.9 billion adults aged ≥18 years were overweight, and of them more than 650 million were obese [1]. Since 1975 the worldwide prevalence of obesity has tripled, thus leading to a major public health problem. Obesity is associated with a large number of comorbidities, such as type 2 diabetes mellitus, hypertension, dyslipidemia, non-alcoholic fatty liver disease and cardiovascular diseases, among others. Furthermore, according to the WHO, people with obesity have a 50 to 150% increased risk of death from any cause compared to individuals with normal weight [2].

Hypocaloric diets and physical activity based treatments for overweight and obesity represent the first line of therapy. Due to the difficulty in achieving and maintaining an adequate adherence to this treatment, many people often turn to nutritional supplements that promise to help them lose weight in the short run or at least, to maintain it. For many people it is an easy solution, which enables them not to modify their lifestyle too much. In many instances, the motivation for the use of these supplements is due to aesthetic reasons, since the beauty standards that mark our society today encompass being slim.

These supplements work through five basic mechanisms, these being stimulation of thermogenesis, reduction in lipogenesis, increase in lipolysis, suppression of appetite, and decrease in lipid absorption. Among these nutritional supplements we find those based on plant extracts, which have been used for many centuries in the Eastern world. Nowadays, their use has become more and more prevalent throughout the world. Among them, *Garcinia cambogia* is one of the most promoted as a potential anti-obesity agent and has received a lot of attention in the media [3].

Although the current consumption of herbs and dietary supplements is unknown, in a study carried out in six European countries, it was estimated that 18.8% of the 2358 consumers selected for the study consumed one or more dietary supplements, not taking into account herbal products. The percentages of plant food supplement consumers were 9.6% in Finland, 16.9% in Germany, 22.7% in Italy, 17.9% in Romania, 18.0% in Spain and 19.1% in the United Kingdom [4].

The aim of this review is to gather the reported information concerning both the effectiveness and the side-effects of nutritional supplements based on *Garcinia cambogia* to promote weight loss. In addition, the efficacy of other *Garcinia* species is also presented. For this purpose, a selective literature search in PubMed and Cochrane databases was performed. To search for the effects on body weight loss, the terms Garcinia, HCA, weight loss, fat mass and obesity were introduced, adding "and", "or", "not" to improve the classification. In addition, the filters "clinical trial" and "meta-analysis" were used. To search for toxic effects, the terms Garcinia, HCA and toxic effects were used. In the Cochrane Library database we sought the term "Garcinia" in the title or the abstract in meta-analyses and clinical trials. The search was extended to a total of 25 years (Figure 1).

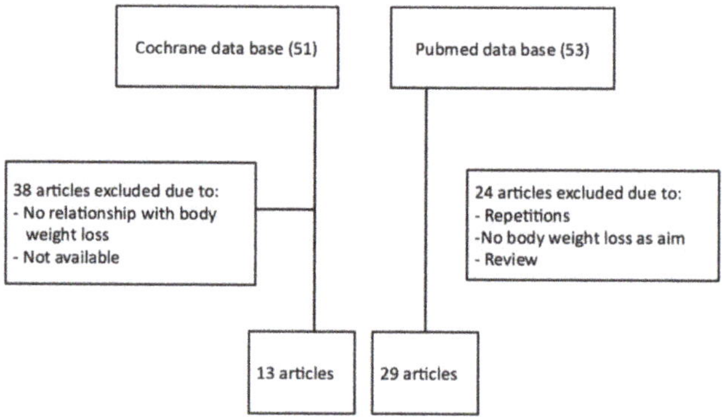

Figure 1. Flow chart showing the process for the inclusions of articles.

2. Active Principle and Mechanisms of Action

The genus *Garcinia,* native to Asia and Africa, belongs to the *Clusiaceae* family and includes more than 300 species, such as *Garcinia cambogia* (Figure 2), *Garcinia mangostana* and *Garcinia atroviridis* [5]. Potential therapeutic effects, such as anti-obesity, anti-ulcerogenic, antioxidant, anti-diabetes, anti-fungal, anti-inflammatory and anti-neoplasic [6], have been attributed to this genus. This has led to multiple investigations by pharmaceutical companies [5]. Some of the bioactive compounds isolated from Garcinia are garcinol, isogarcin, (−)-hydroxycitric acid (HCA), mangostin, and xanthoquimol [6]. It should be noted that out of the species mentioned above, *Garcina cambogia* is the most used as a nutritional supplement for weight loss or maintenance. The anti-obesity properties have been attributed to HCA, which is present in the rind or epicarp of the fruit and represents 20–30% of the dry weight [7]. Many food supplements containing HCA are currently marketed for weight reduction.

Figure 2. Tree, branch, and fruit of *Garcinia cambogia*.

The effects of HCA are associated with a reduction in food intake via serotonin level regulation and metabolic modifications, such as an increase in fat oxidation, a decrease in de novo lipogenesis and the stimulation of hepatic glycogenesis, thus promoting energy expenditure. HCA is a competitive inhibitor of adenosine triphosphate (ATP)-citrate lyase, an enzyme that catalyzes the extramitochondrial breakdown of citrate into oxalacetate and acetyl-CoA, thus limiting the availability of acetyl-CoA, a compound that plays a key role in the synthesis of fatty acids in diets rich in carbohydrates (Figure 3).

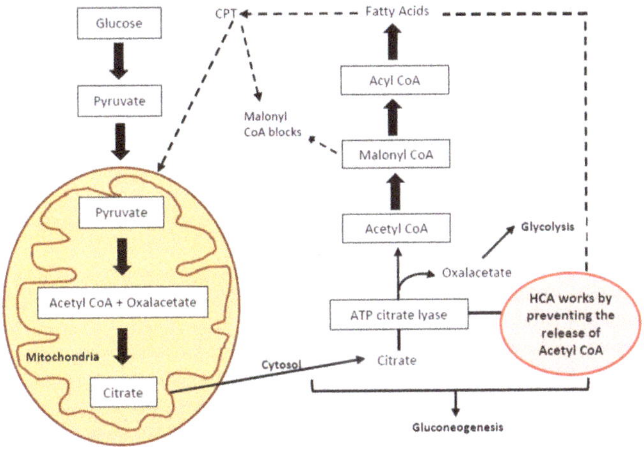

Figure 3. Mechanisms of action of hydroxycitric acid (HCA). CPT: Carnitine palmitoyltransferase.

3. Effectiveness of Garcinia to Lose and Maintain Body Weight

In in vitro studies, HCA has been shown to inhibit fatty acid synthesis [8]. Specifically, in isolated hepatocytes, HCA inhibits the synthesis of fatty acids from glucose, but not from acetate. Therefore, HCA is an inhibitor of lipogenesis only if cytoplasmic acetyl-CoA is produced by ATP-citrate lyase. Nevertheless, fatty acid synthesis is able to continue as long as acetate, another acetyl-CoA precursor, is available. As the synthesis of acetyl-CoA is reduced, that of malonyl-CoA is reduced too, thus decreasing the negative feedback of carnitine acyltransferase. This produces an increase in lipid transport in the mitochondria and inefficient oxidation, that promotes the formation of ketone bodies. These molecules can pass into the bloodstream and reach the brain, where they constitute an energy reserve in the event of fasting [8].

In preclinical studies using animal models, chronic oral administration of HCA to rats significantly reduces food intake in the first hour after administration, together with body weight and concentrations of cholesterol, triglycerides and fatty acids. When evaluating the acute and chronic effects of HCA on energy metabolism in mice, it was observed that oral administration of 10 mg increased serum concentration of free fatty acids and glycogen concentration in skeletal muscle [9].

Leonhardt et al. (2002) studied the long-term effect of HCA in male Sprague–Dawley rats after a notable body weight loss in two different experiments. Each experiment had 23 or 24 rats, respectively [10]. The rats were fed a 1% fat diet or a 12% fat diet, depending on the experiment. Both diets were supplemented with 3% HCA. HCA produced a long-term reduction in body weight recovery in both groups (in both rats fed 1% fat or 12% fat diets). However, only HCA produced a long-term suppressive effect in the case of the group of rats fed the 12% fat diet. No effect on plasma β-hydroxybutyrate levels was observed, so the hypothesis that increased fatty acid oxidation in the liver is involved in suppressing food intake by HCA was not supported by these results [11]. In another study, young lean and obese female Zucker rats were fed a diet (70% glucose, 1% corn oil) supplemented with HCA (52.6 mmol/kg diet) for 39 days. Amongst the lean rats, HCA decreased food intake, body weight, the percent of body fat, and fat cell size. Amongst the obese rats, food intake and body weight were lowered, but body fat percentage remained unchanged [10].

In other studies, instead of isolated HCA, extracts of Garcinia were used for supplementation. Saito et al. (2005) studied the ability of a *Garcinia cambogia* extract, containing HCA, to suppress the accumulation of body fat in growing male obese Zucker rats (6 weeks) [12]. They were given diets containing different amounts of the extract, which provided 10, 51, 102 or 154 mmol of HCA/kg of diet, respectively, for 92–93 days. Rats fed the highest amount of extract, and therefore receiving the highest dose of HCA (154 mmol HCA/kg diet), significantly decreased epididymal fat accumulation, compared to the other groups.

Table 1 presents a summary of the results obtained in 20 reported intervention studies in humans [13–32]. In 12 studies, statistically significant reductions in body weight were observed [14,15,18–22,26–28,30,31]. Conversely, the supplementation turned out to be ineffective in the remaining studies [13,16,17,23–25,29,32]. In the studies where significant weight loss was observed, the average value was between 2 and 6 kg in 2–4 months, with the exception of one where a loss of 14 kg was induced in 6 months [31]. Interestingly, in 11 studies [19–21,23,24,26,27,29–32], significant reductions in other parameters, such as waist and hip circumference, triglycerides, cholesterol and glucose were also observed regardless of body weight loss. It was specified in the vast majority of the studies that the supplement should be consumed approximately half an hour before meals. Splitting the doses of HCA has also been shown to be more effective than utilizing the same amount given as a single dose.

Table 1. Characteristics and results of published intervention studies in humans.

Reference	Type of Study	Participants	Diet and Physical Activity	Treatment	Treatment Duration	Results	
						Weight loss and Related Parameters	Other Results Observed after the Intervention
Heymsfield et al., 1998 [13]	Randomized, double-blind, placebo-controlled study.	135 subjects BMI 27–38 aged 18 to 65 years. Control: 69 Intervention: 66.	1200 Kcal/d diethigh in fibre: -Proteins: 30% -Lipids: 20% -Carbohydrates: 50% Regular physical activity	Pills: 500 mg *Garcinia cambogia* (50% HCA), 2 pills, 3 times/day	3 months	No significant differences.	
Mattes et al., 2000 [14]	Double-blind, placebo-controlled parallel group study.	89 women mean BMI of 28.6 aged 18 to 65 years. Control: 42 Intervention: 47	1200 Kcal/diet (30% lipids) Exercise was encouraged, but no formal regimen was prescribed.	Capsules: 400 mg of *Garcinia cambogia* (50% HCA), 3 times/day	3 months	Significant weight loss. Reduction in waist circumference.	
Thom et al., 2000 [15]	Randomized double-blind study	40 subjects BMI 27.5–39.0 aged ≥18 years. Control: 20 Intervention: 20.	Participants were given diet lists with advice on low-fat foods, supplying an energy intake of approximately 1200 Kcal/d, and were recommended to use this diet during the study.	Suco-Blo™ (tablets): 200 mg *Phaseolus vulgaris* extract, 200 mg inulin and 50 mg *Garcinia cambogia* extract 3 times/day.	3 months	Significant reductionin body weight andbody mass.	

Table 1. *Cont.*

Reference	Type of Study	Participants	Diet and Physical Activity	Treatment	Treatment Duration	Results	
						Weight loss and Related Parameters	Other Results Observed after the Intervention
Kovacs et al., 2001 [16]	Double-blind, placebo-controlled, randomized, and cross-over study.	11 obese men mean BMI of 27.4 mean age of 47 years.	Diet was divided in 3 meals without restrictions on the type and quantity of food and a maximum of one glass of alcohol drink per day.	Isoenergetic snack (cereal bar): 4 times/d Intervention 1:500 mg HCA Intervention 2:500 mg HCA + 3 g MCT	3 Intervention periods of 2 weeks separated by washout periods of 4 weeks.	No significant differences.	
Kovacs et al., 2001 [17]	Double-blind, placebo-controlled, randomized, and crossover study.	21 obese subjects mean BMI of 27 mean of age 43 years.	Diet was divided in 3 meals without restrictions on the type and quantity of food and a maximum of one glass of alcoholic drink per day.	Isoenergetic snack (cereal bar): 4 times/d Control: no supplementation Intervention 1:500 mg HCA Intervention 2:500 mg HCA + 3 g MCT	3 intervention periods of 2 weeks separated by washout periods of 2 or 6 weeks	No significant differences.	
Hayamizu et al., 2003 [18]	Double blind, randomized, placebo-controlled, parallel-group study.	44 subjects aged 20 to 65 years visceral fat area >90 cm^2 Control: 21 Intervention: 23	Maximun 2250 Kcal/d for men and 1800 Kcal/d for women.	Tablets: 185.25 mg of *Garcinia cambogia* extract (60% HCA) 3 tablets before each meal (9 tablets/day)	3 months + 1 month of placebo in both groups at the end.	Reduction in visceral fat area, subcutaneous fat area and total fat area.	
Preuss et al., 2004 [19]	Randomized, double-blind, placebo-controlled study.	30 subjects BMI > 26 aged 21 to 50 years Control:10 Intervention 1:10 Intervention 2:10	2000 Kcal/d divided in 3 meals: - Proteins: 17% - Lipids: 25% - Carbohydrates: 58% 30 min supervised walking exercise program (5 days a week).	HCA-SX (4667 mg) divided in 3 doses: Intervention 1:2800 mg/d of HCA Intervention 2:2800 mg/d of HCA, 4 mg niacin-bound chromium and 400 mg *Gymnema sylvestre* extract.	2 months	Significant weight loss in both intervention groups. Reduction in food intake.	Reduction in total cholesterol, LDL-c, TG, and leptin levels. Increase in fat oxidation, HDL levels and serotonin levels.
Preuss et al., 2004 [20]	Randomized, double-blind, placebo-controlled study.	60 subjects BMI > 26 aged 21 to 50 years Control: 20 Intervention 1:20 Intervention 2:20	2000 Kcal/d - Proteins: 17% - Lipids: 25% - Carbohydrates: 58% 30 min supervised walking exercise program (5 days a week).	HCA-SX (4667 mg) divided in 3 doses: Intervention 1:2800 mg/d of HCA Intervention 2:2800 mg/d of HCA, 4 mg niacin-bound chromium and 400 mg *Gymnema sylvestre* extract.	2 months	Significant weight lossand reduction in food intake in both intervention groups.	Reduction in total cholesterol, LDL-c, TG and serum leptin levels in both intervention groups. Increase in HDL-c and excretion of urinary fat metabolites in both intervention groups.
Roongpisuthipong et al., 2007 [21]	Randomized, double-blind, placebo-controlled study	50 women BMI 25–30 aged 18 to 75 years Control: 25 Intervention: 25	1000 Kcal/d - Proteins: 50 g - Fats: 33 g - Carbohydrates: 125 g	Sachets: 1.15 g of *Garcinia artroviridis* (HCA) 3 times/day	2 months	Significant weight loss during the first 4 weeks. No significant differences over the following 4 weeks. Decrease in fat mass, bicipital, subscapular and suprailiac folds and upper arm circumference.	Increase in lean mass and body water. Decrease in TG.

Table 1. *Cont.*

Reference	Type of Study	Participants	Diet and Physical Activity	Treatment	Treatment Duration	Results	
						Weight loss and Related Parameters	Other Results Observed after the Intervention
Toromanyan et al., 2007 [22]	Double blind, randomized, parallel group, placebo-controlled study.	60 subjects BMI 25–44 aged 25 to 65 years Control: 30 Intervention: 30.	Diet and exercise performed regularly.	Slim339™ (tablets): 132 mg of *Garcinia cambogia* (HCA) + *Matricaria chamomilla, Rosa damascena, Lavandula officinalis and Cananga odorata* 3 times/d.	2 months	Significant weight reduction.	
Vasques et al., 2008 [23]	Randomized double-blind study	58 subjects BMI 30–39.9 aged 25 to 60 years Control: 26 Intervention: 32		Capsules: 800 mg of *Garcinia cambogia* (HCA) + 500 mg de *Amorphophallus konjac* → 3 times/day	3 months	No significant reduction in body weight.	Reduction in total cholesterol and LDL-c.
Kim et al., 2011 [24]	Randomized, double-blind, placebo-controlled study	86 subjects BMI 23–29 aged 20 to 60 years. Control: 29 *Glycine max* leaves (GML): 28 *Garcinia cambogia* (GC): 29	Diet and habitual physical activity	Pills: 2 g/d of the substances corresponding to each group. In the case of the placebo and GML 4 pills/d and for *Garcinia cambogia* 8 pills/day.	2.5 months	No significant reduction in body weight.	GML reduced total cholesterol and increased HDL-c (significant differences compared to the placebo group and the GC and placebo group, respectively).
Lu et al., 2012 [25]	Randomized double-blind study	114 overweight subjects	Nutritional education	Super CitriMax™ (HCA) 2800 mg/day	2 months	No significant reduction in body weight.	
Stern et al., 2013 [26]	Randomized, double-blind, placebo-controlled clinical study	60 subjects BMI 30–40 aged 21 to 50 years Control: 30 Intervention: 30	Participants were given free prepared meals. 2000 Kcal/d - Proteins: 14% - Lipids: 25% - Carbohydrates: 61% Physical activity (walking) 30 min, 5 times/d.	Capsules: 400 mg of *Sphaeranthus indicus* + *Garcinia mangostana*, ratio 3:1 2 times/day	2 months	Significant reduction in body weight, BMI and waist circumference.	Decrease in total cholesterol and TG and increase in adiponectin.
Stern et al., 2013 [27]	Randomized, double-blind, placebo-controlled clinical study	95 subjects BMI 30–40 aged 36 to 40 years Control: 46 Intervention: 49	Same diet and physical activityas in reference 26. In this case the diet is divided into 3 intakes.	Same treatment as in reference 26.	2 months	Reduction in body weight, BMI, waist and hip circumferences.	Decrease in total cholesterol, TG and fasting glucose. Increase in adiponectin. Improvement in physical function and self-esteem (IWQOL questionnaire).
Chong et al., 2014 [28]	Randomized, placebo-controlled, double-blind parallel group study	91 caucasian subjects BMI 25–32 aged 18 to 60 years Control: 45 Intervention: 46	Dietary advice + balanced diet with a deficit of 500 Kcal. −30% lipids.	Tablets: 850 mg 3 tablets 2 times/day. Composition: 650 mg of *Garcinia cambogia* (HCA) + 100 mg of *Camellia sinensis* + 75 mg of *Coffea arabica* + 25 mg of *Lagerstroemia speciosa*	3.5 months	Significant weight loss and reduction in BMI, body fat, waist and hip circumferences	

Table 1. *Cont.*

Reference	Type of Study	Participants	Diet and Physical Activity	Treatment	Treatment Duration	Results	
						Weight loss and Related Parameters	Other Results Observed after the Intervention
Vasques et al., 2014 [29]	Randomized double-blind study	43 women BMI > 25) aged 25 to 60 years Control: 13 Intervention: 30	Individualized diet, with an average caloric restriction of 1523 ± 185 Kcal/day Regular physical activity.	Capsules: 800 mg of *Garcinia cambogia* (HCA) 3 times/day	2 months	No statistically significant differences.	Reduction in TG level.
Kudiganti et al., 2016 [30]	Randomized, double-blind, placebo-controlled clinical study	60 subjects mean BMI of 28.3 aged 21 to 50 years Control: 30 Intervention: 30.	2000 Kcal/d - Proteins: 17% - Lipids: 25% - Carbohydrates: 58%	Capsules: 400 mg of Meratrim™: extracts from the flower heads of *Sphaeranthus indicus* and the fruit rinds of *Garcinia mangostana* 2 times/day	4 months	Significant weight loss and reduction in BMI, waist and hip circumferences.	Reduction in TG and LDL-c cholesterol. Increase in HDL-c.
Maia-Landim et al., 2018 [31]	Non-randomized prospective controlled intervention study	214 subjects BMI > 25 older than 18 years	Balanced diet and regular physical activity, smoking not permitted and controlof alcohol intake.	Capsules: 500 mg of *Garcinia cambogia* (HCA) + 500 mg of *Amorphophallus konjac* 2 times/day	6 months	Reduction in total fat mass and visceral fat mass after 3 and 6 months of intervention.	Increase in basal metabolic rate Reduction in glucose, total cholesterol and TG.
Watanabe et al., 2018 [32]	Prospective, randomized, controlled, parallel study	22 obese women with insulin resistance aged 18 to 65 years Control: 11 Intervention: 11	Hypocaloric diet (300 Kcal restriction) + physical activity of moderate intensity - Proteins: 20–25% - Lipids: 30% - Carbohydrates: 45–50%	Capsules: 400 mg of *Garcinia mangostana* 1 time/day	6.5 months	No significant reduction in body weight.	Reduction in insulin concentration and HOMA-IR.

BMI: body mass index; HCA: hydroxycitric acid; HOMA-IR: insulin resistance index, MCT: medium chain triglyceride; TG: triglycerides; IWQOL: impact of weight on quality of life—lite.

Of note, important differences in the experimental design can be observed among the reported studies. The number of participants ranged from 11 to 214. The fact that the majority of the studies had a small sample size limits the reliability of the results. In addition, in some cases there is a lack of proportion between men and women. The inclusion of both genders is important in order to determine whether there is sexual dimorphism in the effectiveness of the nutritional supplement. Conversely, in all the selected studies, the participants were overweight or obese (based on body mass index) and the duration of treatments ranged from 2 to 4 months in the majority studies, with the exception of two longer ones, in which the duration was 6 or 6.5 months.

The variability is also observed in the type of supplement administered. Upon that, *Garcinia cambogia*/HCA was used in 15 studies, *Garcinia mangostana* in four studies and *Garcinia artroviridis* in one study. Lastly, in some cases HCA or Garcinia appeared to be combined with other ingredients, such as glucomannan (fiber), *Sphaeranthus indicus* or *Coffea arabica*. Consequently, it is not possible to determine whether the effect produced was due to the HCA content, to other ingredients, or to a combination of the effects of the different ingredients. The dose administered is another differential aspect among the studies. Last of all, the presentation of the supplement is different (pills, capsules or sachets). It is indicated in all cases that the supplement should be consumed before the meal (approximately half an hour before).

Differences in other aspects of the experimental design, such as diet and physical activity can also be found. In some studies, the participants were instructed to continue with their usual diet and physical activity, while in others they were given specific instructions. Lastly, in others they were prescribed a hypocaloric diet along with specific practice of regular physical activity.

Some of these differential aspects could explain the lack of effect observed in several studies, for instance, the lowest sample size in the studies reported by Kovacs et al. (2001) [16], Kovacs et al. (2001) [17] and Watanabe et al. (2018) [32] In the study reported by Heymsfield et al. (1998) [13], *Garcinia cambogia* was not effective since the diet was low in energy and high in fiber. The amount of fiber could have inhibited the gastrointestinal absorption of HCA (active compound in *Garcinia cambogia*) and the low energy supplied (approximately 1200 Kcal) could have affected the usefulness of HCA. In addition, the dose of HCA was lower than that used in other studies. Another parameter that could have an influence was the excess of calcium used to stabilize HCA, that could have reduced the solubility of this bioactive compound, and therefore its bioavailability [8]. In a similar way, the lack of effect in the studies carried out by Roongpisuthipong et al. (2007) [21] and Vasques et al. (2014) [29] could be related to the low amount of energy provided by the diet (1000 Kcal/d and 1500 Kcal/d, respectively). In the study conducted by Vasques et al. (2008) [23], the only apparent potential reason could be that neither dieting, nor practicing physical activity were recommended. In contrast, in all the studies in which positive results were obtained, a healthy lifestyle was either prescribed or recommended. Last of all, in the studies conducted by Kim et al. (2011) [24] and Hayamizu et al. (2003) [18], in an Asian population, it is believed that the fact that the fruit of the Garcinia is of common use as part of the traditional diet, could have led to a reduced susceptibility to its effects.

Onakpoya et al. (2011) published a systematic review and meta-analysis of randomized clinical trials devoted to evaluating the effectiveness of Garcinia extracts as weight reduction agents [33]. The authors concluded that these extracts generated short-term weight loss. However, the scale of this effect was small, since it was not statistically significant when only rigorous randomized clinical trials were considered. Therefore, the clinical relevance of these products appears to be questionable and they do not represent an altogether effective measure of the treatment of overweight and obesity.

4. Negative Effects on Health

4.1. Animal Toxicity Studies

Studies of acute, short-term, sub-chronic, and chronic toxicity, as well as studies of genotoxicity, cytotoxicity and toxicity in reproduction, have been conducted in different animal species, although mainly in rats and rabbits. These studies have shown that Garcinia/HCA have good safety profiles, so that they may be used as nutritional supplements for the treatment of obesity [34–36].

Ohia et al. (2002) evaluated the effects of Super Citri-Max™, a novel calcium/potassium-HCA extract (HCA-SX), containing 60% HCA, administered for 14 days. This extract is considerably more soluble and bioavailable than calcium-based HCA ingredients [37]. The study was conducted in Albino rats (males and females) fed ad libitum, treated with a dose of 5000 mg/kg (through a gastric probe), which is equivalent to 350 g or 233 times the maximum dose of 1.5 g/day of HCA in humans. The authors did not report any death or significant clinical changes. Furthermore, no significant tissue injuries were observed during the necropsy, which led them to suggest that the LD50 oral administration of HCA-SX in rats was over 5000 mg/kg. Similarly, in another study carried out by the Wil Research Laboratories, it was shown that 5000 mg HCA/kg of body weight did not produce visible symptoms of toxicity or death in animal models. In line with these studies, Clouatre et al. (2013) defended that HCA was extremely safe and this was corroborated by various reviews where it was claimed that HCA from *Garcinia cambogia* had a protective effect on the liver [38].

Shara et al. (2003) analyzed the effects of HCA intake on weight, testicular and liver lipid peroxidation, and DNA fragmentation, in addition to possible histopathological changes in Sprague–Dawley rats [39]. The animals received 0.2, 2.0, or 5.0% HCA (100–2500 mg/kg) in their diet, equivalent to approximately 100, 1000, and 2500 mg/kg/day, respectively, in humans. The lowest dose was equivalent to the daily recommended dosage in humans, but the doses of 2.0 and 5.0% are 10 to 25 times higher doses. The rats were euthanized at 30, 60, and 90 days of treatment. After 90 days of HCA administration, rats showed decreased body weight, without changes in liver or testicular lipid peroxidation or in DNA fragmentation In a follow-up study, the same authors did not find differences in the weight of various organs. Moreover, no haematological or biochemical disorders or significant histopathological changes or mortality differences were found [40].

As an exception, Kim et al. (2013) in a study addressed in rats fed a high-fat diet (45% of total energy), it was shown that after 16 weeks of treatment with *Garcinia cambogia* (1%, *w/w*, 60% HCA) oxidative stress, inflammation and liver fibrosis were triggered [41]. Consequently, it appears that the form of HCA regarding its extraction process and the residual compounds, among other factors, may spur differences between study outcomes [42].

Toxicological studies have been also addressed with other Garcinia species. Farombi et al. (2013) carried out a study in adult male Wistar rats randomly assigned to four groups of 10 rats each group given *Garcinia kola* orally at different doses (0, 250, 500 and 1000 mg/kg) for 6 weeks [43]. After conducting the study, it was concluded that the administration of *Garcinia kola* increased the antioxidant status and did not produce adverse effects on the liver, testicles and sperm. Saiyed et al. (2015) performed various toxicological studies both in vitro and in animals to evaluate the safety of Meratrim™, a supplement that contains *Garcinia mangostana* [44]. Meratrim™ was determined to be non-irritating to the skin, mildly irritating to the eyes, not mutagenic, and the no-observed-adverse-effect level (NOAEL) for this supplement was 1000 mg/kg body weight/day (in Sprague–Dawley rats). The authors concluded that the safety of Meratrim was demonstrated given the results observed in this study, added to the clinical trials of tolerability already carried out.

4.2. Clinical Toxicity

Based on toxicological studies, Soni et al. (2004) noted that there was sufficient quantitative and qualitative scientific evidence, from both animal and human data, to suggest that HCA intake up to 2800 mg/day is safe for human consumption [35]. As a result, a NOAEL of 2800 mg/d was established [34,35].

Accordingly, none of the studies included in Table 1, devoted to analyzing the effects of Garcinia on body weight reductions at doses below this value have found serious adverse effects. Some of the minor side effects observed were leg cramps, heartburn, diarrhoea, increased gas, higher appetite, headaches, heartburn, rash, menstrual bleeding, and general weakness. In spite of this, other authors have reported toxic manifestations, including hepatotoxicity, acute pancreatitis, serotonin toxicity and psychosis after the consumption of Garcinia-containing products (either as an extract containing other components or pure). In 16 of the 21 cases described in Table 2 [45–60], these adverse effects have occurred after the intake of formulations which contained other ingredients besides Garcinia. One of these formulations is Hydroxycut™. Fourteen different products are marketed under this name, but only eight of them contain *Garcinia cambogia*. In addition, all of them are polyherbal products that can contain up to 20 different substances [34]. When toxic effects are induced by supplements that, in addition to Garcinia, contain other components, it has not been possible to confirm that Garcinia is the agent responsible for the side-effects. In spite of this, the FDA (US Food and Drug Administration) issued a warning in 2009 on Hydroxycut™ products related to hepatotoxicity, which led to the recall of these products.

Table 2. Cases of liver damage associated with the consumption of *Garcinia cambogia* supplements in humans.

Reference	Ageyears	Sex	Type of Supplement	Duration	Symptoms	Test Performed	Diagnosis/Type of Liver Injury
Stevens et al., 2005 [45]	27	M	Hydroxycut™ 3 capsules, 3 times/d.	5 weeks	Fatigue and jaundice.	Laboratory analysis: elevated AST, ALT, AF and PT. Serological study: negative.	Hepatotoxicity Cholestatic liver injury pattern.

Table 2. *Cont.*

Reference	Ageyears	Sex	Type of Supplement	Duration	Symptoms	Test Performed	Diagnosis/Type of Liver Injury
Stevens et al., 2005 [45]	30	M	Hydroxycut™ 9 capsules/d.	5 days	Fever, vomiting, fatigue, and jaundice.	Laboratory analysis: AST, ALT, AF and PT elevated and low albumin. Serological study: negative. CT and cholangiography: normal.	Hepatotoxicity Hepatocyte necrosis was the likely pattern of injury.
Dara et al., 2008 [46]	40	W	Hydroxycut™ 6 capsules/d.	1 week	Mid-epigastric abdominal pain, non-bloody diarrhea, fevers, chills, nausea, vomiting, anorexia and profound fatigue.	Laboratory analysis: acute hepatitis (elevated AST, ALT and AF) Serological study: negative.	Acute hepatitis.
Dara et al., 2008 [46]	33	W	Hydroxycut™	2 weeks	Nausea, crampy abdominal pain, jaundice, acholic stools, dark-colored urine, pruritus and profound fatigue.	Physical examination: jaundice and scleral icterus. Laboratory analysis: elevated AST, ALT, TB and DB. Serological study: negative.	Acute hepatitis.
Shuster et al., 2010 [47]	25	M	Exilis: *Garcinia cambogia*, *Garcinia sylvestre*, L-carnintine and chrome	3 weeks	Two weeks after starting treatment: fatigue and dark urine. In the third week: fever, nausea, vomiting and pain.	Laboratory analysis: elevated ALT, AST, TB and INRA comprehensive study was conducted to determine the etiology of liver damage, but all tests were negative.	Hepatic encephalopathy Liver transplantation required.
Sharma et al., 2010 [48]	19	M	Hydroxycut™	1 week	Fever, severe fatigue, myalgia, arthralgia, and erythematous rash over in lower extremities.	Physical examination: toxic appearance, marked jaundice and fever (39.4 °C). Laboratory analysis: elevated ALT, AF, bilirubin and PT. low blood cell count and hemoglobin. Blood culture, urinalysis, X-rays, abdominal ultrasound, CT and MRCP: normal. Serological study: negative. Hepatic biopsy: acute cholangitis.	Acute cholangitis
Mancano et al., 2015 [49]	42	W	*Garcinia cambogia* pure	1 week	Right upper quadrant abdominal pain and nausea (without emesis).	Laboratory analysis: elevated ALT, AST, AF, ferritin and INR Serological study: negative. Abdominal ultrasound: normal.	Acute hepatitis
Melendez-Rosado et al., 2015 [50]	42	W	*Garcinia cambogia* pure	1 week	Abdominal pain in the right upper quadrant, nausea without emesis and clamminess.	Laboratory analysis: elevated ALT, AST, AF and ferritin. Serological study: negative. Abdominal ultrasound: mildy coarse hepatic echotexture. CT: normal.	Acute hepatitis

Table 2. *Cont.*

Reference	Ageyears	Sex	Type of Supplement	Duration	Symptoms	Test Performed	Diagnosis/Type of Liver Injury
Araujo et al., 2015 [51]	41	M	Hydroxycut SX-7 Clean Sensory™ 2 capsules/day 4 times/week	2 months	Malaise, jaundice, fatigue, nausea, vomiting and asterixis	Physical examination: jaundice and liver edge percussed. Laboratory analysis: elevated AST, ALT, TB, DB, PT and creatinine. Serological study: negative. Abdominal ultrasound: increased liver echogenicity and liver length. CIOMS/RUCAM: 9	Acute hepatocellular liver injury
Smith et al., 2016 [52]	26	M	Multi-ingredient protein supplement with *Garcinia cambogia* (70%).	1 week	Jaundice, fatigue and asterixis.	Laboratory analysis: elevated ALT, AST, AFand bilirubin. Serological study: negative. Hepatic biopsy: liver necrosis. CIOMS: 6	Hepatotoxicity Liver transplantation required.
Corey et al., 2016 [53]	52	W	*Garcinia cambogia* supplement: *Garcinia cambogia* extract (936 mg, 60% HCA), calcium, chromium, potassium 2 capsules/d (1000 mg/day)	3.5 weeks	Decreased appetite, worsening fatigue, and intermittent confusion.	Physical examination: abdominal distention and jaundice. Laboratory analysis: elevated ALT, AST, AF, TB, DB and INR, and low platelet count. CT: nodular liver compatible with necrosis and ascites. Serological study: negative. Biopsy: severe acute hepatitis with necrosis and parenchymal collapse. MELD: it was evolving until it reached a score of 40. CIOMS: 7	Acute liver failure. Liver transplantation required.
Lunsford et al., 2016 [54]	34	M	*Garcinia cambogia* pure 2 capsules of 80 mg, 3 times/day	6.5 months.	Nausea, vomiting, abdominal pain, and dark urine.	Laboratory analysis: elevated transaminases and bilirubin. Asterixis, jaundice, and confusion. Elevated transaminases, bilirubin, elevated INR. Images: cirrhosis or hepatocellular carcinoma. MR: no tumor process.Serological study: positive antinuclear antibody.Hepatic biopsy: necrosis with collapse of the liver architecture.	Severe liver injury. Liver transplantation required.
Crescioli et al., 2018 [55]	61	W	SUPER ANANAS SLIM™: *Garcinia cambogia* (60%), *Ananas comosus* and *Ilex paraguariensis*.	2 months	Abdominal pain, nausea, progressive weakness, jaundice, dark stools, and acholic stools.	Laboratory analysis: ALT, AST, TB, DB, albumin, AF, GGT out of normal range. Serological study: negative. Abdominal ultrasound, MRI. Doppler: normal. CT: small peritoneal effusion and perihepatic lymphadenopathy. Biopsy: cholestatic hepatitis. CIOMS: 7.	Herbal-induced liver damage.

Table 2. *Cont.*

Reference	Ageyears	Sex	Type of Supplement	Duration	Symptoms	Test Performed	Diagnosis/Type of Liver Injury
Crescioli et al., 2018 [55]	39	W	Two sipplements: OBLESS™: *Garcinia cambogia* (72 mg of HCA) and other components: 1 capsule/day and	1 month	Jaundice, asthenia, loss of appetite, and right hypochondrial pain.	Laboratory analysis: elevated ALT, AST, TB, DB, AF, GGT, CRP and lactate dehydrogenase. Serological study: nonspecific antinuclear antibodies and positive bile antibodies. Abdominal ultrasound: normal. CIOMS: 6	Acute cholestatic
			Magistral preparation of different herbs extracts: 1 capsule/day	15 days			
Crescioli et al., 2018 [55]	47	W	THERMO GIALLO™: *Garcinia cambogia* (200 mg HCA) and chromium: y 2 capsules/da.	1 month	Severe abdominal pain.	Laboratory analysis: elevated AST, ALT and TB. Serological study: negative. CIOMS: 6.	Acute hepatitis.
Crescioli et al., 2018 [55]	52	W	2 JILL COOPER BE SLIM™: *Garcinia cambogia* (240 mg) and *Green Coffee* extract 1 capsule / d of each product	1 month		Laboratory analysis: elevated AST, ALT, BT, GGT and AF. Serological test: negative. CIOMS: 6.	Acute hepatitis.
Sharma et al., 2018 [56]	57	W	*Garcinia cambogia* (100%) and vitamin A and D supplement 2 capsules/d (2800 mg/d)	1 month	Abdominal pain (more intense in the right upper quadrant) and vomiting.	Laboratory analysis: elevated ALT, AST, TB, DB, INR, PT. Normal vitamin A and D levels. Serological study: negative Abdominal ultrasound: normal liver CIOMS/RUCAM: 11	Hepatitis secondary to the consumption of *Garcinia cambogia*. After withdrawal of the supplement the levels of the altered enzymes normalized. After six months they elevated again, coinciding with the reintroduction of the supplement.

Table 2. *Cont.*

Reference	Ageyears	Sex	Type of Supplement	Duration	Symptoms	Test Performed	Diagnosis/Type of Liver Injury
Philips et al., 2018 [57]	33	W	Safe Lean™: *Garcinia cambogia* (600 mg), *Allium sativum* (250 mg) and *Trigonella foenum graecum* (100 mg) 1 capsule, 2 times/day	1 month	Nausea, loss of appetite	Laboratory analysis: elevated AST, ALT, AF, TB, gamma-glutamyl transferase, albumin and INR. Serological study: negative. CT: hepatomegaly. RUCAM: 8	Drug induced liver injury, secondary to Safe Lean™.
			Calcium, vitamin A and folic acid supplement 1 time/day	3 months			
Yousaf et al., 2019 [58]	21	W	*Garcinia cambogia* 1400 mg/day	4 weeks	Abdominal pain for 1 wk associated with nausea, vomiting, anorexia and myalgias.	Abdominal ultrasound Laboratory analysis: elevated ALT, AST, alkaline phosphatase	Hepatomegaly Acute liver failure.
Khetpal et al., 2020 [59]	22	W	Hydroxycut™ 2 capsules/day	3 months	Chest pain, fatigue, palpitations and shortness of breath	Physical examination: tachycardia, low oxygen saturation and asterixis. Laboratory analysis: elevated AST, ALT, INR, leukocytes and white blood cells. Serological study: negative. Abdominal ultrasound: hepatomegaly. RUCAM: 9.	Acute drug-induced liver injury likely due to Hydroxycut™.
Ferreira et al., 2020 [60]	26	W	1800 mg of *Garcinia cambogia* (900 mg HCA), 1275 mg of green tea extract with 450 mg of Veldt raisin and 1200 mg of *Coffea arabica* daily.	7 months	Fatigue, nausea and jaundice.	Laboratory analysis: acute hepatitis (elevated AST, ALT, TB and INR). Abdominal Ultrasound: normal. MRCP: normal. Transjugular liver biopsy: acute hepatitis. Serological study: negative. RUCAM: 6.	Subacute liver failure secondary to the consumption of *Garcinia cambogia*. Liver transplantation required.

M: men; W: women; ALT: alanine aminotransferase; AST: aspartate aminotransferase; AF: alkaline phosphatase; GGT: gamma-glutamyl transferase; TB: total bilirrubin; DB: direct bilirrubin; INR: international normalized ratio; PT: prothrombin time; CRP: C-reactive protein; CT: computed tomography; MRCP: magnetic resonance cholangiopancreatography; MR: magnetic resonance; CIOMS/RUCAM: scale for diagnosing drug-induced liver damage; MELD: scale to measure the severity of chronic liver disease. In cases where information about dose is not provided, it is because it was not indicated in the article.

The FDA warned consumers about the serious adverse effects associated with the consumption of Hydroxycut™. This recommendation was based on 23 cases of liver damage, including one death and a liver transplant. All this led to the withdrawal of the supplement from the market. Later on, García-Cortés et al. (2016) reported 29 cases of liver damage induced by this supplement [61].

Taking this into account, as it can observed in Table 2, hepatotoxic effects have also been reported when using pure *Garcinia cambogia* extracts [49,50,54] or supplements that only contained *Garcinia cambogia* with minerals or vitamins [53,55,56]. The reason that justifies the occurrence of hepatotoxic effects in these cases, but not in all the studies gathered in Table 1, is not clear. Genetic interindividual variability leading to different susceptibility to the action of Garcinia cannot be ruled out. Nevertheless, it is important to emphasize that counting all the subjects that participated in the studies described in

Table 1, the lack of toxic effects refers to a quite big sample, whereas toxic effects have only been described in a reduced number of subjects.

The common pattern of symptoms, observed in all these cases, consisted of abdominal pain (predominantly in the right upper quadrant), vomiting, nausea, fatigue and alterations in liver parameters such as transaminases, alkaline phosphatase and bilirubin. Serological tests were performed to rule out other possible causes of liver damage and/or infection, such as hepatitis, Epstein–Barr virus, cytomegalovirus, etc. In eight cases the CIOMS/RUCAM scale was used. This scale is a scoring system used to establish the etiology of drug-induced liver damage, and depending on the score obtained, the substance is classified as a highly probable cause (≥9), probable cause (6–8), possible cause (3–5), unlike cause (1–2) or excluded cause (0) of liver injury. The scores obtained for the supplements in these studies ranged from 6 to 11 points; in other words, probable to highly probable cause. Importantly, after Garcinia supplement withdrawal, the symptoms subsided and all the altered parameters returned to normal levels, although in four cases the patient finally required liver transplantation.

Regarding acute pancreatitis, one case has been reported in an 82-year-old man with past medical history of obesity and two previous episodes of acute pancreatitis in the past. He denied any alcohol use and reported no recent changes in his medications, as well as the intake of *Garcinia cambogia* recently as an appetite suppressant. He was treated with bowel rest and intravenous fluid hydration, providing a significant improvement in his symptoms [62].

Other adverse effects associated with *Garcinia cambogia* ingestion are mania and psychosis, as shown in Table 3 [63–66]. Nevertheless, currently the existing scientific evidence is limited and a causal association has not yet been established with certainty. In some cases, the participants had a previous psychiatric history and/or were treated with selective serotonin reuptake inhibitors (SSRI). HCA acts as a selective serotonin reuptake inhibitor, thus increasing serotonin levels and increasing the risk of toxicity due to this neurotransmitter [63,66]. The most relevant symptoms were irritability and agitation. After the withdrawal of the supplement, the symptoms remitted, and all the altered parameters returned to normal levels.

To conclude, another type of toxicity associated with the consumption of *Garcinia cambogia* can be observed in patients with pre-existing metabolic disorders. In this line, Bystrak et al. (2017) described the case of a 56-year-old insulin-dependent, hypertensive woman with chronic hepatitis C, who developed diabetic ketoacidosis, pancreatitis, and cardiomyopathic stress after consuming a *Garcinia cambogia* supplement (1400 mg HCA/day) to lose weight [67]. Applying the algorithm described by Naranjo et al. (1981) [68] to estimate the causality of an adverse drug reaction, a value of five was obtained, meaning a probable adverse reaction to the use of *Garcinia cambogia* [67].

Table 3. Cases of mania and serotonin toxicity associated with the intake of *Garcinia cambogia* supplements in humans.

Reference	Ageyears	Sex	Previous Psychiatric History	Type of Supplement and Treatment Duration	Psychotropic/ Antidepressant Drugs	Symptoms	Diagnosis
Lopez et al., 2014 [63]	35	W	No	1000 mg *Garcinia cambogia* (60% HCA), chromium, potassium and calcium 2 capsules, 3 times/day. 2–3 months.	Yes: escitalopram (SSRI). 1 year.	Stuttering speech pattern, spontaneous ankle clonus, bilateral ocular clonus, rhythmic jaw movements, profuse sweating, hypertension, tachycardia, and hyperreflexia.	Serotonin toxicity associated with *Garcinia cambogia* ingestion.

Table 3. *Cont.*

Reference	Ageyears	Sex	Previous Psychiatric History	Type of Supplement and Treatment Duration	Psychotropic/ Antidepressant Drugs	Symptoms	Diagnosis
Hendrickson et al., 2016 [64]	50	M	Type I bipolar disorder	*Garcinia cambogia*: 2 capsules/day. 2 months.	No: he had been stable off medications for 6 years.	Irritability, pressured speech, grandiosity, excessive spending, increased social activity and decreased need for sleep.	Bipolar I disorder, manic, severe.
	25	M	No	*Garcinia cambogia*: 1–2 capsules/day.2 months.	No	Inflated self-esteem, grandiosity, decreased need for sleep, increased activity, excessive spending, pressured speech, paranoia and religious delusions.	Bipolar I disorder, manic, severe with psychosis.
	34	W	Type II bipolar disorder and past SSRI-induced hypomania	*Garcinia cambogia* For 1–2 months.	Yes: aripiprazole, bupropion and topiramate.	Irritability, pressured speech, decreased need for sleep and agitation.	Recurrence of bipolar disorder type II, hypomania, moderate.
Cotovio et al., 2017 [65]	51	W	Type I bipolar disorder	*Garcinia cambogia*, calcium, chromium and potassium.	Yes: paroxetine (SSRI) and valproic acid	Irritability, agitation, increased energy and decreased need for sleep.	Hypomanic episode associated with ingestion of *Garcinia cambogia*.
Nguyen et al., 2019 [66]	22	W	No	*Garcinia cambogia* Plus^{TM}: 500 mg *Garcinia cambogia* per capsule (60% HCA) 1 capsule/day during de first 5 days and then 3 capsules/d during the next 5 days.	No	Expansive mood, psychomotor agitation, disorganized and pressured speech, flight of ideas, grandiosity, delusions and auditory hallucinations.	Mania and psychosis secondary to *Garcinia cambogia* ingestion.

M: men; W: women; SSRI: selective serotonin reuptake inhibitor.

5. Concluding Remarks

The reported scientific literature shows that nutritional supplements based on Garcinia extracts are effective in just over half of the reported studies. In these cases, the supplements should not constitute a treatment per se but, they should represent complementary tools to the conventional treatment of excess body fat. Moreover, due to their positive effects on lipid and glycemic profile, these supplements could be useful for the management of the co-morbidities associated with obesity.

Garcinia-based supplements have been shown to be safe in numerous human experiments, but the growing number of cases that report significant adverse effects, mainly associated with liver damage, and to a lesser extent with serotonin toxicity and mania, may lead to reconsideration of the safety of them. Although very often HCA has been signaled as the main element responsible for the toxic effects of Garcinia supplements, it should be noted that the fruit of *Garcinia cambogia*, an important source of HCA, has been consumed for centuries in Southeast Asia and has been generally recognized as safe

(GRAS) [69]. On the other hand, as explained before in this review, in many cases Garcinia supplements contain a great number of components. Consequently, the toxicity cannot be reliably attributable to Garcinia, and it is difficult to make conclusions without giving rise to doubts or objections. Furthermore, potential negative effects due to the combination of the Garcinia supplement with other dietary supplements included in the consumer diet, or even with several drugs, cannot be discarded.

Importantly, adverse effect case reports usually reflect the associations between the observed toxicity and the intake of the dietary supplement, rather than causality. These associations need to be rigorously examined and, if finally, the supplements are found to be the causative factors for the alterations observed, the true agents need to be firmly identified, along with the dose at which the negative effects are induced [69]. In this regard, an important problem in diagnosing the cause that produces the adverse effects, is that many people perceive these type of products as not harmful or as "natural" products, thus, they tend to forget to mention them when they are asked about the foods, beverages and medications that they have consumed. It is therefore likely, that there are more cases than those diagnosed, and consequently, the magnitude of the problem may be underestimated. When the dose administered or the exact content of HCA are not specified, it is not possible to identify whether the dose exceeds the value established as NOAEL. Another important aspect, is that the issue of obesity-related liver co-morbidities as a cause of liver alterations, has been poorly handled in toxic effect reports. Much more attention should be paid to obesity-associated liver diseases in the causality assessment of dietary supplements used for weight reduction [70].

In this scenario, more studies are needed to evaluate the efficacy and safety of these products, using larger sample sizes and longer follow-up periods. Finally, it should be pointed out that there are certain population groups in which the use of these supplements should be discouraged. This is the case for pregnant and lactating women. HCA can affect the production of fatty acids and cholesterol, and can directly influence the production of sterols and steroid hormones. Pregnancy is a time of extreme sensitivity to steroid hormones, therefore, these products are not recommended. In the case of children—although it has not been possible to prove that they are dangerous—the advice is not to consume them in large doses and for long periods of time. On the other hand, the evolution of patients with mild depression or with occasional episodes of hypomania who consume Garcinia should be monitored, since their situation may worsen.

Author Contributions: Conceptualization, M.P.P., Writing-Original Draft Preparation, N.A.; Writing-Review & Editing, M.P.P. and R.M.G.; Funding Acquisition, M.P.P. All authors have read and agreed to the published version of the manuscript.

Funding: This study was funded by Instituto de Salud Carlos III (CIBERobn) under Grant CB12/03/3000 and University of the Basque Country under Grant GIU18-173.

Institutional Review Board Statement: Not applicable.

Informed Consent Statement: Not applicable.

Data Availability Statement: Not applicable.

Conflicts of Interest: The authors declare no conflict of interest.

References

1. The heavy burden of obesity and the economics of prevention. In *The Heavy Burden of Obesity*; OECD Publishing: Paris, France, 2019.
2. World Health Organization. Obesity. Preventing and Managing the Global Epidemic. In *Report of a WHO Consultation on Obesity*; WHO/NUT/NCD/981; WHO: Geneva, Switzerland, 1998.
3. Haber, S.; Awwad, O.; Phillips, A.; Park, A.; Pham, T. Garcinia cambogia for weight loss. *Am. J. Health-Syst. Pharm.* **2018**, *75*, 17–22. [CrossRef] [PubMed]

4. Garcia-Alvarez, A.; Egan, B.; de Klein, S.; Dima, L.; Maggi, F.M.; Isoniemi, M.; Ribas-Barba, L.; Raats, M.M.; Meissner, E.M.; Badea, M.; et al. Usage of plant food supplements across six European countries: Findings from the PlantLIBRA consumer survey. *PLoS ONE* **2014**, *9*, 537. [CrossRef]
5. Hemshekhar, M.; Sunitha, K.; Santhosh, M.; Devaraja, S.; Kemparaju, K.; Vishwanath, B.; Niranjana, S.R.; Girish, K.S. An overview on genus garcinia: Phytochemical and therapeutical aspects. *Phytochem. Rev.* **2011**, *10*, 325–351. [CrossRef]
6. Chuah, L.O.; Ho, W.Y.; Beh, B.K.; Yeap, S.K. Updates on antiobesity effect of Garcinia Origin (-)-HCA. *Evid. Based Complement. Altern. Med.* **2013**, *2013*, 751658. [CrossRef]
7. Semwal, R.B.; Semwal, D.K.; Vermaak, I.; Viljoen, A. A comprehensive scientific overview of *Garcinia cambogia*. *Fitoterapia* **2015**, *102*, 134–148. [CrossRef]
8. Jena, B.S.; Jayaprakasha, G.K.; Singh, R.P.; Sakariah, K.K. Chemistry and biochemistry of (-)-hydroxycitric acid from Garcinia. *J. Agric. Food Chem.* **2002**, *50*, 10–22. [CrossRef] [PubMed]
9. Ishihara, K.; Oyaizu, S.; Onuki, K.; Lim, K.; Fushiki, T. Chronic (-)-Hydroxycitrate Administration Spares Carbohydrate Utilization and Promotes Lipid Oxidation during Exercise in Mice. *J. Nutr.* **2000**, *130*, 2990–2995. [CrossRef] [PubMed]
10. Greenwood, M.R.; Cleary, M.P.; Gruen, R.; Blasé, D.; Stern, J.S.; Triscari, J.; Sullivan, A.C. Effect of (-)-hydroxycitrate on development of obesity in the Zucker obese rat. *Am. J. Physiol. Endocrinol. Metab.* **1981**, *240*, 72–78. [CrossRef]
11. Leonhardt, M.; Langhans, W. Hydroxycitrate has long-term effects on feeding behavior, body weight regain and metabolism after body weight loss in male rats. *J. Nutr.* **2002**, *132*, 1977–1982. [CrossRef] [PubMed]
12. Saito, M.; Ueno, M.; Ogino, S.; Kubo, K.; Nagata, J.; Takeuchi, M. High dose of Garcinia cambogia is effective in suppressing fat accumulation in developing male Zucker obese rats, but highly toxic to the testis. *Food Chem. Toxicol.* **2005**, *43*, 411–419. [CrossRef]
13. Heymsfield, S.B.; Allison, D.B.; Vasselli, J.R.; Pietrobelli, A.; Greenfield, D.; Nunez, C. Garcinia cambogia (Hydroxycitric Acid) as a potential antiobesity agent: A randomized controlled trial. *JAMA* **1998**, *280*, 1596–1600. [CrossRef] [PubMed]
14. Mattes, R.D.; Bormann, L. Effects of (-)-hydroxycitric acid on appetitive variables. *Physiol. Behav.* **2000**, *71*, 87–94. [CrossRef]
15. Thom, E. A randomized, double-blind, placebo-controlled trial of a new weight-reducing agent of natural origin. *J. Int. Med. Res.* **2000**, *28*, 229–233. [CrossRef] [PubMed]
16. Kovacs, E.M.R.; Westerterp-Plantenga, M.S.; Saris, W.H.M. The effects of 2-week ingestion of (–)-hydroxycitrate and (–)-hydroxycitrate combined with medium-chain triglycerides on satiety, fat oxidation, energy expenditure and body weight. *Int. J. Obes.* **2001**, *25*, 1087–1094. [CrossRef] [PubMed]
17. Kovacs, E.M.R.; Westerterp-Plantenga, M.S.; de Vries, M.; Brouns, F.; Saris, W.H.M. Effects of 2-week ingestion of (−)-hydroxycitrate and (−)-hydroxycitrate combined with medium-chain triglycerides on satiety and food intake. *Physiol. Behav.* **2001**, *74*, 543–549. [CrossRef]
18. Hayamizu, K.; Ishii, Y.; Kaneko, I.; Shen, M.; Okuhara, Y.; Shigematsu, N.; Tomi, H.; Furuse, M.; Yoshino, G.; Shimasaki, H. Effects of *Garcinia cambogia* (Hydroxycitric Acid) on visceral fat accumulation: A double-blind, randomized, placebo-controlled trial. *Curr. Ther. Res.* **2003**, *64*, 551–567. [CrossRef]
19. Preuss, H.G.; Bagchi, D.; Bagchi, M.; Rao, C.V.S.; Satyanarayana, S.; Dey, D.K. Efficacy of a novel, natural extract of (–)-hydroxycitric acid (HCA-SX) and a combination of HCA-SX, niacin-bound chromium and *Gymnema sylvestre* extract in weight management in human volunteers: A pilot study. *Nutr. Res.* **2004**, *24*, 45–58. [CrossRef]
20. Preuss, H.G.; Bagchi, D.; Bagchi, M.; Rao, C.V.S.; Dey, D.K.; Satyanarayana, S. Effects of a natural extract of (-)-hydroxycitric acid (HCA-SX) and a combination of HCA-SX plus niacin-bound chromium and *Gymnema sylvestre* extract on weight loss. *Diabetes Obes. Metab.* **2004**, *6*, 171–180. [CrossRef]
21. Roongpisuthipong, C.; Kantawan, R.; Roongpisuthipong, W. Reduction of adipose tissue and body weight: Effect of water soluble calcium hydroxycitrate in *Garcinia atroviridis* on the short term treatment of obese women in Thailand. *Asia Pac. J. Clin. Nutr.* **2007**, *16*, 25.
22. Toromanyan, E.; Aslanyan, G.; Amroyan, E.; Gabrielyan, E.; Panossian, A. Efficacy of Slim339®in reducing body weight of overweight and obese human subjects. *Phytother. Res.* **2007**, *21*, 1177–1181. [CrossRef]
23. Vasques, C.A.R.; Rossetto, S.; Halmenschlager, G.; Linden, R.; Heckler, E.; Fernandez, M.S.P.; Alonso, J.L.L. Evaluation of the pharmacotherapeutic efficacy of *Garcinia cambogia* plus *Amorphophallus konjac* for the treatment of obesity. *Phytother. Res.* **2008**, *22*, 1135–1140. [CrossRef] [PubMed]
24. Kim, J.; Jeon, S.; Park, K.H.; Lee, W.S.; Jeong, T.; McGregor, R.A.; Choi, M.-S. Does Glycine max leaves or Garcinia Cambogia promote weight-loss or lower plasma cholesterol in overweight individuals: A randomized control trial. *Nutr. J.* **2011**, *10*, 94. [CrossRef] [PubMed]
25. Lu, C.H.; Yang, T.H.; Wu, C.C.; Doong, J.Y.; Lin, P.Y.; Chiang, C.M.; Lin, C.L.; Hsieh, S.L. Clinical evaluation of *Garcinia cambogia* and *Phaseolus vulgaris* extract for obese adults in taiwan. *Nutr. Sci. J.* **2012**, *37*, 75–84. [CrossRef]
26. Stern, J.S.; Peerson, J.; Mishra, A.T.; Sadasiva Rao, M.V.; Rajeswari, K.P. Efficacy and tolerability of a novel herbal formulation for weight management. *Obesity* **2013**, *21*, 921–927. [CrossRef] [PubMed]
27. Stern, J.S.; Peerson, J.; Mishra, A.T.; Mathukumalli, V.S.R.; Konda, P.R. Efficacy and tolerability of an herbal formulation for weight management. *J. Med. Food* **2013**, *16*, 529–537. [CrossRef]
28. Chong, P.; Beah, Z.; Grube, B.; Riede, L. IQP-GC-101 reduces body weight and body fat mass: A randomized, double-blind, placebo-controlled study. *Phytother. Res.* **2014**, *28*, 1520–1526. [CrossRef]

29. Vasques, C.A.R.; Schneider, R.; Klein-Júnior, L.C.; Falavigna, A.; Piazza, I.; Rossetto, S. Hypolipemic Effect of Garcinia cambogia in Obese Women. *Phytother. Res.* **2014**, *28*, 887–891. [CrossRef]
30. Kudiganti, V.; Kodur, R.R.; Kodur, S.R.; Halemane, M.; Deep, D.K. Efficacy and tolerability of Meratrim for weight management: A randomized, double-blind, placebo-controlled study in healthy overweight human subjects. *Lipids Health Dis.* **2016**, *15*, 136. [CrossRef]
31. Maia-Landim, A.; Ramírez, J.M.; Lancho, C.; Poblador, M.S.; Lancho, J.L. Long-term effects of *Garcinia cambogia*/Glucomannan on weight loss in people with obesity, PLIN4, FTO and Trp64Arg polymorphisms. *BMC Complement. Altern. Med.* **2018**, *18*, 26. [CrossRef]
32. Watanabe, M.; Gangitano, E.; Francomano, D.; Addessi, E.; Toscano, R.; Costantini, D.; Tuccinardi, D.; Mariani, S.; Basciani, S.; Spera, G.; et al. Mangosteen Extract shows a potent Insulin sensitizing effect in Obese female patients: A prospective randomized controlled pilot study. *Nutrients* **2018**, *10*, 586. [CrossRef]
33. Onakpoya, I.; Hung, S.K.; Perry, R.; Wider, B.; Ernst, E. The use of Garcinia extract (hydroxycitric acid) as a weight loss supplement: A systematic review and meta-analysis of randomised clinical trials. *J. Obes.* **2011**, *2011*, 509038. [CrossRef]
34. Chuah, L.O.; Yeap, S.K.; Ho, W.Y.; Beh, B.K.; Alitheen, N.B. In vitro and in vivo toxicity of Garcinia or hydroxycitric acid: A review. *Evid. Based Complement. Altern. Med.* **2012**, *2012*, 197920. [CrossRef]
35. Soni, M.G.; Burdock, G.A.; Preuss, H.G.; Stohs, S.J.; Ohia, S.E.; Bagchi, D. Safety assessment of (−)-hydroxycitric acid and Super CitriMax®, a novel calcium/potassium salt. *Food Chem. Toxicol.* **2004**, *42*, 1513–1529. [CrossRef]
36. Márquez, F.; Babio, N.; Bulló, M.; Salas-Salvadó, J. Evaluation of the safety and efficacy of hydroxycitric acid or *Garcinia cambogia* extracts in humans. *Crit. Rev. Food Sci. Nutr.* **2012**, *52*, 585–594. [CrossRef]
37. Ohia, S.; Opere, C.; LeDay, A.; Bagchi, M.; Bagchi, D.; Stohs, S. Safety and mechanism of appetite suppression by a novel hydroxycitric acid extract (HCA-SX). *Mol. Cell. Biochem.* **2002**, *238*, 89–103. [CrossRef]
38. Clouatre, D.L.; Preuss, H.G. Hydroxycitric acid does not promote inflammation or liver toxicity. *World J. Gastroenterol.* **2013**, *19*, 8160–8162. [CrossRef]
39. Shara, M.; Ohia, S.E.; Yasmin, T.; Zardetto-Smith, A.; Kincaid, A.; Bagchi, M.; Chatterjee, A.; Bagchi, D.; Stohs, S.J. Dose- and time-dependent effects of a novel (−)-hydroxycitric acid extract on body weight, hepatic and testicular lipid peroxidation, DNA fragmentation and histopathological data over a period of 90 days. *Mol. Cell. Biochem.* **2003**, *254*, 339–346. [CrossRef]
40. Shara, M.; Ohia, S.E.; Schmidt, R.E.; Yasmin, T.; Zardetto-Smith, A.; Kincaid, A.; Bagchi, M.; Chatterjee, A.; Bagchi, D.; Stohs, S.J. Physico-chemical properties of a novel (-)-hydroxy- citric acid extract and its effect on body weight, selected organ weights, hepatic lipid peroxidation and DNA fragmentation, hematology and clinical chemistry, and histopathological changes over a period of 90 days. *Mol. Cell. Biochem.* **2004**, *260*, 171–186. [CrossRef]
41. Kim, Y.; Choi, M.; Park, Y.B.; Kim, S.R.; Lee, M.; Jung, U.J. *Garcinia cambogia* attenuates diet-induced adiposity but exacerbates hepatic collagen accumulation and inflammation. *World J. Gastroenterol.* **2013**, *19*, 4689–4701. [CrossRef]
42. Kothadia, J.P.; Kaminski, M.; Samant, H.; Olivera-Martinez, M. Hepatotoxicity associated with use of the weight loss supplement Garcinia cambogia: A case report and review of the literature. *Case Rep. Hepatol.* **2018**, *2018*, 6483605. [CrossRef]
43. Farombi, E.O.; Adedara, I.A.; Oyenihi, A.B.; Ekakitie, E.; Kehinde, S. Hepatic, testicular and spermatozoa antioxidant status in rats chronically treated with *Garcinia kola seed*. *J. Ethnopharmacol.* **2013**, *146*, 536–542. [CrossRef]
44. Saiyed, Z.M.; Sengupta, K.; Krishnaraju, A.V.; Trimurtulu, G.; Lau, F.C.; Lugo, J.P. Safety and toxicological evaluation of Meratrim®: An herbal formulation for weight management. *Food Chem. Toxicol.* **2015**, *78*, 122–129. [CrossRef]
45. Stevens, T.; Qadri, A.; Zein, N.N. Two patients with acute liver injury associated with use of the herbal weight-loss supplement hydroxycut. *Ann. Intern. Med.* **2005**, *142*, 477–478. [CrossRef]
46. Dara, L.; Hewett, J.; Lim, J.K. Hydroxycut hepatotoxicity: A case series and review of liver toxicity from herbal weight loss supplements. *World J. Gastroenterol.* **2008**, *14*, 6999–7004. [CrossRef]
47. Shuster, J. Priapism and risperidone; Psoriasis after certolizumab therapy; bruxism with two different SSRIs; Nasal septum deviation associated with bevacizumab treatment; Hepatic failure with another herbal weight loss supplement; Bupropion-related dystonia and parkinsonism; bendamustine-associated hemolytic anemia. *Hosp. Pharm.* **2010**, *45*, 100–109. [CrossRef]
48. Sharma, T.; Wong, L.; Tsai, N.; Wong, R.D. Hydroxycut(®) (herbal weight loss supplement) induced hepatotoxicity: A case report and review of literature. *Hawaii Med. J.* **2010**, *69*, 188–190.
49. Mancano, M.A. *Garcinia cambogia*–Induced Acute Hepatitis; Varenicline-Induced Parkinsonism; Resistant Hypocalcemia after Zoledronic Acid Administration; Zonisamide-Induced Acute Kidney Injury; Psychosis Associated with Guanfacine. *Hosp. Pharm.* **2015**, *50*, 564–568. [CrossRef]
50. Melendez-Rosado, J.; Snipelisky, D.; Matcha, G.; Stancampiano, F. Acute hepatitis induced by pure Garcinia cambogia. *J. Clin. Gastroenterol.* **2015**, *49*, 449–450. [CrossRef]
51. Araujo, J.L.; Worman, H.J. Acute liver injury associated with a newer formulation of the herbal weight loss supplement Hydroxycut. *BMJ Case Rep.* **2015**, *2015*, bcr2015210303. [CrossRef]
52. Smith, R.J.; Bertilone, C.; Robertson, A.G. Fulminant liver failure and transplantation after use of dietary supplements. *Med. J. Aust.* **2016**, *204*, 30–32. [CrossRef]
53. Corey, R.; Werner, K.T.; Singer, A.; Moss, A.; Smith, M.; Noelting, J.; Rakela, J. Acute liver failure associated with *Garcinia cambogia* use. *Ann. Hepatol.* **2016**, *15*, 123. [CrossRef]

54. Lunsford, K.E.; Bodzin, A.S.; Reino, D.C.; Wang, H.L.; Busuttil, R.W. Dangerous dietary supplements: *Garcinia cambogia*-associated hepatic failure requiring transplantation. *World J. Gastroenterol.* **2016**, *22*, 10071–10076. [CrossRef]
55. Crescioli, G.; Lombardi, N.; Bettiol, A.; Marconi, E.; Risaliti, F.; Bertoni, M.; Ippolito, F.M.; Maggini, V.; Gallo, E.; Firenzuoli, F.; et al. Acute liver injury following *Garcinia cambogia* weight-loss supplementation: Case series and literature review. *Intern. Emerg. Med.* **2018**, *13*, 857–872. [CrossRef]
56. Sharma, A.; Akagi, E.; Njie, A.; Goyal, S.; Arsene, C.; Krishnamoorthy, G.; Ehrinpreis, M. Acute hepatitis due to *Garcinia cambogia* extract, an herbal weight loss supplement. *Case Rep. Gastrointest. Med.* **2018**, *2018*, 9606171. [CrossRef]
57. Philips, C.A.; Augustine, P. Chemical analysis of weight loss herbal supplement safe lean™ associated with acute liver injury—A concern for spurious drug, misbranding and adulteration. *J. Clin. Exp. Hepatol.* **2018**, *8*, 471–473. [CrossRef]
58. Yousaf, M.N.; Chaudhary, F.S.; Hodanazari, S.M.; Sittambalam, C.D. Hepatotoxicity associated with *Garcinia cambogia*: A case report. *World J. Hepatol.* **2019**, *11*, 735–742. [CrossRef]
59. Khetpal, N.; Mandzhieva, B.; Shahid, S.; Khetpal, A.; Jain, A.G. Not All Herbals are Benign: A Case of Hydroxycut-induced Acute Liver Injury. *Cureus* **2020**, *12*, e6870. [CrossRef]
60. Ferreira, V.; Mathieu, A.; Soucy, G.; Giard, J.M.; Erard-Poinsot, D. Acute severe liver injury related to long-term *Garcinia cambogia* intake. *ACG Case Rep. J.* **2020**, *7*, e00429. [CrossRef]
61. García-Cortés, M.; Robles-Díaz, M.; Ortega-Alonso, A.; Medina-Caliz, I.; Andrade, R.J. Hepatotoxicity by dietary supplements: A tabular listing and clinical characteristics. *Int. J. Mol. Sci.* **2016**, *17*, 537. [CrossRef]
62. Iqbal, U.; Anwar, H.; Siddiqui, H.U.; Mehmood, A. Acute Pancreatitis Secondary to Use of Appetite Suppressant: *Garcinia cambogia*. *Cureus* **2019**, *11*, e4676. [CrossRef]
63. Lopez, A.; Kornegay, J.; Hendrickson, R. Serotonin toxicity associated with *Garcinia cambogia* over-the-counter supplement. *J. Med. Toxicol.* **2014**, *10*, 399–401. [CrossRef]
64. Hendrickson, B.P.; Shaikh, N.; Occhiogrosso, M.; Penzner, J.B. Mania induced by *Garcinia cambogia*: A case series. *Prim. Care Companion CNS Disord.* **2016**, 18. [CrossRef]
65. Cotovio, G.; Oliveira-Maia, A.J. Hypomania induced by a *Garcinia cambogia* supplement. *Aust. N. Z. J. Psychiatry* **2017**, *51*, 641–642. [CrossRef]
66. Nguyen, D.C.; Timmer, T.K.; Davison, B.C.; McGrane, I.R. Possible *Garcinia cambogia*-induced mania with psychosis: A case report. *J. Pharm. Pract.* **2019**, *32*, 99–102. [CrossRef]
67. Bystrak, T.; Cervera-Hernandez, M.E.; Reddy, N.; King, Z.; Bratberg, J. *Garcinia cambogia*, Diabetic Ketoacidosis, and Pancreatitis. *RI Med. J.* **2017**, *100*, 48–50.
68. Naranjo, C.A.; Busto, U.; Sellers, E.M.; Sandor, P.; Ruiz, I.; Roberts, E.A.; Janecek, E.; Domecq, C.; Greenblatt, D.J. A method for estimating the probability of adverse drug reactions. *Clin. Pharmacol. Ther.* **1981**, *30*, 239–245. [CrossRef]
69. Stohs, S.J.; Preuss, H.G.; Ohia, S.E.; Kaats, G.R.; Keen, C.L.; Williams, L.D.; Burdock, G.A. No evidence demonstrating hepatotoxicity associated with hydroxycitric acid. *World J. Gastroenterol.* **2009**, *15*, 4087–4089. [CrossRef]
70. Teschke, R.; Wolff, A.; Eickhoff, A.; Danan, G. Is obesity rather than the dietary supplement used for weight reduction the cause of liver injury? *JGH Open* **2018**, *2*, 152–157. [CrossRef]

MDPI
St. Alban-Anlage 66
4052 Basel
Switzerland
Tel. +41 61 683 77 34
Fax +41 61 302 89 18
www.mdpi.com

Nutrients Editorial Office
E-mail: nutrients@mdpi.com
www.mdpi.com/journal/nutrients

www.ingramcontent.com/pod-product-compliance
Lightning Source LLC
Chambersburg PA
CBHW040833250926
51850CB00007BA/1302